AF360734

Functional Cereal Foods for Health Benefits: Genetic and/or Processing Strategies to Enhance the Quali-Quantitative Composition of Bioactive Components

Functional Cereal Foods for Health Benefits: Genetic and/or Processing Strategies to Enhance the Quali-Quantitative Composition of Bioactive Components

Editors

Barbara Laddomada
Weiqun Wang

MDPI • Basel • Beijing • Wuhan • Barcelona • Belgrade • Manchester • Tokyo • Cluj • Tianjin

Editors
Barbara Laddomada
CNR, Institute of Sciences of
Food Production (ISPA),
Lecce, Italy

Weiqun Wang
Department of Food,
Nutrition, Dietetics & Health,
Kansas State University,
Manhattan, KS 66506, USA

Editorial Office
MDPI
St. Alban-Anlage 66
4052 Basel, Switzerland

This is a reprint of articles from the Special Issue published online in the open access journal *Foods* (ISSN 2304-8158) (available at: https://www.mdpi.com/journal/foods/special_issues/cereal_foods_bioactive).

For citation purposes, cite each article independently as indicated on the article page online and as indicated below:

LastName, A.A.; LastName, B.B.; LastName, C.C. Article Title. *Journal Name* **Year**, *Volume Number*, Page Range.

ISBN 978-3-0365-4781-7 (Hbk)
ISBN 978-3-0365-4782-4 (PDF)

Contents

About the Editors

Barbara Laddomada

Barbara Laddomada has been a researcher at the Institute of Sciences of Food Production of the Italian Research Council since 2000. Her major research focuses include the genetic and biochemical characterization of bioactive components of wheat grains, and the improvement of the quality of wheat-based foods. She graduated in Agricultural Sciences in 1993 at the University of Bari, Italy. In 1997, she received her PhD in Plant Genetics from the University of La Tuscia, Viterbo, Italy. In 1995/96, she was a visiting scientist at the Department of Plant Pathology, KSU (Manhattan, KS, USA) working on the genetic dissection of protein content in durum wheat. In 1997, she earned a one-year fellowship from FAO at the University of Basilicata, Italy, identifying molecular markers associated with yield components in durum wheat. In 1999, she worked as a post-doctoral fellow at the University of Bari, Italy, working on the genetic analysis of quantitative traits in durum wheat. In 2010/11, she was visiting scientist at the Plant Pathology Department, KSU, USA. Since 2016, she has been a member of the Wheat Initiative expert working group for improving wheat quality for processing and health, and durum wheat genomics and breeding.

Weiqun Wang

Weiqun Wang (Professor of Food, Nutrition, Dietetics and Health, and Director of the Graduate Program at Kansas State University) received his BS in Biochemistry and his Ph.D. in Animal Physiology and Biochemistry in China. His past research experience includes working as a postdoctoral fellow at the University of Hawaii, a junior researcher at the Cancer Research Center of Hawaii, and a senior scientist at Iowa State University. His research focuses on the molecular mechanisms of cancer prevention by dietary calorie restriction, exercise, and phytochemicals. As a PI or co-investigator on 28 funded research grants from NIH, USDA, and AHA, Dr. Wang has laid groundwork for research and demonstrated a record of productive research in more than 120 scholarly publications. He has also mentored 13 postdoctoral and 83 graduate students as a major professor and committee chair. Along with serving on several editorial boards of journals as Editor-in-Chief or Associate Editor, Dr. Wang has also participated in NIH and USDA grant review panels as a panelist or panel manager.

foods

Editorial

Multiple Approaches to Improve the Quality of Cereal-Based Foods

Barbara Laddomada [1,*] and Weiqun Wang [2]

1 Institute of Sciences of Food Production (ISPA), National Research Council (CNR), Via Monteroni, 73100 Lecce, Italy
2 Department of Food, Nutrition, Dietetics and Health, Kansas State University, Manhattan, KS 66506, USA; wwang@ksu.edu
* Correspondence: barbara.laddomada@ispa.cnr.it; Tel.: +39-0832-422-613

Citation: Laddomada, B.; Wang, W. Multiple Approaches to Improve the Quality of Cereal-Based Foods. *Foods* **2022**, *11*, 1849. https://doi.org/10.3390/foods11131849

Received: 16 June 2022
Accepted: 22 June 2022
Published: 23 June 2022

Publisher's Note: MDPI stays neutral with regard to jurisdictional claims in published maps and institutional affiliations.

The interest in improving the health benefits of cereal foods is continuously increasing. This is essentially due to the high frequency of their consumption worldwide, and to the chance of using them to vehicle health promoting components in the diet that may counteract the occurrence of non-communicable diseases (NCDs). This would contribute to improving health maintenance and disease prevention on line with aims of the 2030 Agenda for Sustainable Development that recognizes NCDs as a "major challenge for sustainable development". Incorrect dietary behaviors are among the major "behavioral risk factors" in NCDs incidences. In fact, the increasing consumption of ready meals that are rich in sugars and lipids, and poor in bioactive compounds and fibers, is contributing to overweight and obesity worldwide, with an increasing trend in low-, middle-, and even high-income countries.

Improving the functional and technological properties of cereal wholemeal flours is a first and direct goal in favor of the production and spread of cereal-based foods with improved functional skills. The challenge is to improve the cereal foods not only for health, but also for taste in order to guarantee consumer's acceptance, and the economic potential of the new products. This aim can be achieved using different strategies, some being based on the selection of the best genetic materials and on breeding programs, some others focusing on the upgrade of milling processes or on using new foods formulations.

The Special Issue "Functional Cereal Foods for Health Benefits: Genetic and/or Processing Strategies to Enhance the Quali-Quantitative Composition of Bioactive Components" collects 12 original research articles on different innovative approaches to improve the health potential of cereal foods contributing to a Mediterranean diet-based lifestyle. Some of the research articles address the issue of finding the best milling strategies to produce wholemeal flour with improved contents of bioactive components, but also reduced contents of contaminants and toxic components. Above all, the aim is to improve the health value of the end-products while maintaining good technological and sensorial quality. The latter issues seemed to be not completely overcome when different pilot-scale milling methods were tested to produce whole-wheat flour to be used for Chinese Steamed Bread (CSB) and Chinese leavened pancakes (CLP) [1]. In fact, dietary fibers increased by 1.6 fold, and ferulic acid by 1.9 fold, but such improvements were accompanied by an increase of damaged starch percentage, and overall affected the quality of end-products [1]. The development of an upgraded micronization plant and of a modified air classification plant able to produce different types of durum wheat milling fractions enriched in bran particles and consequently of phenolic bioactive compounds was also presented in this special issue [2]. This innovative solution offered the advantage of directly producing different unrefined milling products with peculiar features, making them suitable to produce diverse types of unrefined end-products (pasta, bread, biscuits, etc.). Due to the presence of bran particles in different rates within each air-classified fraction, the millings had a higher technological quality compared to that of semolina supplemented with the addition of bran aliquots. Nevertheless, the alvegraphic behavior of the unrefined millings displays a significant reduction in the alveographic parameters, especially with regard to a P/L increase [3]. This technology has the advantage of reducing the levels of organic and inorganic contaminants, thus reducing health hazards to consumers [4].

Another aspect of the special issue is about the effects of seed fermentation or dehulling on improving the concentration of nutritional and bioactive compounds. As an example, xylanase and cellulose, produced by *Aspergillus awamori* and *Thricoderma reesei*, were used on rice seeds [5]. In the selected conditions, such treatment facilitated a consistent increase of Ca, P, Iron, free amino acids, phenolic compounds and proteins. Another application of fermentation regarded coix seeds by using *Monascus purpureus* [6]. As a result, the contents of tocols, γ-oryzanol, and coixenolide increased improving also the associated antioxidant and anticancer activity on laryngeal carcinomatous HEp2 cells [6].

Dehulling also could have an indirect effect on the improvement of phenolic compounds with health benefits [7]. In fact, the germination of dehulled buckwheat seeds resulted in 1.8-fold and 1.9-fold higher phenolics and antioxidant activity compared to hulled seeds [7].

Ultimately, as a third topic, this special issue includes studies investigating the effect of extreme environmental conditions on the accumulation of phenolic acids in the wholemeal flour of different durum wheat cultivars [8]. Good yield performances and high accumulation of bioactive compounds across highly differing growing conditions could be significant to improve both the durum wheat resilience and health-promoting value. This in agreement with Goal 13 of the 2030 Agenda requiring urgent actions to combat climate change and its impacts.

Notably, cereal polyphenols can be found associated with cancer prevention, though more studies are needed in order to receive specific health claims by FDA and EFSA. To this aim, phenolic extracts from sorghum with black pericarp were tested on the inhibition of hepatocarcinoma HepG2 cells and colorectal adenocarcinoma Caco-2 cells [9]. The cell growth inhibition by the sorghum phenolic extracts was significantly associated with their phenolic content and the inhibition appeared to be mediated by cytostatic and apoptotic mechanisms rather than cytotoxicity [9].

The quality of the majority of cereal foods depends largely on protein composition. In wheat, especially the high and low molecular weight glutenins have a great role on that. The effect of these proteins was analyzed gene-by-gene or as haplotypes providing new insights in the importance of common and rare allelic variants to improve gluten quality [10]. As a matter of fact, the biological value of cereal proteins is limited by the deficiency of essential amino acids. To counteract malnutrition due to a poorly balanced diet, it is urgent to select cereals that, besides having high yield potential, contain higher amounts of essential amino acids. Such a work was carried out in maize and is particularly valuable in view of future hybridization to produce highly nutritious maize hybrids to address malnutrition caused by a maize-based diet [11].

Finally, this special issue includes three reviews discussing the nutritional significance of wheat proteins on human health [12] and of their evolution from wheat domestication to modern cultivars [13]. The reviews particularly address common misconceptions that are associated with wheat consumption in relation to health; the latter provides a holistic view of the temporal and proteogenomic evolution of wheat from its domestication to the massively produced high-yield crop of our day [12,13].

In relation to the celiac disease and gluten sensitivity, sorghum grains can enter gluten-free diets being also interesting for their potential phenolic-induced health benefits [14]. The third review covers aspects of sorghum health benefits and explores their mechanisms of action [14].

In conclusion, this Special Issue offers an interesting contribution to the field providing readers information on multiple approaches to improve the cereal-based foods. The research articles and the reviews can be useful both for researchers and for industry operators.

In most cases, these strategies may alter the technological and sensory properties, but confer different quality characteristics that, when explained, can be accepted by consumers and producers.

Author Contributions: B.L. writing—original draft preparation, B.L. and W.W. writing—review and editing, B.L. funding acquisition. All authors have read and agreed to the published version of the manuscript.

Funding: This work was funded by PRIMA SECTION 1 2020 AGROFOOD VALUE 997 CHAIN IA TOPIC: 1.3.1-2020 (IA), MEDWHEALTH, project grant no. 2034.

Conflicts of Interest: The authors declare no conflict of interest.

References

1. Tian, W.; Tong, J.; Zhu, X.; Martin, F.P.; Li, Y.; He, Z.; Zhang, Y. Effects of Different Pilot-Scale Milling Methods on Bioactive Components and End-Use Properties of Whole Wheat Flour. *Foods* **2021**, *10*, 2857. [CrossRef] [PubMed]
2. Cammerata, A.; Laddomada, B.; Milano, F.; Camerlengo, F.; Bonarrigo, M.; Masci, S.; Sestili, F. Qualitative Characterization of Unrefined Durum Wheat Air-Classified Fractions. *Foods* **2021**, *10*, 2817. [CrossRef] [PubMed]
3. Cammerata, A.; Sestili, F.; Laddomada, B.; Aureli, G. Bran-Enriched Milled Durum Wheat Fractions Obtained Using Innovative Micronization and Air-Classification Pilot Plants. *Foods* **2021**, *10*, 1796. [CrossRef] [PubMed]
4. Cammerata, A.; Marabottini, R.; Del Frate, V.; Allevato, E.; Palombieri, S.; Sestili, F.; Stazi, S.R. Use of Air-Classification Technology to Manage Mycotoxin and Arsenic Contaminations in Durum Wheat-Derived Products. *Foods* **2022**, *11*, 304. [CrossRef] [PubMed]
5. Kothakota, A.; Pandiselvam, R.; Siliveru, K.; Pandey, J.P.; Sagarika, N.; Srinivas, C.H.S.; Kumar, A.; Singh, A.; Prakash, S.D. Modeling and Optimization of Process Parameters for Nutritional Enhancement in Enzymatic Milled Rice by Multiple Linear Regression (MLR) and Artificial Neural Network (ANN). *Foods* **2021**, *10*, 2975. [CrossRef] [PubMed]
6. Zeng, H.; Qin, L.; Liu, X.; Miao, S. Increases of Lipophilic Antioxidants and Anticancer Activity of Coix Seed Fermented by Monascus purpureus. *Foods* **2021**, *10*, 566. [CrossRef] [PubMed]
7. Živkovic, A.; Polak, T.; Cigic, B.; Požrl, T. Germinated Buckwheat: Effects of Dehulling on Phenolics Profile and Antioxidant Activity of Buckwheat Seeds. *Foods* **2021**, *10*, 740. [CrossRef] [PubMed]
8. Laddomada, B.; Blanco, A.; Mita, G.; D'Amico, L.; Singh, R.P.; Ammar, K.; Crossa, J.; Guzmán, C. Drought and Heat Stress Impacts on Phenolic Acids Accumulation in Durum Wheat Cultivars. *Foods* **2021**, *10*, 2142. [CrossRef] [PubMed]
9. Chen, X.; Shen, J.; Xu, J.; Herald, T.; Smolensky, D.; Perumal, R.; Wang, W. Sorghum Phenolic Compounds Are Associated with Cell Growth Inhibition through Cell Cycle Arrest and Apoptosis in Human Hepatocarcinoma and Colorectal Adenocarcinoma Cells. *Foods* **2021**, *10*, 993. [CrossRef]
10. Roncallo, P.F.; Guzmán, C.; Larsen, A.O.; Achilli, A.L.; Dreisigacker, S.; Molfese, E.; Astiz, V.; Echenique, V. Allelic Variation at Glutenin Loci (Glu-1, Glu-2 and Glu-3) in a Worldwide Durum Wheat Collection and Its Effect on Quality Attributes. *Foods* **2021**, *10*, 2845. [CrossRef]
11. Amegbor, I.; van Biljon, A.; Shargie, N.; Tarekegne, A.; Labuschagne, M. Identifying Quality Protein Maize Inbred Lines for Improved Nutritional Value of Maize in Southern Africa. *Foods* **2022**, *11*, 898. [CrossRef]
12. Sabença, C.; Ribeiro, M.; de Sousa, T.; Poeta, P.; Bagulho, A.S.; Igrejas, G. Wheat/Gluten-Related Disorders and Gluten-Free Diet Misconceptions: A Review. *Foods* **2021**, *10*, 1765. [CrossRef] [PubMed]
13. De Sousa, T.; Ribeiro, M.; Sabença, C.; Igrejas, G. The 10,000-Year Success Story of Wheat! *Foods* **2021**, *10*, 2124. [CrossRef] [PubMed]
14. Xu, J.; Wang, W.; Zhao, Y. Phenolic Compounds in Whole Grain Sorghum and Their Health Benefits. *Foods* **2021**, *10*, 1921. [CrossRef]

 *foods*

Article

Effects of Different Pilot-Scale Milling Methods on Bioactive Components and End-Use Properties of Whole Wheat Flour

Wenfei Tian [1,2,†], **Jingyang Tong** [1,†], **Xiaoyue Zhu** [3], **Philipp Fritschi Martin** [3], **Yonghui Li** [4], **Zhonghu He** [1,2] and **Yan Zhang** [1,*]

1   National Wheat Improvement Centre, Institute of Crop Sciences, Chinese Academy of Agricultural Sciences, Beijing 100081, China; tianwenfei@caas.cn (W.T.); 82101192196@caas.cn (J.T.); hezhonghu02@caas.cn (Z.H.)
2   International Maize and Wheat Improvement Center (CIMMYT) China Office, Chinese Academy of Agricultural Sciences, Beijing 100081, China
3   Grain Technology Center, Buhler (Wuxi) Commercial Co., LTD., Wuxi 214142, Jiangsu, China; arena.zhu@buhlergroup.com (X.Z.); philipp.fritschi@buhlergroup.com (P.F.M.)
4   Department of Grain Science and Industry, Kansas State University, Manhattan, KS 66506, USA; yonghui@ksu.edu
*   Correspondence: zhangyan07@caas.cn; Tel.: +86-10-8210-8741
†   These authors contributed equally to this work.

Abstract: The health benefits from consumption of whole wheat products are widely recognized. This study investigated the effects of different pilot-scale milling methods on physicochemical properties, bioactive components, Chinese steamed bread (CSB), and Chinese leavened pancakes (CLP) qualities of whole wheat flour (WWF). The results indicated that WWF-1 from the reconstitution of brans processed by a hammer mill had the best CSB and CLP quality overall. WWF from entire grain grinding by a jet mill (65 Hz) contained the highest concentration of bioactive components including dietary fibers (DF) and phenolic acids. A finer particle size did not necessarily result in a higher content of phenolic antioxidants in WWF. DF contents and damaged starch were negatively correlated with CSB and CLP quality. Compromised reduced quality observed in CLP made from WWF indicated its potentially higher acceptance as a whole-grain product.

Keywords: milling methods; dietary fiber; phenolic acid; steamed bread; leavened pancake

Citation: Tian, W.; Tong, J.; Zhu, X.; Martin, P.F.; Li, Y.; He, Z.; Zhang, Y. Effects of Different Pilot-Scale Milling Methods on Bioactive Components and End-Use Properties of Whole Wheat Flour. *Foods* **2021**, *10*, 2857. https://doi.org/10.3390/foods10112857

Academic Editors: Laura Gazza, Barbara Laddomada and Weiqun Wang

Received: 8 October 2021
Accepted: 5 November 2021
Published: 18 November 2021

Publisher's Note: MDPI stays neutral with regard to jurisdictional claims in published maps and institutional affiliations.

1. Introduction

Bread wheat (*Triticum aestivum* L.) is among the most widely planted cereal crops [1]. Consumption of whole-grain products has been associated with reduced risk of chronic diseases, such as cardiovascular disease, type 2 diabetes, obesity, and some types of cancer [2,3]. The health benefits of whole-grain products can be partially attributed to their contents of dietary fiber (DF) and phytochemicals, such as phenolic acids and phytosterols [4,5]. Phenolic acids have promising antioxidative, anti-inflammatory, and antimicrobial activities [6,7].The content of phenolic compounds is becoming another parameter evaluating the quality of the whole wheat flour [8,9]. Although whole-wheat flour (WWF) can provide high amounts of DF and other bioactive components, products made from it remain underutilized globally. This can be partially attributed to poorer sensory properties, a shorter shelf life, and a higher risk of rancidity [10,11]. The proper use of enzymes, emulsifiers, and hydrocolloids can improve the sensory properties of the whole wheat products [12]. However, the common whole-wheat improvers, such as diacetyl tartaric acid ester of mono- and diglycerides (DATEM) and sodium stearoyl lactylate (SSL), cannot pass the clean label test which may limit their application in food industry in the future. Different milling techniques can also influence the technical and end-use properties of WWF, possibly by modifying its particle sizes and chemical composition [13]. A good amount of study have indicated the effects of bran particle sizes on the quality of WWF [14–18]. Obviously, different bran particle sizes come from different milling methods including different types of the mills and rotor speeds. Therefore, it is necessary to develop proper milling methods to produce WWF products with enhanced sensory properties. Currently, most commercial WWF is produced using roller mills. The roller mill separates the endosperm fraction from the coarse bran and germ fraction. Coarse brans are usually micronized and then re-mixed with the endosperm for the production of WWF. Hammer mills (HM) and jet mills (JM) can be used for bran micronization [19]. WWF can also be prepared by direct pulverization of entire wheat grains using jet mills without separation and reconstitution of the components [19,20]. However, this method may result in a higher content of damaged starch. It is important to investigate the effects of this entire grain grinding method on the quality of the resulting WWF.

Chinese steamed bread (CSB) is a traditional fermented and steamed food in China with distinctive cultural features across the country [21]. It is a staple food in northern China with increasing popularity beyond [22]. There are many significant differences between CSB and western-style baked bread owing to the different production procedures. For instance, the lower processing temperature (100 °C) for CSB might permit higher preservation of diverse endogenous nutrients and reduce the production of toxic Maillard reaction products, such as acrylamide and furans [23]. Further, the general ingredients of CSB only include flour, water, and yeast. The absence of added sugar, oil, and salt renders CSB additional health benefits. Chinese leavened pancake (CLP) is another popular wheat product consumed as a staple food in some parts of China. However, previous literature on the quality of CLP made from WWF is not available. Compared to western-style whole wheat bread, factors affecting quality of whole wheat CSB and CLP are under-investigated. It was reported that finer particle sizes increased the specific volume and improved the crumb texture of CSB [15]. Bran dietary fibers and phenolics also influenced the CSB quality. As for the effect of milling process, to our knowledge, there was only one study whole results suggested that CSB from bran reconstitution had larger loaf volume, but the entire grain grinding improved the color of the CSB [19]. More studies are needed to gain further understanding on the quality of CSB and CLP. Furthermore, the effects of different milling methods on the bioactive components, such as dietary fiber and phenolic acids of WWF, are also not thoroughly investigated. Due to their potential health benefits [24,25], the content of total phenolics is becoming another factor determining wheat market preference [26]. In addition, previous studies were conducted using lab-scale mills. Information at pilot-scale level is very limited. Therefore, the objective of this study was to systematically investigate the effects of different pilot-scale milling methods on bioactive components (including dietary fiber and phenolic acids), physicochemical properties of WWF, as well as its effect on CSB and CLP quality. The outcome of the study provides useful information for the quality improvement of WWF products and contributes to the facilitation of high-DF food consumption in enhancing health benefits for consumers.

2. Materials and Methods

2.1. Wheat Grains and Chemicals

A sound grain mixture (1:1) of wheat cultivars Zhongmai 578 (1000 kg) and Zhongmai 175 (1000 kg) was used for preparation of WWF. Zhongmai 578 (farinograph stability time = 23.6 min, wet gluten = 34.2%) is an important cultivar in the Yellow-Huai river winter wheat region of China. Zhongmai 175 (farinograph stability time = 2.0 min, wet gluten = 27.4%) is the most widely grown cultivar in the northern China plain winter wheat region. General chemicals, HPLC grade water, HPLC grade acetonitrile, and analytical standards (including vanillic acid, syringic acid, *para*-coumaric acid, *trans*-ferulic acid, and sinapic acid) were purchased from Aladdin Biochemical Technology Co., Ltd. (Shanghai, China). The instant dry yeast was purchased from Lesaffre (Marcq-en-Barœul, France). The colza oil was purchased from Luhua Group Co., Ltd. (Yantai, China).

2.2. Pilot-Scale Milling Methods for Producing Whole Wheat Flour

After cleaning and tempering to 16.0% moisture for 28 h, the mixed grain sample of 2000 kg was milled on a Buhler roller mill located at Buhler Commercial Co. Ltd. (Wuxi, China). The mill was equipped with four break sections (B1, B2, B3, B4) and four reduction sections (R1, R2, R3, R4). During the process, a portion of crushed wheat from B1 stream (200 kg) were taken out, divided equally into two parts and then directly pulverized by jet mill (AHFL, Buhler Commercial Co. Ltd., Wuxi, China) with the rotor speeds of 45 Hz (JM-45 Hz) and 65 Hz (JM-65 Hz). The WWF from JM-45 Hz and JM-65 Hz were designated as WWF-4 and WWF-5, respectively. The other grains of 1800 kg went through the whole roller mill process and the yield of the refined flour was 72%. The bran and germ fractions from the roller mill were collected, equally divided into three parts, and then micronized using a hammer mill (AHZCφ1 mm, Buhler Commercial Co. Ltd., Wuxi, China) and a jet mill (AHFL, Buhler Commercial Co. Ltd., Wuxi, China) with the rotor speeds of 45 Hz (JM-45 Hz) and 65 Hz (JM-65 Hz), respectively. The micronized wheat bran was remixed with the refined flour at a ratio of 28:72 (bran of 28 kg, refined flour of 72 kg) to obtain WWF samples designated as WWF-1 (bran micronized by HM), WWF-2 (bran micronized by JM45-Hz), and WWF-3 (bran micronized by JM-65 Hz). For each WWF, two batches were prepared. (Supplementary Document, Table S1).

2.3. Analysis of Particle Size Distribution

Particle size distribution was determined using an on-line particle analyzer (MYTA, Buhler, Switzerland) equipped with a dry sample delivery and measurement system that combined laser diffraction and image processing. The system was equipped with a series of sieves with particle sizes ranging from 10 to 5000 μm. During the measurement, the percentage of flours passing through the certain sieve size was determined. Mean particle sizes (MPS) were determined as:

MPS = $\sum$ sieve average particle size (μm) × percentage of particles that pass through the sieve (Through X%). Original data are provided in the Supplementary Documents.

2.4. Analysis of Chemical Composition

Starch damage was determined using a rapid method with a SDmatic analyzer (Chopin Technologies, Paris, France) according to AACC International Approved Method 76-33.01 [27]. The method measured the amount of iodine absorbed by the tested sample in a liquid suspension. Higher content of damaged starch led to more fixed iodine. Protein content was recorded with a Foss-Tecator 1241 (Foss, Högänas, Sweden) NIR analyzer. Contents of soluble dietary fiber (SDF) and insoluble dietary fiber (IDF) were analyzed according to the Chinese national standard method GB 5009.88-2014 [28] which was very similar to the AACC approved method 32-07.01 [29]. Briefly, the flour was treated by sequentialenzymatic hydrolysis using heat-stable alpha-amylase, protease, and amyloglucosidase. IDF was filtered, and the residue was washed with warm distilled water and weighed. For SDF, the filtrate and water washing from the above step were precipitated with 95% ethanol, and the residue was then filtered, dried, and weighed. Determination of fatty acids contents followed Chinese national standard GB/T 5510-2011 [30]. Briefly, the free fatty acids were extracted with benzene and titrated with KOH solution.

2.5. Phenolic Acid Composition of Whole Wheat Flour

The phenolic acids in WWF were extracted according to a method reported previously [31] with some modifications. Briefly, WWF (1 g) was hydrolyzed with 2 M of NaOH (10 mL) for 4 h under nitrogen protection in darkness. The mixture was then acidified to pH 1 using 6 M of HCl and extracted three times with ethyl acetate. The combined extract was evaporated to dryness and re-dissolved in HPLC grade methanol (2 mL). The final extract was filtered through a 0.22-μm filter and analyzed within 12 h.

A UPLC-PDA system from Waters Corporation (Milford, MA, USA) equipped with an ACQUITY UPLC BEH C18 (2.1 mm × 50 mm) column was used to analyze phenolic acids following a previously reported gradient protocol [32] with some modifications. Briefly, mobile phase A comprised HPLC-grade water containing 0.1% trifluoroacetic acid (TFA), and mobile phase B was HPLC-grade acetonitrile with 0.1% TFA. The PDA detector was set to record the spectra information from 210 nm to 400 nm and also the absorption at 280 nm. The flow rate of the mobile phase was kept at 0.4 mL/min and the percentage of mobile phase B was kept at 6% from 0–1.0 min, which then increased linearly to 14% from 1.0 min to 3.0 min and increased linearly to 18% from 3.0 min to 5.5 min. The column was re-equilibrated for 2 min between each injection. Phenolic acids were identified by comparison with the retention time of analytical standards and quantified using external calibration curves using the absorbance at 280 nm. *Cis*-ferulic acid was identified according to a previous study [31] and tentatively quantified according to the calibration of *trans*-ferulic acid since the analytical standard of *cis*-ferulic acid was not available.

2.6. Analysis of Dough Properties

Dough rheological properties were determined by Farinograph and Extensograph (Brabender, Duisburg, Germany), following the AACC International Approved Methods 54-21.02 [33] and 54-10.01 [34], respectively. Water absorption (%), dough development time (min), and stability (min) were measured by Farinograph. Dough energy (cm^2), extensibility (mm), and maximum resistance (BU) were determined by Extensograph.

2.7. Preparation and Sensory Evaluation of Chinese Steamed Bread

CSB was prepared and evaluated according to a previous study [35], with minor modifications. In brief, WWF (100 g) was mixed with yeast (2 g) and water for two minutes in a National mixer (National MFG. Co, Lincoln, NL, USA). The mixer speed was set at 90 rpm and the water addition was set at 80% of optimal water absorption (%) pre-determined by Farinograph analysis in preliminary tests. For example, if the optimal water absorption was determined as 70% by the Farinograph, we would use 56% as the optimal water level for CSB making. Fairnograph is the technique developed for western-style bread making use of refined flours. The dough was then sheeted by passing through a pair of rollers for 10 times. After each pass, the sheeted dough was folded along the side and rotated through 90°. The dough piece was then rounded five times with a suitably sized bowl. The rounded dough was proofed for 20 min in a fermentation cabinet (35 °C, 85% RH) and steamed for 25 min in a steamer initially containing cold water.

The CSB score was a weighted value based on specific volume (20), skin color (10), smoothness (10), shape (10), structure (15), and stress relaxation (35), according to the method proposed previously. Specific volume, skin color, and stress relaxation were based on objective measurements with sufficient details provided in the Supplementary Documents. Specific volume indicated the loaf volume and weight ratio of CSB. Loaf volume was determined by the rapeseed displacement method, according to AACC Approved Method 10-05.01. Skin color was measured with a Minolta CR 310 chromameter (Minolta Camera, Osaka, Japan). Skin color score was calculated according to the method reported previously [36]. Stress relaxation was measured with a TA-XT2i texture analyzer (Stable Micro Systems, Surrey, Godalming, UK). Briefly, after 15 min of steaming, two slices (3 cm thick) were cut from the center of the CSB. Each slice was placed on a flat metal plate and compressed twice to 50% of its original thickness at a speed of 1.0 mm/s with a cylindrical P35 probe. The calculation and scoring of stress relaxation was conducted according to a previous report [36]. The smoothness, shape, and structure of steamed bread were scored subjectively by five trained panelists. A high score for smoothness was given for very smooth skin; freedom of wrinkles, dimples, blisters, or gelatinized spots; and a round shape and fine crumb structure combined with uniform porosity contributed to high scores for shape and structure, respectively.

2.8. Preparation and Sensory Evaluation of Chinese Leavened Pancakes

To prepare CLP, WWF (100 g) was mixed with 120 mL of water (35 °C), 5 g sugar, 1 g of salt, 2 g of yeast, and 16 g of colza oil in a National mixer for 1.5 min (National MFG. Co, Lincoln, NE, USA). The mixer speed was set at 90 rpm. The dough was proofed for 1 h in a fermentation cabinet (35 °C, 85% RH). The fermented dough was gently kneaded by hand to form a rounded piece with a smooth upper surface. The rounded piece was flattened using a rolling pin to produce a 1-cm-thick dough sheet and then rested for 15 min at room temperature. The dough sheet was finally cooked for 8 min in an electric baking pan (MC-JS 3406, Midea Life Appliance Manufacturing Co., Ltd., Foshan, China). To our knowledge, there is not an international method on sensory evaluation of CLP. In this study, the weighted CLP score included appearance (20), inner structure (20), stress relaxation (35), stickiness (15), and taste and flavor (10). Stress relaxation was measured with a TA-XT2i texture analyzer (Stable Micro Systems, Surrey, Godalming, UK) according to the same method as CSB. Appearance, inner structure, stickiness, and taste and flavor of CLP were scored subjectively by five trained panelists. A high score for appearance was given to pale yellow skin, round shape, and freedom of large black spots. A fine crumb with uniform porosity and non-sticking contributed to high scores for inner structure and stickiness, respectively. Freedom from bad smell and a rich wheat flavor related to high scores for taste and flavor.

2.9. Short-Term Storage Quality of Chinese Steamed Bread

Texture profiles of CSB from WWF were determined after storage for 6, 12, 24, 48, and 72 h at room temperature using a TA-XT2i texture analyzer. Briefly, two slices (3 cm thick) were cut from the center of the CSB. Each slice was placed on a flat metal plate and compressed twice to 50% of its original thickness at a speed of 1.0 mm/s with a cylindrical P35 probe. Firmness and resilience were recorded.

2.10. Statistical Analyses

For chemical analyses, such as determination of phenolic acid composition, the results were reported as mean values from three replicates. For dough property parameters, the results were reported as mean values from two replicates. Evaluation of CSB and CLP were conducted by five trained panelists and reported as mean scores from the five panelists. The results were reported as mean $\pm$ standard deviation (SD). Results were subjected to one-way analysis of variance (AVONA) and Turkey's test using SAS software, version 9.3 (Cary, NC, USA). $p < 0.05$ was considered as significantly different.

3. Results and Discussions

The quality of wheat flour is evaluated according to their chemical composition including protein content, damaged starch, and bioactive components (such as phenolic acids and dietary fibers). Dough rheological properties can be evaluated with a farinograph and an extensograph. Particle size is an important factor affecting the end-use properties of whole wheat flour.

3.1. Particle Sizes and Chemical Compositions of Whole Wheat Flours

Particle size distribution is an important parameter affecting end-use quality of whole wheat flour [17]. Particle size distributions of WWFs from different milling process are plotted in Figure S1 in the Supplementary Documents. Mean particle sizes (MPS) for WWF-5 were 236 μm, 146 μm, 124 μm, 191 μm and 146 μm, respectively. The jet milling was more effective than the hammer milling in reducing bran particle size. Chemical compositions of the WWFs are presented in Table 1. WWF-3 and WWF-5 from the JM-65 Hz contained higher contents of total dietary fibers (TDF) and damaged starch than the other WWFs. The higher content of damaged starch can be attributed to the higher rotor speed. Interestingly, there was a strong positive correlation between the content of damaged starch content and total DF ($R^2 = 0.7974$). WWF-5 had a significantly lower

($p < 0.05$) protein content than other WWFs. WWF-3 contained the highest content of fatty acids, a factor that can lead to rancidity of food products during storage [37].

Table 1. Chemical composition of WWFs obtained from different milling methods.

Sample	SDF (g/100 g)	IDF (g/100 g)	TDF (g/100 g)	Fatty Acids (mg/100 g)	Damaged Starch (%)	Protein (%)
WWF-1	7.35 ± 0.16c	1.50 ± 0.24b	8.85 ± 0.07c	45.3 ± 2.26c	4.12 ± 0.06e	14.6 ± 0.06bc
WWF-2	7.27 ± 0.02c	1.73 ± 0.02ab	9.00 ± 0.05c	53.9 ± 2.61b	5.06 ± 0.01c	15.0 ± 0.04a
WWF-3	8.64 ± 0.30b	2.12 ± 0.12a	10.76 ± 0.17b	64.4 ± 2.85a	5.71 ± 0.07b	14.8 ± 0.14ab
WWF-4	6.18 ± 0.28d	1.28 ± 0.14b	7.46 ± 0.27d	42.1 ± 2.10c	4.56 ± 0.08d	14.1 ± 0.11d
WWF-5	10.24 ± 0.40a	2.03 ± 0.05a	12.27 ± 0.45a	53.0 ± 1.98b	6.35 ± 0.01a	13.5 ± 0.07e

The results were reported as mean values from three replicates. Within each column, values followed by different letters are significantly different ($p < 0.05$). WWF: whole-wheat flour; SDF: soluble dietary fiber; IDF: insoluble dietary fiber; TDF: total dietary fiber. WWF-1: total reconstitution of brans grinded by hammer mill; WWF-2: total reconstitution of brans grinded by jet mill at 45 Hz. WWF-3: total reconstitution of brans grinded by jet mill at 65 Hz; WWF-4: entire grains grinded by jet mill at 45 Hz; WWF-5: entire grains grinded by jet mill at 65 Hz.

3.2. Phenolic Acid Compositions of Whole Wheat Flour

Phenolic acids are a major type of bioactive phytochemical in whole wheat [38,39]. Antioxidant and anti-inflammatory activities of phenolic acid are widely recognized [40,41]. Vanillic acid, syringic acid, *para*-coumaric acid, *trans*-ferulic acid, sinapic acid, and *cis*-ferulic acid were isolated and quantified. A typical UPLC spectra was presented in Figure 1. The retention time for vanillic acid, syringic acid, *para*-coumaric acid, *trans*-ferulic acid, sinapic acid, and *cis*-ferulic acid were 1.60, 1.84, 2.20, 2.95, 3.20, and 3.55, respectively. *Trans*-ferulic acid was the predominant phenolic acid (Table 2).

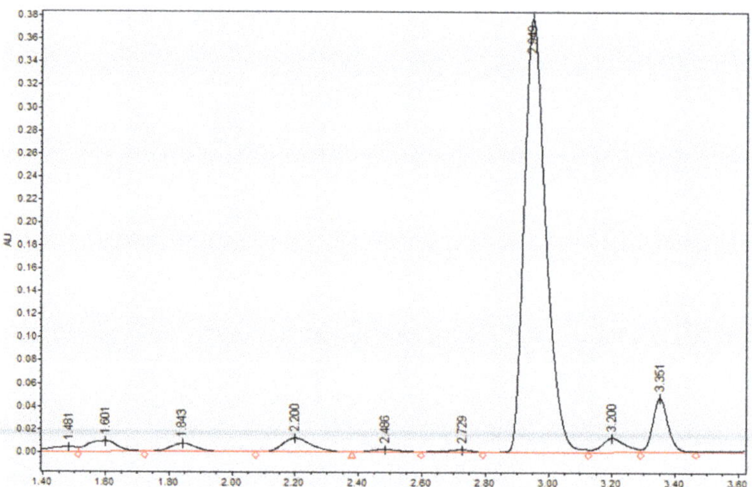

Figure 1. A typical UPLC spectra of the tested sample at 280 nm.

WWF-1 and WWF-2 contained similar concentrations of *trans*-ferulic acids (720.16 and 742.15 µg/g, respectively), but their particle sizes were significantly different ($p < 0.05$). Although WWF-3 and WWF-5 were both processed by the JM-65 Hz mill, they had comparable particle size distributions and dietary fiber contents; WWF-3 contained the lowest (534.90 µg/g) and WWF-5 contained the highest (1002.11 µg/g) concentrations of *trans*-ferulic acid. The high rotor speed of the JM-65 Hz on wheat bran in the bran reconstitution method (WWF-3) led to significant loss of *trans*-ferulic acid during processing. This suggests that, in the entire grain grinding method using the JM-65 Hz mill (WWF-5), the starchy endosperm physically prevents the loss of ferulic acid during reduction in particle size. As a result, WWF-5 contained the highest concentration of *trans*-ferulic acid. Similarly, WWF-5 also contained a higher content of vanillic acid, syringic acid, and *para*-coumaric acid than WWF-4. This explanation is partially supported by the fact that WWF-4 prepared by entire grain grinding using the JM 45 Hz mill had a significantly lower ($p < 0.05$) concentration of *trans*-ferulic acid than WWF-5. This was probably due to smaller particle size in WWF-5 that enhanced the extractability of phenolic acids as generally reductions in particle size enhance the extractability of phenolic acids.

Table 2. Phenolic acid composition of WWFs obtained by different milling methods.

Sample	Vanillic Acid (μg/g)	Syringic Acid (μg/g)	*para*-Coumaric Acid (μg/g)	*trans*-Ferulic Acid (μg/g)	Sinapic Acid (μg/g)	*cis*-Ferulic Acid (μg/g)
WWF-1	29.57 ± 0.80b	17.20 ± 0.68b	18.50 ± 0.81ab	720.16 ± 4.58bc	77.47 ± 1.22abc	75.08 ± 0.63b
WWF-2	31.22 ± 0.83ab	16.76 ± 0.87bc	17.82 ± 0.89ab	742.15 ± 14.62b	80.97 ± 2.10a	73.48 ± 1.40b
WWF-3	30.18 ± 1.54b	14.47 ± 0.53c	13.30 ± 0.65c	534.90 ± 6.83d	78.49 ± 2.18ab	50.81 ± 1.41c
WWF-4	27.72 ± 1.43b	14.75 ± 0.99bc	15.81 ± 0.57bc	647.33 ± 23.64c	70.65 ± 2.94c	65.77 ± 1.81b
WWF-5	34.35 ± 3.97a	19.64 ± 2.46a	19.87 ± 4.37a	1002.11 ± 101.04a	73.57 ± 6.81bc	111.25 ± 16.03a

The results were reported as mean values from three replicates. Within each column, values followed by different letters are significantly different ($p < 0.05$). WWF: whole-wheat flour. WWF-1: total reconstitution of brans grinded by hammer mill; WWF-2: total reconstitution of brans grinded by jet mill at 45 Hz. WWF-3: total reconstitution of brans grinded by jet mill at 65 Hz; WWF-4: entire grains grinded by jet mill at 45 Hz; WWF-5: entire grains grinded by jet mill at 65 Hz.

Previous studies reported that reductions in bran particle size led to enhanced antioxidant capacity of wheat samples [42,43]. However, those results were obtained by in vitro methods, such as total phenolic content (TPC) and 2,2'-azino-bis(3-ethylbenzothiazoline-6-sulfonic acid) (ABTS) assays. The non-specificity and limitations of those methods, which could lead to incorrect data interpretations, were recognized in recent years [44,45]. Our analysis using a UPLC instrument on phenolic acids, i.e., the major antioxidants in WWF, indicated that a reduction in particle size does not necessarily lead to enhanced phenolic acid concentration. Different milling strategies, types of mills, and rotor speeds can produce WWFs with similar particles size but significantly different phenolic acid compositions. In summary, the milling method had a great impact on phenolic acid compositions of WWF and improved its availability with grain grinding using a JM-65 Hz procedure. To our knowledge, this is the first study reporting the high variation of total dietary fibers (1.6-fold) and ferulic acid content (1.9-fold) of WWF from different milling methods.

3.3. Rheological Properties of Doughs Made from Whole Wheat Flour

The effects of different milling methods on dough properties of WWF are shown in Table 3. WWF-5 had the highest water absorption (73.0%) and WWF-4 had the lowest (64.8%). This could be explained by the high DF and damaged starch contents in WWF-5, leading to higher water affinity and hydration properties [46–48]. Dough from WWF-4 exhibited significantly shorter ($p < 0.05$) development time than other WWF doughs. Generally, moderate development time is considered a favorable dough property. Stability time is an essential factor in evaluating dough properties. Dough from WWF-1 had the longest stability time. This can be attributed to its lowest contents of DF and damaged starch. Dough from WWF-5 had the shortest stability time due to the high content of damaged starch and DF that can lead to the formation of a weakened gluten network [49]. WWF-4 and WWF-5 from entire grain milling involved significantly less energy than the recombined WWF–3. Extensibility is another important parameter for the preparation of western-style bread. Higher extensibility is considered as a favorable dough property. Generally, entire grain milling also had negative effects on extensibility of WWF. To conclude, different milling methods had significant effects on dough rheological properties of WWF. WWF from bran reconstitution, especially WWF-1 from HM, exhibited better dough properties than WWF from entire grain grinding.

Table 3. Dough rheological properties of WWF obtained by different milling methods.

Sample	Water Absorption (%)	Development time (min)	Stability (min)	Energy (cm^2)	Extensibility (mm)	Maximum Resistance (BU)
WWF-1	68.4 ± 0.1c	4.7 ± 0.0b	6.0 ± 0.1a	47.0 ± 0.3b	132.9 ± 0.5b	229.8 ± 1.8b
WWF-2	69.8 ± 0.1b	4.9 ± 0.0ab	4.7 ± 0.0b	52.8 ± 0.2a	140.0 ± 0.8a	249.0 ± 3.9a
WWF-3	68.0 ± 0.1c	5.1 ± 0.1a	4.8 ± 0.1b	49.0 ± 0.3b	142.1 ± 0.9a	227.6 ± 2.3b
WWF-4	64.8 ± 0.3d	3.8 ± 0.1c	3.8 ± 0.1c	39.5 ± 0.3c	133.6 ± 0.5b	193.0 ± 1.9c
WWF-5	73.0 ± 0.3a	4.8 ± 0.1b	2.1 ± 0.0d	38.3 ± 0.3c	105.9 ± 0.4c	243.8 ± 3.1a

The results were reported as mean values from two replicates. Within each column, values followed by different letters are significantly different ($p < 0.05$). WWF: whole-wheat flour. WWF-1: total reconstitution of brans grinded by hammer mill; WWF-2: total reconstitution of brans grinded by jet mill at 45 Hz. WWF-3: total reconstitution of brans grinded by jet mill at 65 Hz; WWF-4: entire grains grinded by jet mill at 45 Hz; WWF-5: entire grains grinded by jet mill at 65 Hz.

3.4. Sensory Properties of Chinese Steamed Bread and Chinese Leavened Pancakes

The CSB quality evaluations from the different WWFs presented in Table 4 and illustrated in Figure 2A,B showed that the CSB from WWF-4 had the highest total score (59), although that score was not significantly higher from the CSB score from WWF-1 (57) and WWF-2 (58). WWF-4 exhibited inferior dough properties compared to WWF-1 and WWF-2 in terms of stability, energy, and extensibility, suggesting that inferior dough properties may not affect the quality of final products. WWF-3 showed a significantly higher score than WWF-5, probably attributable to the higher damaged starch and DF contents of WWF-5 interfering with the formation of the gluten network [14,50]. Compared to CSB from WWF-1 and WWF-4, CSB from WWF-2, WWF-3, and WWF-5 received higher scores in smoothness but lower scores in specific volume. This could be due to the relatively smaller particle sizes of WWF-2, WWF-3, and WWF-5 [19]. All CSBs from WWF obtained a 0 score color as the white color is currently an essential factor that influences acceptance by consumers [21].

The scores were reported as mean values from five panelists. Within each column, values followed by different letters are significantly different ($p < 0.05$).

Sensory qualities of CLP prepared from the different flours are summarized in Table 5, and Figure 2C,D illustrates the appearance of CLP and its internal structures, respectively. CLP from WWF-1 received the highest total score (70) followed by CLP from WWF-4 (66) and WWF-2 (63). All CLP showed flat and tidy surfaces, except for that from WWF-5, which showed broken edges and very low specific volume. This was likely caused by higher DF and damaged contents [14,18].

Table 4. Sensory evaluation for Chinese steamed bread produced from WWF obtained by different milling methods.

Sample	Specific Volume (20)	Stress Relaxation (35)	Skin Color (10)	Smoothness (10)	Shape (10)	Structure (15)	Total Score (100)
WWF-1	14a	21a	0a	5d	8a	9b	57a
WWF-2	14a	19b	0a	7b	7b	11a	57a
WWF-3	9b	19b	0a	8a	7b	9b	52b
WWF-4	14a	21a	0a	6c	7b	11a	59a
WWF-5	9b	13c	0a	7b	5c	8c	42c
RF	20	33	9	9	8	14	93

The scores were reported as mean values from five panelists. Within each column, values followed by different letters are significantly different ($p < 0.05$). WWF: whole-wheat flour, RF: refined flour. WWF-1: total reconstitution of brans grinded by hammer mill; WWF-2: total reconstitution of brans grinded by jet mill at 45 Hz. WWF-3: total reconstitution of brans grinded by jet mill at 65 Hz; WWF-4: entire grains grinded by jet mill at 45 Hz; WWF-5: entire grains grinded by jet mill at 65 Hz.

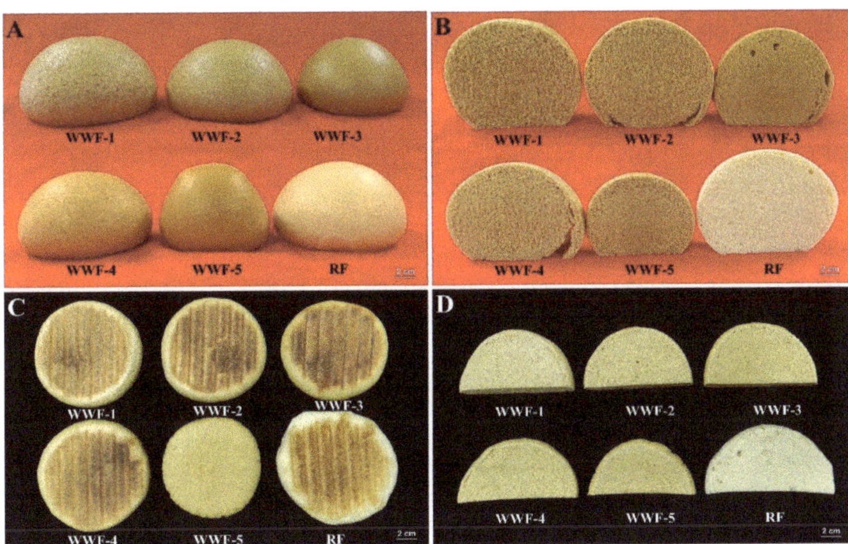

Figure 2. Appearance of CSB (**A**), internal structures of CSB (**B**), appearance of CLP (**C**) and internal structures of CLP (**D**) produced from different whole wheat flours. CSB: Chinese steamed bread; CLP: Chinese leavened pancakes. WWF-1: total reconstitution of brans grinded by hammer mill; WWF-2: total reconstitution of brans grinded by jet mill at 45 Hz. WWF-3: total reconstitution of brans grinded by jet mill at 65 Hz; WWF-4: entire grains grinded by jet mill at 45 Hz; WWF-5: entire grains grinded by jet mill at 65 Hz.

Table 5. Sensory evaluation of Chinese leavened pancakes produced from WWF obtained by different milling methods.

Sample	Appearance (20)	Stress Relaxation (35)	Structure (20)	Stickiness (15)	Taste and Flavor (10)	Total Score (100)
WWF-1	16a	20c	16a	13a	5b	70a
WWF-2	14b	20c	14b	11bc	4c	63b
WWF-3	12c	20c	12c	8d	3d	55c
WWF-4	16a	20c	14b	10c	6a	66b
WWF-5	6d	20c	10d	7d	2e	45d
RF	18	32	16	12	9	87

The scores were reported as mean values from five panelists. Within each column, values followed by different letters are significantly different ($p < 0.05$). WWF: whole-wheat flour, RF: refined flour. WWF-1: total reconstitution of brans grinded by hammer mill; WWF-2: total reconstitution of brans grinded by jet mill at 45 Hz. WWF-3: total reconstitution of brans grinded by jet mill at 65 Hz; WWF-4: entire grains grinded by jet mill at 45 Hz; WWF-5: entire grains grinded by jet mill at 65 Hz.

CSB made from refined flour received a score of 93 which was 34 points higher than CSB from WWF-4 with a score of 59. In contrast, CLP from refined flour received a score of 87 which was only 17 points higher than CLP from WWF-1 with a score of 70. This result indicated a potentially higher consumer acceptance for whole-grain CLP products. Whole-grain CLP can be a promising contributor for increased whole-grain consumption in the future due to its favorable sensory properties.

3.5. Correlation of Sensory Quality and Chemical Composition

The general effects of damaged starch and DFs on sensory properties of whole-grain products have been extensively documented. However, to our knowledge, correlations between sensory properties and damaged starch and DF have not been widely investigated, possibly due to the lack of a comprehensive scoring system for evaluation of sensory properties. In this study, total scores for sensory evaluation as well as DF and damaged starch contents were obtained, thus paving the way for correlation analysis between certain components and end-use quality (Supplementary Document, Figure S2). Negative correlations were found between (a) the CSB score and damaged starch (R^2 = 0.7553); (b) the CSB score and total dietary fiber (R^2 = 0.8931); (c) the CLP score and damaged starch (R^2 = 0.9623); and (d) the CLP score and total dietary fiber (R^2 = 0.8375). These correlations clearly demonstrate that end-use quality of WWFs can be enhanced by reduced damaged starch content.

3.6. Storage Quality of CSB

Changes on texture profiles, including hardness and resilience, were measured at 6, 12, 24, 48, and 72 h storage at room temperature to assess the storage properties of CSB (Supplementary Document, Figure S3). A lower value of hardness is considered as a favorable property. At 0 h, CSB from WWF-1 had the lowest hardness (4240.5 g) while CSB from WWF-5 had the higher hardness (11,040.8 g). After 72 h of storage, the hardness of CSB from WWF-1 and WWF-5 increased to 10,519.7 g and 17,068.0 g, respectively. Generally, as shown in Figure S3A, the final hardness (72 h) of all samples showed a consistent relationship with initial hardness (0 h) as CSB from WWF-3 and WWF-5 had both high initial and final hardness, whereas that from WWF-1 had both low initial and final hardness. Compared to WWF-1 (236 μm) and WWF-4 (191 μm), WWFs with finer particle sizes, i.e., WWF-3 (124 μm) and WWF-5 (146 μm), led to the CSB products with higher hardness values. The increase in hardness during storage is probably due to recrystallization of amorphously melted starch [49]. The lower hardness of WWF-1 might result from the coarse bran particles slowing down the formation of double helix starch and retrogradation [51].

Changes in resilience are presented in Figure S3B. Resilience reflects how well a CSB slice can restore to the original height after application of pressure. A higher resilience value is generally considered preferable. CSB from WWF-1 had the highest initial resilience of 0.44, followed by CSB from WWF-4 (0.41). WWF-3 and WWF-5 had lower initial resilience, at 0.35 and 0.34, respectively. After 72 h, WWF-3 and WWF-5 maintained relatively higher resilience, at 0.24 and 0.23, respectively, while the resilience of WWF-1 and WWF-4 decreased to 0.19 and 0.19, respectively. This observation was possibly due to the fact that the degree of change in resilience is positively related to the original CSB volume [52]. Therefore, the resilience of CSB from WWF-1, WWF-2, and WWF-4 with larger original volumes decreased more dramatically than that from WWF-3 and WWF-5, both of which contained slightly higher contents of soluble dietary fiber (SDF) than other WWFs (Table 1). Rezaei et al. [53] reported that a moderately enhanced content of SDF improves the elasticity of the gluten–starch matrix and reduces starch retrogradation. The resilience of CSB from WWF-1 was comparable to that of CSB from WWF-3 and WWF-5, especially during storage of 0–48 h. By combing the data of hardness and resilience, it can be concluded that CSB from WWF-1 processed by hammer mill had the best overall storage quality over 0 to 72 h.

4. Conclusions

This study used five different pilot-scale milling methods to prepare whole-wheat flour.

The different milling methods had significant effects on physiochemical properties, bioactive components, and end-use properties of the WWFs. Particle size distribution was important for the quality of CSB and CLP. The content of damaged starch and dietary fiber was negatively correlated with an evaluation score of CSB and CLP. WWF-1 constituted from total bran reconstitution processed by a hammer mill exhibited best end-use properties, especially for Chinese leavened pancakes. WWF-5 from the entire grain ground by JM-65 Hz had the highest content of bioactive dietary fiber and phenolic acids, rendering its superiority in nutraceutical values. CSB from WWF-1 by total bran reconstitution using the hammer mill had the best short-term storage property. In general, WWF with better end-use properties can be obtained by the milling processes, resulting in appropriate particle size distributions and a low content of damaged starch. Although DF and phenolic acids had negative effects on sensory property of WWF, their health-promoting effects must be recognized. Different milling methods can result in WWF with significantly different contents of phenolic acids and dietary fibers. Further studies will be worthwhile for developing novel techniques that produce whole-wheat products with both high consumer acceptance and superior nutraceutical value.

Supplementary Materials: The following are available online at https://www.mdpi.com/2304-8158/10/11/2857/s1, Table S1. Different milling methods used for preparing WWF, Figure S1. Particle size distributions of WWFs, Figure S2. Correlation analyses of sensory quality and chemical composition, Figure S3. Storage quality of CSB.

Author Contributions: W.T.: conceptualization, methodology, formal analysis, investigation, writing the original drafts, writing, review & editing; J.T.: sample preparation, formal analysis, writing the original draft; X.Z.: investigation, sample preparation; P.F.M.: investigation, methodology; Y.L.: writing, review & editing; Z.H.: writing, review & editing, funding acquisition; Y.Z.: conceptualization, methodology, formal analysis, investigation, resources, writing, review & editing, project administration. All authors have read and agreed to the published version of the manuscript.

Funding: This research was financially supported by the National Key Research and Development Program of China (2020YFE0202300) and the Agricultural Science and Technology Innovation Program of CAAS (CAAS-XTCX20190025-2 and CAAS-XTCX2018020-7).

Institutional Review Board Statement: The study was conducted according to the guidelines of the Declaration of Helsinki, and approved by the Institutional Review Board of Institute of Crop Sciences, Chinese Academy of Agricultural Sciences.

Informed Consent Statement: Informed consent was obtained from all subjects involved in the study.

Data Availability Statement: The datasets generated for this study are available on request to the corresponding author.

Acknowledgments: We are grateful to R.A. McIntosh, Plant Breeding Institute, the University of Sydney, for reviewing the manuscript.

Conflicts of Interest: The authors declare no competing financial interests.

References

1. Bakke, A.; Vickers, Z. Consumer Liking of Refined and Whole Wheat Breads. *J. Food Sci.* **2007**, *72*, S473–S480. [CrossRef] [PubMed]
2. Jonnalagadda, S.S.; Harnack, L.; Hai Liu, R.; McKeown, N.; Seal, C.; Liu, S.; Fahey, G.C. Putting the Whole Grain Puzzle Together: Health Benefits Associated with Whole Grains—Summary of American Society for Nutrition 2010 Satellite Symposium. *J. Nutr.* **2011**, *141*, 1011S–1022S. [CrossRef] [PubMed]
3. Sang, S.; Idehen, E.; Zhao, Y.; Chu, Y. Emerging Science on Whole Grain Intake and Inflammation. *Nutr. Rev.* **2020**, *78*, 21–28. [CrossRef]
4. Okarter, N.; Liu, C.-S.; Sorrells, M.E.; Liu, R.H. Phytochemical Content and Antioxidant Activity of Six Diverse Varieties of Whole Wheat. *Food Chem.* **2010**, *119*, 249–257. [CrossRef]
5. Tian, W.; Hu, R.; Chen, G.; Zhang, Y.; Wang, W.; Li, Y. Potential Bioaccessibility of Phenolic Acids in Whole Wheat Products during In Vitro Gastrointestinal Digestion and Probiotic Fermentation. *Food Chem.* **2021**, *362*, 130135. [CrossRef] [PubMed]
6. Călinoiu, L.F.; Vodnar, D.C. Whole Grains and Phenolic Acids: A Review on Bioactivity, Functionality, Health Benefits and Bioavailability. *Nutrients* **2018**, *10*, 1615. [CrossRef]
7. Călinoiu, L.F.; Cătoi, A.-F.; Vodnar, D.C. Solid-State Yeast Fermented Wheat and Oat Bran as A Route for Delivery of Antioxidants. *Antioxidants* **2019**, *8*, 372. [CrossRef] [PubMed]
8. Tian, W.; Chen, G.; Gui, Y.; Zhang, G.; Li, Y. Rapid Quantification of Total Phenolics and Ferulic Acid in Whole Wheat Using UV–Vis Spectrophotometry. *Food Control* **2021**, *123*, 107691. [CrossRef]
9. Shewry, P.R.; Charmet, G.; Branlard, G.; Lafiandra, D.; Gergely, S.; Salgó, A.; Saulnier, L.; Bedő, Z.; Mills, E.N.C.; Ward, J.L. Developing New Types of Wheat with Enhanced Health Benefits. *Trends Food Sci. Technol.* **2012**, *25*, 70–77. [CrossRef]
10. Heiniö, R.L.; Noort, M.W.J.; Katina, K.; Alam, S.A.; Sozer, N.; de Kock, H.L.; Hersleth, M.; Poutanen, K. Sensory Characteristics of Wholegrain and Bran-Rich Cereal Foods—A Review. *Trends Food Sci. Technol.* **2016**, *47*, 25–38. [CrossRef]

11. Li, M.; Ho, K.K.H.Y.; Hayes, M.; Ferruzzi, M.G. The Roles of Food Processing in Translation of Dietary Guidance for Whole Grains, Fruits, and Vegetables. *Annu. Rev. Food Sci. Technol.* **2019**, *10*, 569–596. [CrossRef]
12. Tebben, L.; Shen, Y.; Li, Y. Improvers and Functional Ingredients in Whole Wheat Bread: A Review of Their Effects on Dough Properties and Bread Quality. *Trends Food Sci. Technol.* **2018**, *81*, 10–24. [CrossRef]
13. Gómez, M.; Gutkoski, L.C.; Bravo-Núñez, Á. Understanding Whole-Wheat Flour and Its Effect in Breads: A Review. *Compr. Rev. Food Sci. Food Saf.* **2020**, *19*, 3241–3265. [CrossRef]
14. Noort, M.W.J.; van Haaster, D.; Hemery, Y.; Schols, H.A.; Hamer, R.J. The Effect of Particle Size of Wheat Bran Fractions on Bread Quality—Evidence for Fibre–Protein Interactions. *J. Cereal Sci.* **2010**, *52*, 59–64. [CrossRef]
15. Wang, N.; Hou, G.G.; Dubat, A. Effects of Flour Particle Size on the Quality Attributes of Reconstituted Whole-Wheat Flour and Chinese Southern-Type Steamed Bread. *LWT-Food Sci. Technol.* **2017**, *82*, 147–153. [CrossRef]
16. Coda, R.; Kärki, I.; Nordlund, E.; Heiniö, R.-L.; Poutanen, K.; Katina, K. Influence of Particle Size on Bioprocess Induced Changes on Technological Functionality of Wheat Bran. *Food Microbiol.* **2014**, *37*, 69–77. [CrossRef]
17. Protonotariou, S.; Stergiou, P.; Christaki, M.; Mandala, I.G. Physical Properties and Sensory Evaluation of Bread Containing Micronized Whole Wheat Flour. *Food Chem.* **2020**, *318*, 126497. [CrossRef]
18. Lin, S.; Jin, X.; Gao, J.; Qiu, Z.; Ying, J.; Wang, Y.; Dong, Z.; Zhou, W. Impact of Wheat Bran Micronization on Dough Properties and Bread Quality: Part I—Bran Functionality and Dough Properties. *Food Chem.* **2021**, *353*, 129407. [CrossRef] [PubMed]
19. Liu, C.; Liu, L.; Li, L.; Hao, C.; Zheng, X.; Bian, K.; Zhang, J.; Wang, X. Effects of Different Milling Processes on Whole Wheat Flour Quality and Performance in Steamed Bread Making. *LWT Food Sci. Technol.* **2015**, *62*, 310–318. [CrossRef]
20. Guan, E.; Yang, Y.; Pang, J.; Zhang, T.; Li, M.; Bian, K. Ultrafine Grinding of Wheat Flour: Effect of Flour/Starch Granule Profiles and Particle Size Distribution on Falling Number and Pasting Properties. *Food Sci. Nutr.* **2020**, *8*, 2581–2587. [CrossRef] [PubMed]
21. Zhu, F. Influence of Ingredients and Chemical Components on the Quality of Chinese Steamed Bread. *Food Chem.* **2014**, *163*, 154–162. [CrossRef] [PubMed]
22. Ma, S.; Wang, X.; Zheng, X.; Tian, S.; Liu, C.; Li, L.; Ding, Y. Improvement of the Quality of Steamed Bread by Supplementation of Wheat Germ from Milling Process. *J. Cereal Sci.* **2014**, *60*, 589–594. [CrossRef]
23. Su, D.; Ding, C.; Li, L.; Su, D.; Zheng, X. Effect of Endoxylanases on Dough Properties and Making Performance of Chinese Steamed Bread. *Eur. Food Res. Technol.* **2005**, *220*, 540–545. [CrossRef]
24. Tian, W.; Ehmke, L.; Miller, R.; Li, Y. Changes in Bread Quality, Antioxidant Activity, and Phenolic Acid Composition of Wheats During Early-Stage Germination. *J. Food Sci.* **2019**, *84*, 457–465. [CrossRef]
25. Luna-Vital, D.; Luzardo-Ocampo, I.; Cuellar-Nuñez, M.L.; Loarca-Piña, G.; Gonzalez de Mejia, E. Maize Extract Rich in Ferulic Acid and Anthocyanins Prevents High-Fat-Induced Obesity in Mice by Modulating SIRT1, AMPK and IL-6 Associated Metabolic and Inflammatory Pathways. *J. Nutr. Biochem.* **2020**, *79*, 108343. [CrossRef] [PubMed]
26. Tian, W.; Chen, G.; Zhang, G.; Wang, D.; Tilley, M.; Li, Y. Rapid Determination of Total Phenolic Content of Whole Wheat Flour Using Near-Infrared Spectroscopy and Chemometrics. *Food Chem.* **2021**, *344*, 128633. [CrossRef] [PubMed]
27. AACC Approved Methods of Analysis, 11th Edition—AACC Method 76-33.01. Damaged Starch—Amperometric Method by SDmatic. Available online: https://www.cerealsgrains.org/resources/Methods/Pages/76Starch.aspx (accessed on 21 October 2021).
28. GB 5009.88-2014. Available online: https://www.chinesestandard.net/China/Chinese.aspx/GB5009.88-2014 (accessed on 21 October 2021).
29. AACC Approved Methods of Analysis, 11th Edition—AACC Method 32-07.01. Soluble, Insoluble, and Total Dietary Fiber in Foods and Food Products. Available online: https://www.cerealsgrains.org/resources/Methods/Pages/32Fiber.aspx (accessed on 21 October 2021).
30. GB/T 5510-2011. Available online: http://www.gb-gbt.cn/PDF.aspx/GBT5510-2011 (accessed on 21 October 2021).
31. Tian, W.; Wilson, T.L.; Chen, G.; Guttieri, M.J.; Nelson, N.O.; Fritz, A.; Smith, G.; Li, Y. Effects of Environment, Nitrogen, and Sulfur on Total Phenolic Content and Phenolic Acid Composition of Winter Wheat Grain. *Cereal Chem.* **2021**, *98*, 903–911. [CrossRef]
32. Tian, W.; Chen, G.; Tilley, M.; Li, Y. Changes in Phenolic Profiles and Antioxidant Activities during the Whole Wheat Bread-Making Process. *Food Chem.* **2021**, *345*, 128851. [CrossRef]
33. AACC Approved Methods of Analysis, 11th Edition—AACC Method 54-21.02. Rheological Behavior of Flour by Farinograph: Constant Flour Weight Procedure. Available online: https://www.cerealsgrains.org/resources/Methods/Pages/54PhysicalDoughTests.aspx (accessed on 21 October 2021).
34. AACC Approved Methods of Analysis, 11th Edition—AACC Method 54-10.01. Extensigraph Method, General. Available online: https://www.cerealsgrains.org/resources/Methods/Pages/54PhysicalDoughTests.aspx (accessed on 21 October 2021).
35. Chen, F.; He, Z.; Chen, D.; Zhang, C.; Zhang, Y.; Xia, X. Influence of Puroindoline Alleles on Milling Performance and Qualities of Chinese Noodles, Steamed Bread and Pan Bread in Spring Wheats. *J. Cereal Sci.* **2007**, *45*, 59–66. [CrossRef]
36. Huang, S.; Quail, K.; Moss, R.; Best, J. Objective Methods for the Quality Assessment of Northern-Style Chinese Steamed Bread. *J. Cereal Sci.* **1995**, *21*, 49–55. [CrossRef]
37. De Brier, N.; Delcour, J.A. Pearling Affects the Lipid Content and Composition and Lipase Activity Levels of Wheat (*Triticum aestivum* L.) Roller Milling Fractions. *Cereal Chem.* **2017**, *94*, 588–593. [CrossRef]

38. Adom, K.K.; Sorrells, M.E.; Liu, R.H. Phytochemical Profiles and Antioxidant Activity of Wheat Varieties. *J. Agric. Food Chem.* **2003**, *51*, 7825–7834. [CrossRef] [PubMed]
39. Tian, W.; Li, Y. Phenolic Acid Composition and Antioxidant Activity of Hard Red Winter Wheat Varieties. *J. Food Biochem.* **2018**, *42*, e12682. [CrossRef]
40. Liu, R.H. Whole Grain Phytochemicals and Health. *J. Cereal Sci.* **2007**, *46*, 207–219. [CrossRef]
41. Sang, S.; Landberg, R. The Chemistry behind Health Effects of Whole Grains. *Mol. Nutr. Food Res.* **2017**, *61*, 1770074. [CrossRef]
42. Brewer, L.R.; Kubola, J.; Siriamornpun, S.; Herald, T.J.; Shi, Y.-C. Wheat Bran Particle Size Influence on Phytochemical Extractability and Antioxidant Properties. *Food Chem.* **2014**, *152*, 483–490. [CrossRef] [PubMed]
43. Rosa, N.N.; Barron, C.; Gaiani, C.; Dufour, C.; Micard, V. Ultra-Fine Grinding Increases the Antioxidant Capacity of Wheat Bran. *J. Cereal Sci.* **2013**, *57*, 84–90. [CrossRef]
44. Granato, D.; Shahidi, F.; Wrolstad, R.; Kilmartin, P.; Melton, L.D.; Hidalgo, F.J.; Miyashita, K.; van Camp, J.; Alasalvar, C.; Ismail, A.B.; et al. Antioxidant Activity, Total Phenolics and Flavonoids Contents: Should We Ban in Vitro Screening Methods? *Food Chem.* **2018**, *264*, 471–475. [CrossRef]
45. Harnly, J. Antioxidant Methods. *J. Food Compos. Anal.* **2017**, *64*, 145–146. [CrossRef]
46. Sanz Penella, J.M.; Collar, C.; Haros, M. Effect of Wheat Bran and Enzyme Addition on Dough Functional Performance and Phytic Acid Levels in Bread. *J. Cereal Sci.* **2008**, *48*, 715–721. [CrossRef]
47. Okusu, H.; Otsubo, S.; Dexter, J. Wheat Milling and Flour Quality Analysis for Noodles in Japan. In *Asian Noodles*; John Wiley & Sons, Inc.: Hoboken, NJ, USA, 2010; pp. 57–73, ISBN 9780470179222.
48. Rosell, C.M.; Santos, E.; Collar, C. Mixing Properties of Fibre-Enriched Wheat Bread Doughs: A Response Surface Methodology Study. *Eur. Food Res. Technol.* **2006**, *223*, 333–340. [CrossRef]
49. Xu, X.; Xu, Y.; Wang, N.; Zhou, Y. Effects of Superfine Grinding of Bran on the Properties of Dough and Qualities of Steamed Bread. *J. Cereal Sci.* **2018**, *81*, 76–82. [CrossRef]
50. Wang, C.-C.; Yang, Z.; Guo, X.-N.; Zhu, K.-X. Effects of Insoluble Dietary Fiber and Ferulic Acid on the Quality of Steamed Bread and Gluten Aggregation Properties. *Food Chem.* **2021**, *364*, 130444. [CrossRef] [PubMed]
51. Palacios, H.R.; Schwarz, P.B.; D'Appolonia, B.L. Effect of α-Amylases from Different Sources on the Retrogradation and Recrystallization of Concentrated Wheat Starch Gels: Relationship to Bread Staling. *J. Agric. Food Chem.* **2004**, *52*, 5978–5986. [CrossRef]
52. Liu, W.; Brennan, M.; Serventi, L.; Brennan, C. Effect of Wheat Bran on Dough Rheology and Final Quality of Chinese Steamed Bread. *Cereal Chem. J.* **2017**, *94*, 581–587. [CrossRef]
53. Rezaei, S.; Najafi, M.A.; Haddadi, T. Effect of Fermentation Process, Wheat Bran Size and Replacement Level on Some Characteristics of Wheat Bran, Dough, and High-Fiber Tafton Bread. *J. Cereal Sci.* **2019**, *85*, 56–61. [CrossRef]

 *foods*

Article

Qualitative Characterization of Unrefined Durum Wheat Air-Classified Fractions

Alessandro Cammerata [1], Barbara Laddomada [2,*], Francesco Milano [2], Francesco Camerlengo [3], Marco Bonarrigo [3], Stefania Masci [3] and Francesco Sestili [3,*]

[1] Council for Agricultural Research and Economics, Research Centre for Engineering and Agro-Food Processing, Via Manziana 30, 00189 Rome, Italy; alessandro.cammerata@crea.gov.it
[2] Institute of Sciences of Food Production (ISPA), National Research Council (CNR), Via Monteroni, 73100 Lecce, Italy; francesco.milano@cnr.it
[3] Department of Agriculture and Forest Sciences (DAFNE), University of Tuscia, Via San Camillo de Lellis snc, 01100 Viterbo, Italy; f.camerlengo@unitus.it (F.C.); bonarrigo.marco@yahoo.it (M.B.); masci@unitus.it (S.M.)
* Correspondence: barbara.laddomada@ispa.cnr.it (B.L.); francescosestili@unitus.it (F.S.); Tel.: +39-0832-422613 (B.L.); +39-328-8866276 (F.S.)

Abstract: Durum wheat milling is a key process step to improve the quality and safety of final products. The aim of this study was to characterize three bran-enriched milling fractions (i.e., F250, G230 and G250), obtained from three durum wheat grain samples, by using an innovative micronization and air-classification technology. Milling fractions were characterized for main standard quality parameters and for alveographic properties, starch composition and content, phenolic acids, antioxidant activity and ATIs. Results showed that yield recovery, ash content and particle size distributions were influenced either by the operating conditions (230 or 250) or by the grain samples. While total starch content was lower in the micronized sample and air-classified fractions, the P/L ratio increased in air-classified fractions as compared to semolina. Six main individual phenolic acids were identified through HPLC-DAD analysis (i.e., ferulic acid, vanillic acid, *p*-coumaric acid, sinapic acid, syringic and *p*-hydroxybenzoic acids). Compared to semolina, higher contents of all individual phenolic components were found in all bran-enriched fractions. The highest rise of TPAs occurred in the F250 fraction, which was maintained in the derived pasta. Moreover, bran-enriched fractions showed significant reductions of ATIs content versus semolina. Overall, our data suggest the potential health benefits of F250, G230 and G250 and support their use to make durum-based foods.

Keywords: durum wheat; air-classified fractions; alveographic properties; antioxidants; starch; ATI

Citation: Cammerata, A.; Laddomada, B.; Milano, F.; Camerlengo, F.; Bonarrigo, M.; Masci, S.; Sestili, F. Qualitative Characterization of Unrefined Durum Wheat Air-Classified Fractions. *Foods* **2021**, *10*, 2817. https://doi.org/10.3390/foods10112817

Academic Editor: Silvana Cavella

Received: 16 September 2021
Accepted: 13 November 2021
Published: 16 November 2021

Publisher's Note: MDPI stays neutral with regard to jurisdictional claims in published maps and institutional affiliations.

1. Introduction

Durum wheat (*Triticum turgidum* L. ssp. *durum*) is an important crop, especially in the Mediterranean basin, as it is used for the production of daily foods, such as pasta, couscous, bulgur, and unleavened and leavened bread. The main component of the kernel is starch, followed by storage proteins, lipids and other minor compounds. Durum wheat grain also contains several bioactive compounds of health interest, such as insoluble fiber, phenolic acids, and alkylresorcinols, which are mostly concentrated within the coating structure of the kernel [1]. Phenolic acids are among the most abundant and studied components promoting human health. As dietary antioxidants, they act as free-radical scavengers [2,3], and reduce the inflammatory response in endothelial cells and monocytes [4]. Besides their antioxidant and antiradical activity, phenolic acids participate in plant cell walls as structural components and are involved in plant adaptation to abiotic and biotic stresses [5,6]. Among other wheat constituents that recently have become a major target of research in wheat there are the Amylase/Trypsin Inhibitors (ATIs) [7]. ATIs contribute about 3% of total kernel proteins and, similarly to phenolic acids, they are involved in plant defense mechanisms against pests and pathogens [7]. However, differently from them, ATI proteins are concentrated mainly in the endosperm and are associated with negative effects on human health [7]. ATIs, in fact, are putative factors triggering Non-Celiac Wheat Sensitivity (NCWS), one of the most common adverse reactions to wheat, and are responsible for baker's asthma, the most common occupational Ig-E mediated allergy in Europe [7].

So far, the use of raw materials rich in health-promoting compounds and poor in toxic components constitutes a key point for a proper nutritional approach [8–10]. Diets rich in whole cereals, legumes, vegetables and fruit guarantee greater protection against the onset of age-related metabolic diseases that are mainly spread in developed countries [11,12]. Compared to refined products, whole grain-based foods, besides having higher contents of dietary fiber and bioactive compounds, are characterized by reduced amounts of starch, which is associated with postprandial responses, the risk of type 2 diabetes and the incidence of other non-communicable diseases, such as cardiovascular disorders and colon cancer [13–15].

Several attempts were made to improve the poor quality of refined wheat milling products. Among these, the technology based on micronization and subsequent air-classification of milling particles was among the most promising [1,16]. The air-classification technology is a suitable tool to this purpose giving rise to sub-fractions with diverse particle size and composition that can be selected to enhance the beneficial health effects associated with bran particles, while limiting the content of toxic contaminants [17–21]. Through the use of this latter technology, several types of unrefined milling products can be directly obtained from the air-classified plants and evaluated on the basis of qualitative tests without the need for adding selected fractions enriched with specific outer layers such as aleurone [22,23]. Air-classification is based on pneumatic transportation of micronized samples inside the plant, where a series of ascending airflows are controlled by setting the airflow inlet valve at different opening conditions [17]. At the end of each cycle, G fractions (heavier gross particles) and F fractions (fine particles), are collected. The use of micronization and subsequent air-classification treatment has expanded the possibility of selecting milling fractions with technological and nutritional characteristics that are more suitable to make healthier and safer unrefined wheat-based products, also taking into account good testing results and consumer acceptability [16]. These latter aspects are of great interest because durum end-products (e.g., pasta, bread) must meet good technological requirements and consumers' taste. Moreover, constant upgrading of micronization and air-classification technologies significantly improved their efficiency over the last decade. Among the main innovative solutions, a Programmable Logic Control (PLC) used to manage the airflow inside the separation chamber and a decreasing section chamber were adopted [17]. The latter modifications allowed to better split heavier gross fractions (G) from fine fractions (F). Three air-classified fractions of major interest resulted, namely F250, G230 and G250, showing a good compromise between technological properties and bran enrichment [17].

More in detail, these three fractions proved to be the most interesting on the basis of the following criteria:

- F250 (unrefined product with higher bran content): excellent yield (60–70%), good particle size composition, higher ash percentage than micronized samples, and mycotoxin presence;

- G250 (unrefined product with more semolina particles and less fine middlings): insufficient yield (20–30%) but excellent particle size composition, good ash content (higher than semolina), and strong reduction of the mycotoxin content;

- G230 (unrefined product containing both more semolina and more fine middlings): excellent yield (>60%), balanced particle size composition, good ash content (higher than semolina), and strong reduction of the mycotoxin content.

In the present work, the air-classified fractions F250, G230 and G250 were thoroughly characterized in terms of main quality parameters, alveographic properties, starch composition, phenolic acids profile and antioxidant activity, along with the measure of ATIs amount.

2. Materials and Methods

2.1. Plant Materials

Three grain samples were considered in the present study, each belonging to: (i) cv. Saragolla, grown in the Lazio region (hereafter named: Saragolla_LA); (ii) cv. Antalis, grown in the Basilicata region (hereafter named: Antalis_BA); and (iii) cv. Antalis, grown in the Marche region (hereafter named: Antalis_MA). The plants were grown in experimental fields under conventional farming during the 2018–2019 growing season.

2.2. Grain Characterization

Grain samples were evaluated for test weight, moisture and protein content, gluten percentage and yellow color through near-infrared analysis in transmission mode (NIT) by using *Infratec*™ mod 1241 (FOSS, Hillerød, Denmark). Figure 1 shows the flow chart of the experimental plan.

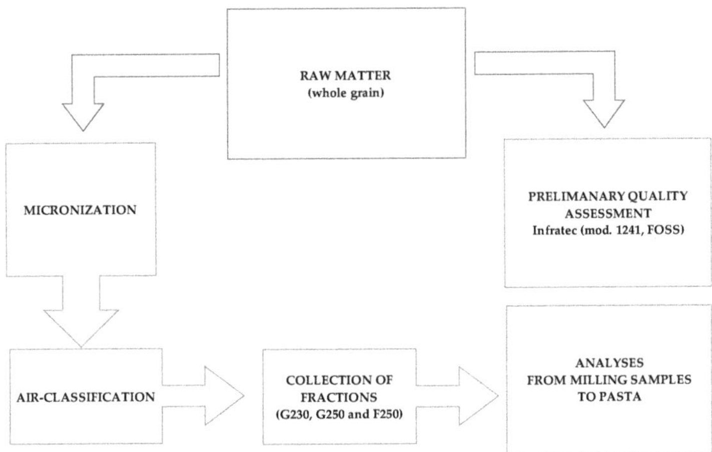

Figure 1. Flow chart of the experimental plan.

2.3. Whole Grain Micronizationand Air Classification

Durum wheat grain samples (11 kg per each grain sample) were micronized using a micronizer pilot plan (mod. 32300, KMXi-300-7,5; Separ Microsystem S.a.s, Brescia, Italia). The micronization step did not require a preventive conditioning of the grains. The micronizing pilot plant was equipped with a hammer crusher impeller with a reduced cross-section inside the grinding chamber. Compared to the traditional type, a sieving grid ($\varnothing$ = 0.7 mm) suitable to obtain a more homogeneous product was added to the plan. Afterwards, the micronized sample was submitted to an air-classifier pilot plant (model SX-LAB; Separ Microsystem S.a.s, Brescia, Italia) suitable for a particle size up to a setting limit ($\varnothing \leq$ 1.5 mm). More in detail, micronized aliquots of 2.0 kg were air-classified for each cycle (in total 3–5 cycles) at a time by setting the airflow inlet valve at 230 and 250. At the end of each cycle the fractions of type G (heavier gross particles) and F (fine particles) were collected but only G230, G250 and F250 were submitted to analysis. All steps of the milling process were carried out both using the plants already updated and following the detailed procedure recently described [17].

2.4. Traditional Roller Milling Process

Grain samples were conditioned by adding water until a moisture value of 17% was achieved and left to rest for 24 h. Such a specific treatment was helpful to favor both the undressing process of kernels and softening the endosperm, thus being suitable to submit grain samples to the traditional roller milling plant (Bühler, model MLU 202, Uzwil,

Switzerland). The main milling products collected were semolina refined through a sieving treatment (sieve types: 38GG, 40GG and 44GG) by the use of a suitable pilot plant sieving system (NAMAD Impianti, Rome, Italy), bran fractions (coarse and refined types) and fine middlings.

2.5. Yield, Ash and Particle Size

Mean yield percentages of the milling fractions were calculated as the weight percentage over the weight of the starting sample. Ash content was determined by following the official method EN ISO 2171:2010 [24]. Particle size analysis of each fraction was carried out using certified test sieves (Giuliani Tecnologie S.r.l., Turin, Italy). Concerning this last point, an electronic sieve (Giuliani Tecnologie S.r.l., Turin, Italy, mod. IG/3) consisting of six stacked sieves of different screens (500-425-355-250-180-125 µm) was used. The milling fraction retained in each sieve was then weighed. The yield parameter was used as a key indicator of the milling quality along with ash and grain size parameters; these last two are expressly mentioned in the Italian legislation [25].

2.6. Alveographic Parameters

The alveographic parameters were evaluated on the basis of the viscoelastic behavior of the dough. The analyses were performed using the *Chopin* alveograph (model NG) AACC Method 54-30.02 and UNI 10453 (Cedex, France) [26].

2.7. Total Starch, Damaged Starch and Amylose

Starch content and damaged starch were determined using the Total Starch Assay Kit (K-TSTA, Megazyme, Irishtown, Ireland) and the Starch Damage Assay Kit (K-SDAM, Megazyme, Irishtown, Ireland), respectively. Total starch contents were determined using the protocol specific for "samples containing also resistant starch". For each fraction, the values represented the mean of four technical replicates.

Amylose content was determined from 15 mg of sample using a colorimetric assay based on the iodine–amylose reaction [27]. The standard curve was created using a mixture of potato amylose (Fluka, Neu-Ulm, Germany) and wheat amylopectin (Sigma Aldrich, St. Louis, MO, USA). Each value represented the mean of four technical replicates.

2.8. Phenolic Acids Analysis

Phenolic acid analysis was performed on the raw materials (micronized samples; F250, G250 and G230 air classification fractions; and semolina) and on uncooked and cooked pasta made using the F250, G250 and G230 air classification fractions and semolina as a control. Total phenolic acids (TPA), comprising the soluble and insoluble fractions, were extracted from samples and analyzed by HPLC according to the procedure detailed in Laddomada et al. [28]. In brief, samples underwent delipidation using hexane, hydrolyzation with 2M NaOH, and acidification with HCl 12 M up to pH 2 prior to ethyl acetate extraction. Ethyl acetate extracts were dried under nitrogen flux, dissolved in 80:20 methanol/water and analyzed using an Agilent 1100 Series HPLC-DAD system (Agilent Technologies, Santa Clara, CA, USA). Individual phenolic acids were identified by comparing their retention times and UV–Vis spectra to those of authentic phenolic standards and quantified via their ratio to the internal standard (3, 5-dichloro-4-hydroxybenzoic acid) added to every sample and using calibration curves for this standard. All analyses were performed on duplicate extracts.

2.9. Pasta making Process

Semolina (control) and the F250, G250 and G230 air-classified fractions (1 kg) were mixed with 280 mL water in a premixing chamber for 15 min; afterwards the dough was transferred to a pilot plan extruder (NAMAD impianti, Rome, Italy) equipped with a 1.6 mm Teflon-coated spaghetti die. Fresh spaghetti were dried using a pilot plan drier

(AFREM, Lyon, France), at 50° C for 18 h. Dried pasta samples were stored in sealed plastic bags at room temperature.

2.10. Antioxidant Activity Assayes

Trolox equivalent antioxidant capacity (TEAC) was measured for all extracts using the ABTS decolorization assay according to Durante et al. [29] with modifications. ABTS$^+$ stock solution was prepared by incubating overnight in the dark 7 mM ABTS (2,2′-azinobis (3-ethylbenzothiazoline-6-sulfonic acid) and 2.45 mM potassium persulfate in water. Trolox standard solutions in the interval of 0–300 µM were prepared by diluting in 80:20 methanol/water the ethanolic 30 mM stock solution. Samples were diluted in 80:20 methanol/water by a factor varying between 10 and 80 according to their TPAcontent and mixed with diluted ABTS$^+$ (A_{734} = 0.7) solution in PBS (Phosphate Buffer Solution) (50 µL samples in duplicate or Trolox standard in 950 µL ABTS$^+$). After 5 min of incubation at 25 °C, the absorbance at 734 nm was measured by means of a spectrometer (Shimadzu UV-1800). TEAC values were calculated from the Trolox standard curve and values in µeq/mL were converted into µeq/g dry matter considering the initial amount of samples used for the TPA extraction.

2.11. Extraction of Albumin and Globulin Fractions (A/G)

The A/G fraction was extracted according to Lupi et al. [30]. Briefly, 1 g of wheat semolina or air-classified fractions were suspended in 27 mL of 0.05 M phosphate buffer/ 0.1 M NaCl (pH = 7.8) and incubated for 2 h at 4 °C. After centrifugation at 8000× *g* for 15 min at 15 °C, the supernatant was recovered, and the proteins (A/G fraction) were precipitated with four volumes of cold (−20 °C) acetone. After 1 h, the supernatant was discarded and the protein pellet was dissolved in 50 mM carbonate buffer (pH = 9.6).

2.12. Indirect Enzyme-Linked Immunosorbent Assay (ELISA) with Anti-ATI Antibodies

The wells of a microtiter plate (ELISA plate 82.1582.100, Sarstedt, Nümbrecht, Germany) were coated with 5 µg/mL of antigen prepared as above in 50 mM carbonate buffer (pH = 9.6) overnight at 4 °C. The plate was washed three times with PBS-0.05% Tween 20, and then the wells were blocked with PBS-BSA 4% for 1 h at 37 °C. After three washes with PBS-0.05% Tween 20, the plate was incubated for 1 h at 37 °C with serial dilutions (from 1:2 to 1:20,000) in PBS-BSA 2% of anti-ATI polyclonal antibodies (developed by BIA-INRA, Nantes, France). Following three washes with PBS-0.05% Tween 20, the secondary antibody (anti-rabbit IgG conjugated with horseradish peroxidase) diluted 1:3000 in PBS-BSA 2% was added to each well and the plate was incubated for 1 h at 37 °C. After three additional washings with PBS-0.05% Tween 20, the plate was incubated at room temperature with the colorimetric substrate, composed of o-phenylenediamine (OPD) in 0.05 M citrate buffer (pH = 5.5) plus hydrogen peroxide. After 30 min, the reaction was stopped with H_2SO_4 4N and the plate was read at 492 nm (Multiskan GO, Thermo Scientific). Three technical replicates were used and the data were subjected to one-way ANOVA followed by Tukey's post hoc test. When one outlier was present, this was excluded from analysis.

2.13. Statistical Analysis

With regards to all data the analysis of variance was performed applying ANOVA (post-hoc: Tukey test) by using the statistical software PAST 2.12 [31].

3. Results

3.1. Grain Characterization

The main qualitative parameters of grain samples used are summarized in Table 1. Test weight values ranged from 81.0 to 84.4 kg/hl, revealing the absence of significant quantities of shriveled kernels, thus falling in the first-class group according to the official standards UNI 10709:1998 [32]. The moisture content of the grains varied from 10.2% to 10.3%, therefore falling below the maximum limit of 14.0%. Except for a low yellow color, protein and gluten percentages were satisfactory, varying from 12.7% to 15.0% (protein content), and from 8.7% to 10.5% (gluten).

Table 1. *Infratec*™ analysis: test weight, moisture, protein content, gluten and yellow color of durum wheat grain samples Saragolla_LA, Antalis_BA, Antalis_MA. Results indicate mean values of replicated analyses (*n* = 10); d.m. = dry matter.

Durum Wheat Grain Sample	Test Weight (kg/hl)	Moisture (%)	Protein Content (% d.m.)	Gluten (% d.m.)	Yellow Color Index
Saragolla_LA	83.4	10.2	12.7	8.7	14.1
Antalis_BA	80.2	10.2	13.9	9.8	15.0
Antalis_MA	81.7	10.3	15.0	10.5	13.6

3.2. Milling Yield and Particle Size Parameters

The average yield recovery of micronized samples exceeded 99%, thus showing a negligible loss of the starting grain samples. Yield percentages of air-classified fractions were high for the F250 and G230 milling fractions (Figure 2). Some significant differences were observed within each F and G fraction as depending on grain samples. In fact, within G milling products, Saragolla_LA yields varied from 42% (G250 setting) to 70% (G230 setting), whereas Antalis_BA yield recovery was lower and comprised between 25% (G250) and 61% (G230), and that of Antalis_MA ranged from 33% (G250) to 66% (G230). Among F fractions, yields varied from a minimum of 56% (Saragolla_LA) to a maximum of 74% (Antalis_BA) (Figure 2).

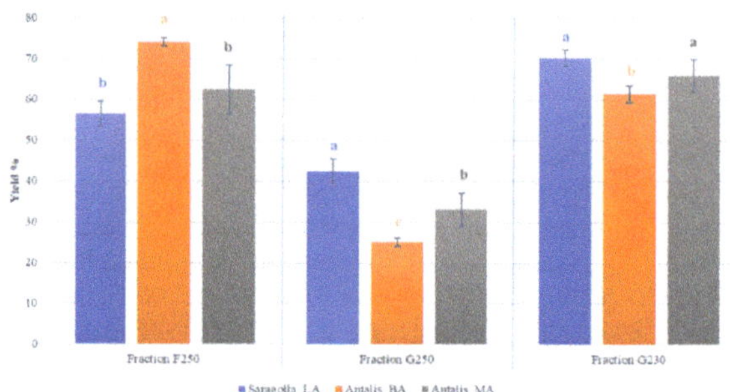

Figure 2. Mean yield (%) of F250, G250 and G230 air-classified fractions obtained from Saragolla_LA, Antalis_BA and Antalis_MA samples. Different letters indicate statistically significant differences (*p* < 0.05, *n* = 3) within each fraction type. F250: Saragolla_LA = 3.4 kg, Antalis_BA = 4.5 kg, Antalis_MA = 3.9 kg; G250: Saragolla_LA = 2.5 kg, Antalis_BA = 1.5 kg, Antalis_MA = 1.9 kg; G230: Saragolla_LA = 3.5 kg, Antalis_BA = 3.0 kg, Antalis_MA = 3.3 kg.

About particle size (Ø), in semolina samples the range was between 425 and 180 μm, whereas in air-classified fractions varied depending on the type of fraction (G or F) (Figure 3).

F fractions had a higher concentration of fine particles (Ø < 180 μm), which was between 49% (Antalis_MA) and 68% (Saragolla_LA). By contrast, G fractions contained more coarse particles (Ø > 425 μm) with contents ranging from 12% to 28% (Antalis_BA). The content of intermediate particles and fine particles of all air-classified fractions (except for G250) was significantly different (*p* < 0.05) from that found in semolina samples (Figure 3).

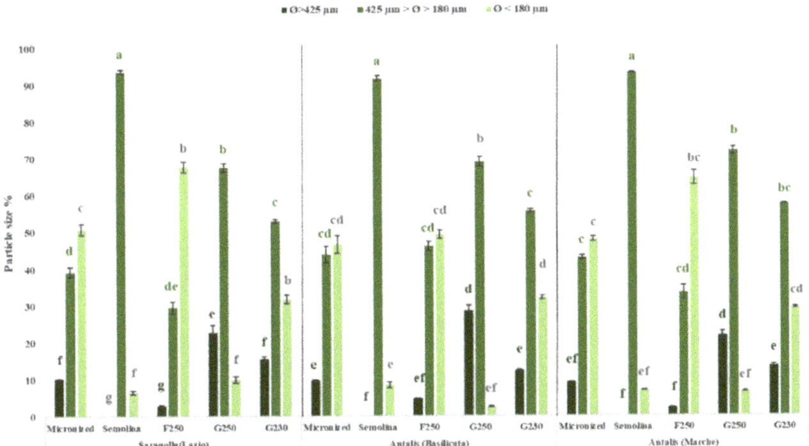

Figure 3. Mean particle size fractions (%) of micronized samples, semolina and air-classified fractions (F250, G250 and G230) obtained from Saragolla_LA, Antalis_BA and Antalis_MA. Different letters indicate statistically significant difference ($p < 0.05$, $n = 2$) within each cultivar sample.

3.3. Ash Content

In general, ash content of micronized samples and bran-enriched fractions was significantly ($p < 0.05$) higher compared to that of semolina (Figure 4). Among F fractions, ash content ranged from 2.46 (Antalis_BA) to 2.10 (Antalis_MA), which was significantly higher ($p < 0.05$) than in G, varying from 1.89 (Antalis BA) to 1.59 (Antalis_MA).

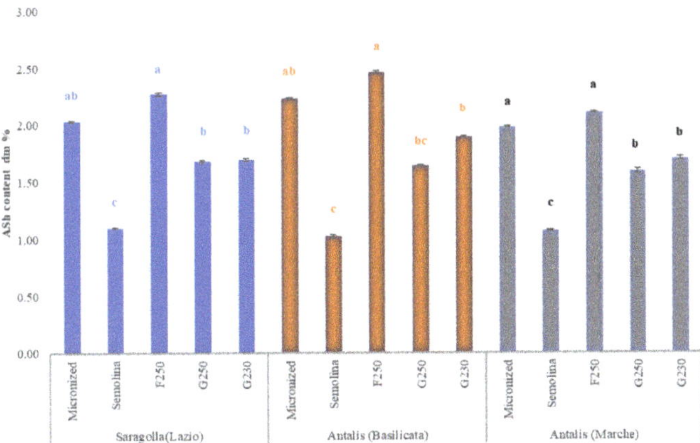

Figure 4. Mean values of ash content (d.m.%) of five milling products: micronized sample, semolina, F250, G250 and G230 air-classified fractions obtained from Saragolla_LA, Antalis_BA and Antalis_MA. Different letters indicate statistically significant differences ($p < 0.05$, $n = 3$) within each cultivar sample.

3.4. Alveographic Properties of Air-Classified Fractions

Alveographic parameters of air-classified fractions, in comparison with the corresponding semolina samples, showed an increased P/L ratio (Table 2). While semolina samples showed P/L ratios ranging from a minimum of 1.49 ± 0.15 (Antalis_BA) to a maximum of 3.96 ± 0.38 (Saragolla_LA), higher values were observed in air-classified fractions, varying from 4.69 ± 0.21 (Antalis_BA G250) to 11.32 ± 0.08 (Antalis_MA F250). The W parameters showed more homogeneous values between air-classified fractions and semolina and a narrower range of variation (i.e., from $133 \pm 20 \, J \cdot 10^{-4}$ (Antalis_BA F250) to $286 \pm 48 \, J \cdot 10^{-4}$ (Antalis_MA semolina).

Table 2. Alveographic parameters of semolina and F250, G250 and G230 air-classified fractions obtained from three durum grain samples (Saragolla_LA, Antalis_BA, Antalis_MA). Different letters indicate significant differences ($p < 0.05$) within each milling products; $n = 5$.

Durum Wheat Grain Samples	Milling Products	W($J \cdot 10^{-4}$)	P/L
Saragolla_LA	Semolina	241 ± 18 [a]	3.96 ± 0.38 [a]
Antalis_BA	Semolina	162 ± 7 [b]	1.49 ± 0.15 [b]
Antalis_MA	Semolina	286 ± 48 [a]	3.85 ± 0.93 [a]
Saragolla_LA	F250	129 ± 13 [a]	10.80 ± 0.37 [a]
Antalis_BA	F250	133 ± 20 [a]	9.32 ± 0.66 [b]
Antalis_MA	F250	156 ± 13 [a]	11.32 ± 0.08 [a]
Saragolla_LA	G250	144 ± 10 [b]	8.35 ± 0.60 [a]
Antalis_BA	G250	137 ± 5 [b]	4.69 ± 0.21 [c]
Antalis_MA	G250	189 ± 4 [a]	7.50 ± 0.31 [b]
Saragolla_LA	G230	144 ± 28 [b]	8.93 ± 0.93 [a]
Antalis_BA	G230	151 ± 6 [b]	6.82 ± 0.58 [b]
Antalis_MA	G230	217 ± 19 [a]	7.59 ± 0.95 [ab]

3.5. Starch Composition of Air-Classified Fractions

Starch, amylose and damaged starch contents were determined in the air-classified fractions (F250, G230 and G250) along with semolina and micronized samples (Table 3). Total starch content was significantly lower in micronized samples and in all air-classified samples as compared to semolina ($p < 0.05$). Saragolla_LA showed the highest amount of total starch (between 61.19% and 74.87%), followed by Antalis_BA (55.09–67.68%) and Antalis_MA (53.01–64.83%). G fractions had superior amounts of total starch than F fractions (Table 3). Interestingly, damaged starch values were not significantly different ($p > 0.05$) between F50 fraction and semolina, whereas significant differences ($p < 0.05$) resulted between each G fraction and the corresponding semolina sample.

Table 3. Total starch, damaged starch and amylose (% d.m.) of five milling products: micronized sample, air-classified fractions (F250, G250, G230) and semolina obtained from three durum grain samples (Saragolla_LA, Antalis_BA, Antalis_MA). Different letters indicate statistically significant differences ($p < 0.05$, $n = 3$) within each milling product ($n = 4$).

Durum Wheat Grain Sample	Milling Product	Total Starch	Damaged Starch	Amylose
Saragolla_LA	Micronized	61.82 ± 0.94 [cd]	4.88 ± 0.21 [d]	15.23 ± 2.99 [hi]
	Semolina	74.87 ± 1.45 [a]	5.96 ± 0.56 [abc]	26.70 ± 3.41 [ab]
	F250	61.19 ± 1.23 [d]	6.41 ± 0.68 [ab]	14.02 ± 1.17 [i]
	G250	67.49 ± 1.20 [b]	2.62 ± 0.27 [e]	21.56 ± 2.29 [defg]
	G230	65.18 ± 1.44 [b]	2.93 ± 0.11 [e]	17.41 ± 1.02 [ghi]
Antalis_BA	Micronized	56.46 ± 1.11 [fg]	5.32 ± 0.39 [cd]	21.67 ± 3.82 [def]
	Semolina	67.68 ± 1.61 [b]	5.80 ± 0.48 [bcd]	28.84 ± 2.18 [a]
	F250	55.09 ± 1.07 [gh]	5.91 ± 0.22 [abc]	18.39 ± 2.05 [fgh]
	G250	60.26 ± 1.78 [de]	2.76 ± 0.13 [e]	23.80 ± 0.76 [bcde]
	G230	59.26 ± 1.58 [def]	3.51 ± 0.10 [e]	25.54 ± 4.11 [abcd]
Antalis_MA	Micronized	54.70 ± 1.36 [gh]	5.38 ± 0.54 [cd]	20.30 ± 1.58 [efg]
	Semolina	64.83 ± 0.93 [bc]	6.36 ± 0.29 [ab]	26.09 ± 1.42 [abc]
	F250	55.14 ± 1.25 [gh]	6.83 ± 0.43 [a]	18.82 ± 1.31 [fgh]
	G250	53.01 ± 1.01 [h]	3.02 ± 0.27 [e]	23.75 ± 3.05 [bcde]
	G230	57.13 ± 0.67 [efg]	3.29 ± 0.08 [e]	22.46 ± 2.43 [cdef]

The amylose content was directly determined on semolina, micronized samples and air-classified fractions. Results showed a significant reduction of amylose content in all air-classified fractions and micronized samples due to the reduced amount of starch as compared to semolina (Table 3). However, this reduction was not significant ($p > 0.05$) in the case of evaluation of the amylose/total starch ratio (data not shown).

3.6. Phenolic Acids Analysis

Phenolic acids content and composition of micronized samples, F250, G250 and G230 air classified fractions, and semolina from three grain samples are shown in Table 4. Overall, each milling product exhibited a typical phenolic acid profile, differing significantly ($p < 0.05$) for almost all individual components. Independently of the grain sample and milling type, ferulic acid was the most abundant phenolic acid being comprised between 63.52 µg/g dry matter (Antalis_BA, semolina) and 569.60 µg/g dry matter (Antalis_BA, F250). Second in abundancy was sinapic acid, ranging from 8.28 µg/g dry matter (Saragolla_LA, semolina) to 68.26 µg/g dry matter (Antalis_MA, F250).

Table 4. Phenolic acid profiles (µg/g dry matter) and antioxidant activity (µeq Trolox/g dry matter) of micronized samples, air-classified fractions (F250, G250, G230) and semolina obtained from three durum grain samples (Saragolla_LA, Antalis_BA, Antalis_MA). Different letters within columns indicate significant differences ($p < 0.05$).

Durum Grain Sample	Milling Product	*p*-Hydroxy Benzoic Acid	Syringic Acid	Vanillic Acid	*p*-Coumaric Acid	Ferulic Acid	Sinapic Acid	TEAC
Saragolla_LA	Micronized	4.37 ± 0.50 [ab]	5.98 ± 0.21 [abc]	8.51 ± 0.48 [bc]	7.96 ± 0.52 [cd]	502.19 ± 8.05 [b]	51.04 ± 1.64 [abc]	8.10 ± 1.54 [cde]
	Semolina	1.12 ± 0.01 [cd]	1.43 ± 0.04 [fg]	1.50 ± 0.42 [f]	0.14 ± 0.01 [f]	80.32 ± 2.79 [e]	8.28 ± 0.66 [g]	2.67 ± 0.18 [gh]
	F250	6.22 ± 0.36 [a]	8.12 ± 0.4 [a]	11.83 ± 0.35 [a]	11.60 ± 0.73 [ab]	498.04 ± 5.08 [b]	61.72 ± 3.83 [ab]	15.22 ± 0.27 [a]
	G250	2.96 ± 0.47 [bcd]	3.23 ± 0.57 [defg]	5.07 ± 0.83 [de]	3.52 ± 0.62 [ef]	383.78 ± 7.31 [c]	32.38 ± 4.01 [cde]	7.05 ± 0.58 [cdef]
	G230	3.07 ± 0.44 [bcd]	4.12 ± 0.51 [cde]	5.59 ± 0.65 [cd]	4.33 ± 0.47 [e]	408.18 ± 10.25 [c]	30.50 ± 3.49 [cdef]	8.58 ± 0.38 [bcd]
Antalis_BA	Micronized	3.36 ± 0.75 [bc]	5.12 ± 0.19 [bcd]	5.92 ± 0.26 [cd]	8.59 ± 0.11 [bc]	409.65 ± 4.24 [c]	41.47 ± 4.12 [bcde]	9.45 ± 0.05 [bc]
	Semolina	0.76 ± 0.15 [d]	0.76 ± 0.18 [g]	0.73 ± 0.09 [f]	0.09 ± 0.00 [f]	63.52 ± 3.32 [e]	9.81 ± 0.47 [fg]	2.47 ± 0.06 [gh]
	F250	4.72 ± 0.56 [ab]	8.15 ± 0.20 [a]	7.53 ± 0.62 [bc]	13.02 ± 0.59 [a]	569.60 ± 5.41 [a]	44.21 ± 4.77 [bcd]	12.26 ± 0.17 [ab]
	G250	2.07 ± 0.08 [bcd]	1.60 ± 0.75 [fg]	2.12 ± 0.70 [f]	2.17 ± 1.00 [ef]	240.90 ± 6.59 [d]	25.36 ± 1.38 [defg]	5.03 ± 0.15 [efgh]
	G230	1.53 ± 0.12 [cd]	1.64 ± 0.23 [efg]	2.22 ± 0.28 [f]	2.81 ± 0.74 [ef]	203.66 ± 5.87 [d]	20.59 ± 2.11 [efg]	5.48 ± 0.17 [defg]
Antalis_MA	Micronized	3.12 ± 0.01 [bc]	4.69 ± 0.33 [cd]	6.76 ± 0.15 [bcd]	8.49 ± 0.30 [bc]	394.44 ± 4.44 [c]	54.71 ± 5.03 [ab]	9.07 ± 0.39 [bc]
	Semolina	0.83 ± 0.01 [cd]	1.01 ± 0.01 [fg]	1.19 ± 0.11 [f]	0.16 ± 0.00 [f]	67.51 ± 3.48 [e]	10.99 ± 0.05 [fg]	2.07 ± 0.10 [h]
	F250	4.04 ± 0.11 [ab]	7.73 ± 0.58 [ab]	9.09 ± 0.49 [ab]	12.40 ± 0.40 [a]	512.48 ± 4.37 [b]	68.26 ± 3.61 [a]	11.90 ± 0.67 [ab]
	G250	1.91 ± 0.01 [bcd]	4.49 ± 0.53 [cd]	3.42 ± 0.03 [def]	4.57 ± 0.21 [de]	241.48 ± 5.66 [d]	29.74 ± 2.59 [def]	4.26 ± 0.02 [fgh]
	G230	1.88 ± 0.01 [cd]	3.49 ± 0.04 [cdef]	3.18 ± 0.27 [def]	4.28 ± 0.35 [e]	214.11 ± 4.38 [d]	25.47 ± 1.62 [defg]	4.32 ± 0.17 [fgh]

Four minor compounds were also detected, namely sinapic, *p*-coumaric, vanillic, syringic and *p*-hydroxybenzoic acids (Table 4).

Semolina had the lowest concentration of all phenolic acids, whereas the micronized samples and the F250 fraction had the highest (Table 4).

The antioxidant activity associated with phenolic acids was lower in semolina samples (ranging from 2.07 to 2.67 µeq Trolox/g dry matter) compared to that of all air-classified fractions; the highest value was in Saragolla_LA, for the F250 fraction (15.22 µeq Trolox/g dry matter) (Table 4). Considering the average values of TPAs evaluated on the overall data across cultivars and environments, significant differences ($p < 0.05$) were found among the different milling products (Figure 5). The observed variation ranged from a minimum of 82.48 µg/g dry matter (semolina) to a maximum of 619.57 µg/g dry matter (F250). Compared to semolina, the F250 fraction had the highest TPAs content (7.5-fold vs. semolina), whereas the G250 and G230 fractions showed a four-fold higher content. The pasta-making and cooking processes caused a slight decrease of TPAs content compared to that of raw materials (Figure 5). Nevertheless, uncooked pasta made with air-classified fractions had significant enrichments of TPAs compared to traditional pasta made with semolina; such an increase varied from seven-fold (F250) to four-fold (G250 and G230) (Figure 5). A complete and detailed view of the phenolic acids profiles and antioxidant activity of uncooked and cooked pasta is presented in Supplementary Tables S1 and S2.

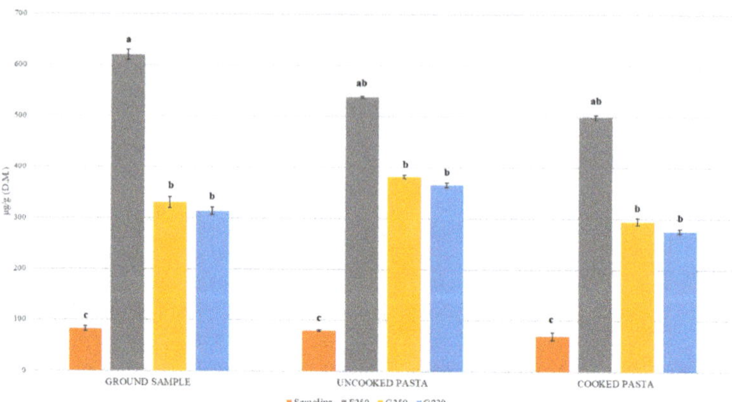

Figure 5. Average contents (*n* = 3) of total phenolic acids (TPAs), expressed as µg/g dry matter (d.m.) of four milling types: semolina, F250, G250 and G230 air classified fractions, obtained from the overall durum grain samples used in this study, and derived uncooked and cooked spaghetti. Different letters show significant differences ($p < 0.05$).

3.7. ATIs Content

An ELISA test was performed for detecting differences in the amount of ATI on the albumin and globulin fractions (A/G) of the different milling types (micronized grains, F250, G250, and G230 air-classified fractions, and semolina). Data relative to the four antibody dilutions used are reported in Supplementary Table S3. Since the dilution 1:200 corresponded to the most linear region, it was used to build the histograms reported in Figure 6. Overall, results showed that semolina had higher ATIs contents compared to the other milling products, excepting for the Antalis_BA G230 air-classified fraction that had slightly higher, though not significant ($p < 0.05$), amounts (Figure 6). Among the five milling types obtained from the Saragolla_LA grain sample, the G230 fraction showed the lowest amount of ATI, whereas both Antalis_BA and Antalis_MA displayed the lowest content of ATI in the F250 air classified fraction (Figure 6).

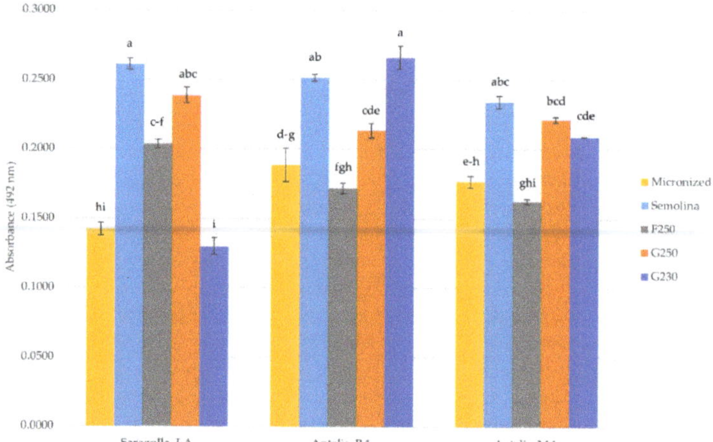

Figure 6. Indirect ELISA performed with anti-ATI polyclonal antibodies dilution 1:200 of micronized samples, semolina and air-classified fractions (F250, G250, G230) obtained from the durum grain samples used in this study. Different letters indicate significant differences ($p < 0.05$).

4. Discussion

One of the key challenges to enhance the health value of durum wheat foods is the use of raw materials rich in health-promoting compounds and poor in toxic components. Bran is the compartment accounting for only 15% of the wheat grain, but it is an important source of components associated with health benefits, such as phenolic acids. To maintain these healthy bran components, and reduce the negative impacts associated with some endosperm proteins, such as ATIs, diverse bran-rich streams can be used [16,17].

In this paper, we evaluated several qualitative features of F250, G230 and G250 air-classified fractions as compared to semolina and micronized samples. The use of this updated technology directly provided several unrefined milling products, each having different characteristics, and without adding fractions rich in bioactive compounds [19]. This fact could lead to time and money savings, especially in a large-scale application.

First, milling yield percentages were assessed as they are among the major milling quality features [33]. The total recovery of milling products (F + G fractions) along the micronizing and air classification processes reached values around 99%, thus ensuring a minimal loss of raw material. The present results were comparable to those observed in previous evaluations [17]; slight differences detected here could depend on diverse grain size and texture [34]. The operating conditions that we used gave rise to G230 and F250 fractions with yield values that were near or above 60%, with some differences depending on grain samples ($p < 0.05$) [17].

The distribution of coarse, intermediate and fine particles within each G and F fraction resulted in agreement with a preliminary study [17]. The variability observed among the overall samples depended on grain samples and the setting valve point, supporting previous findings [17]. The typical particle size distribution characterizing each fraction type suggested that each could be more suitable to make specific unrefined end-products (pasta, bread, biscuits, etc.). The presence of bran particles in different rates within each air-classified fraction could lead to durum end-products (such as pasta) of higher quality than those obtained by adding bran aliquots to semolina [35]. The occurrence of coarse semolina particles in milling products could be important to enhance the gluten index and yellow color of end products [36]. Results regarding ash characterization suggested a higher content of minerals in the F fractions than in G fractions [37]. Overall, ash values were higher in air-classified fractions and micronized samples than in semolina. On the whole, results concerning yield, ash content and particle sizes' composition proved to be in agreement with previous studies [17]. The alvegraphic behavior of the unrefined samples revealed a great reduction in the alveographic parameters, especially with regard to a P/L ratio increase. This latter observation confirms previous results on the quality effects of the addition of bran aliquots to semolina [38]. Total starch content was lower in micronized flour and in all air-classified fractions compared to semolina. G fractions had superior amounts of total starch compared to two out of three F250 samples. These data were expected, since F fractions contain a higher amount of bran and a reduced content of endosperm. Results of amylose content showed a significant reduction in all air-classified fractions and micronized samples compared to semolina due to the reduced amount of starch in these fractions. In this regard, when considering the amylose/total starch, no significant differences were detected. The reduction of starch, observed in the air-classified fraction compared to semolina, is potentially interesting for the realization of low glycemic foods, useful for the prevention of some diet-related diseases (type II diabetes, obesity and cardiovascular disorders) [13–15].

Based on phenolic acids analysis of raw materials and derived uncooked and cooked pasta, six main individual phenolic acids were identified, namely ferulic acid, vanillic acid, *p*-coumaric acid, sinapic acid, and syringic and *p*-hydroxybenzoic acids. These findings were in line with previous works carried out along the durum chain from seed to pasta [22,39]. Literature studies established that phenolic acids have important biological effects, independently if they are easily absorbed by the small intestine as free forms [40] or reach the colon intact as bound forms [41]. Upon these evidences, we estimated the overall content of free and bound forms for each individual phenolic acid [22,39]. Results showed that semolina had the lowest concentration of all individual phenolic components, whereas micronized samples and the F250 fraction had the highest (Table 4). That was expected for the diverse endosperm concentration in semolina and milling in which bran occurred, though at a different rate [38]. Considering the total sum of individual phenolic acids, the F250 fraction had the highest increase over semolina (650%) (Figure 5), suggesting that it would be the best to use for enriching the content of phenolic acids of pasta. In fact, the F250-derived pasta had the highest content of TPAs with an increase by up to 616% over that made with semolina (Figure 5). A slight, though not significant TPAs reduction (between 11% and 19%) resulted in cooked pasta compared to raw materials (Figure 5). These findings were expected and in line with other studies [39,42]. The antioxidant activity of phenolic extracts from raw materials and derived pasta were in good agreement with predictions considering the ferulic acid content with a TEAC of 3.0, indicating that the main source of antioxidant power of tested samples are the extracted phenolic acids.

Regarding ATIs, the highest amounts were found in semolina, which was expected since these components are mostly present in the starchy endosperm as reviewed [7]. So far, it can be concluded that our data show that micronization and air-classification treatment decrease the amount of ATIs, giving a potential added value to derived pasta products. The importance of ATIs in triggering adverse reactions to wheat is increasing, since they seem to be not only involved in allergies and sensitivity, but also in apparently different pathologies, such as Alzheimer's disease, at least in murine models [43]. Thus, increasing attention is paid to the different procedures that can be used to achieve the goal of decreasing ATIs amounts. ATIs can be reduced in wheat flours through food processing, but the results are ambiguous (reviewed in [7]). Moreover, most of these processing procedures include fermentation, which is not typically used for pasta production. A genetic approach could theoretically be the method of choice; for example, transgenic and genome edited plants were produced in previous papers [44,45]. However, due to legislation restrictions, these genotypes cannot be used for commercial purposes. In addition, the production of new plant lines/varieties takes a long time. The air-fractionation procedure, described in this paper, has the advantage that it can be applied to different genotypes without the issues related to the above reported strategies, giving the possibility to produce pasta with high technological quality.

5. Conclusions

In the present work, F250, G230 and G250 air-classified fractions were characterized for standard quality parameters, starch, phenolic acids and ATIs content. The percentage distribution of coarse, intermediate and fine particles within each G and F fraction depended both on grain samples and setting valve points. While the rheological behavior revealed a reduction in the alveographic parameters, all fractions had significant improvements in other qualitative properties. In fact, total starch content diminished in micronized samples and in all air-classified fractions compared to semolina, suggesting their use for the production of low glycemic foods. All air-fractionated millings, especially F250, also showed strong improvements in phenolic acids content and antioxidant activity versus semolina and traditional pasta. Finally, micronization and air-classification treatments decreased the amount of ATIs. Overall, our data suggest the potential use of the F250, G230 and G250 air-classified fractions to make more nutritious, healthier and safer foods.

Supplementary Materials: The following are available online at https://www.mdpi.com/2304-815 8/10/11/2817/s1, Table S1: Phenolic acid profiles (µg/g dry matter) and antioxidant activity (µeq Trolox/g dry mat-ter) of un-cooked pasta made with air-classified fractions (F250, G250, G230) and semolina ob-tained from three durum grain samples (Saragolla_LA, Antalis_BA, Antalis_MA), Table S2: Phe-nolic acid profiles (µg/g dry matter) and antioxidant activity (µeq Trolox/g dry matter) of cooked pasta made with air-classified fractions (F250, G250, G230) and semolina obtained from three durum grain samples (Saragolla_LA, Antalis_BA, Antalis_MA), Table S3: Anti-ATI polyclonal antibodies on micronized wholemeal, F250, G250, G230 air-classified fractions, and semolina from three durum grain samples.

Author Contributions: Conceptualization, A.C., B.L. and F.S.; methodology, A.C., B.L., F.M., M.B. and F.C.; validation, A.C., F.S., B.L., F.M., S.M. and F.S.; formal analysis, A.C., F.S., B.L., F.M. and S.M.; data curation, A.C., F.S., B.L., F.M., M.B., S.M. and F.C.; writing—original draft preparation, F.S., A.C., B.L., S.M. and F.M.; writing—review and editing, B.L., F.S., S.M. and A.C.; supervision, F.S. All authors have read and agreed to the published version of the manuscript.

Funding: This research was partially funded by the Italian Ministry of Education, University, and Research (MIUR) in the frame of the MIUR initiative "Departments of excellence", Law 232/2016.

Data Availability Statement: The data is contained within the article.

Acknowledgments: The authors wish to thank: Gabriella Aureli, Alessandra Arcangeli, Roberto Mortaro and Samuela Palombieri for their precious support. Sandra Denery-Papini, Colette Larré and Roberta Lupi (INRAE, BIA, Nantes, France) are kindly acknowledged for providing the Anti-ATI antibodies used in this paper.

Conflicts of Interest: The authors declare no conflict of interest.

References

1. Hemery, Y.; Rouau, X.; Lullien-Pellerin, V.; Barron, J.C.; Abecassis, J. Dry processes to develop wheat fractions and products with enhanced nutritional quality. *J. Cereal Sci.* **2007**, *46*, 327–347. [CrossRef]
2. Sevgi, K.; Tepe, B.; Sarikurkcu, C. Antioxidant and DNA damage protection potentials of selected phenolic acids. *Food Chem. Toxicol.* **2015**, *77*, 12–21. [CrossRef] [PubMed]
3. Pandey, K.B.; Rizvi, S.I. Plant polyphenols as dietary antioxidants in human health and disease. *Oxidative Med. Cell. Longev.* **2009**, *2*, 270–278. [CrossRef] [PubMed]
4. Calabriso, N.; Massaro, M.; Scoditti, E.; Pasqualone, A.; Laddomada, B.; Carluccio, M.A. Phenolic extracts from whole wheat biofortified bread dampen overwhelming 1 inflammatory response in human endothelial cells and monocytes: Major role of VCAM-1 and CXCL-10. *Eur. J. Nutr.* **2020**, *59*, 2603–2615. [CrossRef] [PubMed]
5. Laddomada, B.; Blanco, A.; Mita, G.; D'Amico LSingh, R.P.; Ammar, K.; Crossa, J.; Guzmán, C. Drought and heat stress impacts on phenolic acids accumulation in durum wheat cultivars. *Foods* **2021**, *10*, 2142. [CrossRef] [PubMed]
6. Nigro, D.; Grausgruberb, H.; Guzmàn, C.; Laddomada, B. Phenolic compounds in wheat kernels: Genetic and genomic studies of biosynthesis and regulation. In *Wheat Quality for Improving Processing and Health*; Igrejas, G., Ikeda, T., Guzmàn, C., Eds.; Springer Sciences + Business Media: Dordrecht, The Netherlands, 2020; pp. 223–251, ISBN 978-3-030-34162-6.
7. Geisslitz, S.; Shewry, P.; Brouns, F.; America, A.H.P.; Caio, G.P.I.; Daly, M.; D'Amico, S.; De Giorgio, R.; Gilissen, L.; Grausgruber, H.; et al. Wheat ATIs: Characteristics and Role in Human Disease. *Front. Nutr.* **2021**, *8*, 667370. [CrossRef]
8. Otles, S.; Ozgoz, S. Health effects of dietary fiber. *Acta Sci. Pol. Technol. Aliment.* **2014**, *13*, 191–202. [CrossRef] [PubMed]
9. Probst, Y.C.; Guan, V.X.; Kent, K. Dietary phytochemical intake from foods and health outcomes: A systematic review protocol and preliminary scoping. *Br. Med. J.* **2017**, *7*, e013337. [CrossRef] [PubMed]
10. Stephen, A.M.; Champ, M.M.J.; Cloran, S.J.; Fleith, M.; van Lieshout, L.; Mejborn, H.; Burley, V.J. Dietary fibre in Europe: Current state of knowledge on definitions, sources, recommendations, intakes and relationships to health. *Nutr. Res. Rev.* **2017**, *30*, 149–190. [CrossRef]
11. Jacobs, D.; Anderson, F.L.; Blomhoff, R. Wholegrain consumption is associated with a reduced risk of non cardiovascular, non cancer death attributed to inflammatory disease in the Iowa Women's Health Study. *Am. J. Clin. Nutr.* **2007**, *85*, 1606–1614. [CrossRef]
12. Kushi, L.H.; Meyer, K.A.; Jacobs, D.R. Cereals, legumes, and chronic disease risk reduction: Evidence from epidemiologic studies. *Am. J. Clin. Nutr.* **1999**, *70*, 451s–458s. [CrossRef]
13. Aune, D.; Keum, N.N.; Giovannucci, E.; Fadnes, L.T.; Boffetta, P.; Greenwood, D.C.; Tonstad, S.; Vatten, L.J.; Riboli, E.; Norat, T. Whole grain consumption and risk of cardiovascular disease, cancer, and all cause and cause specific mortality: Systematic review and dose-response meta-analysis of prospective studies. *Br. Med. J.* **2016**, *353*, i2716. [CrossRef]
14. Călinoiu, L.F.; Vodnar, D.C. Whole grains and phenolic acids: A review on bioactivity, functionality, health benefits and bioavailability. *Nutrients* **2018**, *10*, 1615. [CrossRef]

15. Giacco, R.; Costabile, G.; Della Pepa, G.; Anniballi, G.; Griffo, E.; Mangione, A.; Cipriano, P.; Viscovo, D.; Clemente, G.; Landberg, R.; et al. A whole-grain cereal-based diet lowers postprandial plasma insulin and triglyceride levels in individuals with metabolic syndrome. *Nutr. Metab. Cardiovasc. Dis.* **2014**, *24*, 837–844. [CrossRef] [PubMed]

16. Hemery, Y.; Chaurand, M.; Holopainen, U.; Lampi, A.-M.; Lehtinen, P.; Piironen, V.; Sadoudi, A.; Rouau, X. Potential of dry fractionation of wheat bran for the development of food ingredients, part I: Influence of ultra-fine grinding. *J. Cereal Sci.* **2011**, *53*, 1–8. [CrossRef]

17. Cammerata, A.; Sestili, F.; Laddomada, B.; Aureli, G. Bran-Enriched Milled Durum Wheat Fractions Obtained Using Innovative Micronization and Air-Classification Pilot Plants. *Foods* **2021**, *10*, 1796. [CrossRef] [PubMed]

18. Cammerata, A.; Marabottini, R.; Allevato, E.; Aureli, G.; Stazi, S.R. Content of minerals and deoxynivalenol in the air-classified fractions of durum wheat. *Cereal Chem. J.* **2021**, *98*, 1101–1111. [CrossRef]

19. Ficco, D.B.M.; Borrelli, G.M.; Giovanniello, V.; Platani, C.; De Vita, P. Production of anthocyanin-enriched flours of durum and soft pigmented wheats by air-classification, as a potential ingredient for functional bread. *J. Cereal Sci.* **2018**, *79*, 118–126. [CrossRef]

20. Gomez-Caravaca, A.M.; Verardo, V.; Candigliota, T.; Marconi, E.; Segura-Carretero, A.; Fernandez-Gutierrez, A.; Caboni, M.F. Use of air classification technology as green process to produce functional barley flours naturally enriched of alkylresorcinols, β-glucans and phenolic compounds. *Food Res. Int.* **2015**, *73*, 88–96. [CrossRef]

21. Laudadio, V.; Bastoni, E.; Introna, M.; Tufarelli, V. Production of low-fiber sunflower (*Helianthus annuus* L.) meal by micronization and air classification processes. *CyTA-J. Food* **2013**, *11*, 398–403. [CrossRef]

22. Chen, Z.; Li, W.; Ren, W.; Li, Y.; Chen, Z. Triboelectric separation of aleurone cell-cluster from wheat bran fragments in nonuniform electric field. *Food Res. Int.* **2014**, *62*, 111–120. [CrossRef]

23. Chen, Z.; Zha, B.; Wang, L.; Wang, R.; Chen, Z.; Tian, Y. Dissociation of aleurone cell cluster from wheat bran by centrifugal impact milling. *Food Res. Int.* **2013**, *54*, 63–71. [CrossRef]

24. EN ISO 2171:2010. *Cereals, Pulses and by-Products-Determination of Ash Yield by Incineration (ISO 2171:2007).* European Directive, 2010.

25. Decreto del Presidente della Repubblica 9 febbraio 2001, n. 187. In *Regolamento per la Revisione Della Normativa Sulla Produzione e Commercializzazione di Sfarinati e Paste Alimentari, a Norma Dell'articolo 50 Della Legge 22 feb-Braio 1994, n. 146*; Italian Legislation: Italy, 2001.

26. AACC Method 54-30.02 and UNI 10453. In *Alveograph Method for Soft and Hard Wheat Flour*; AACC Method 54-30.02: International Method, First approval October 3, 1984, Reapproval 3 November 1999; UNI 10453; Italian National Unification Bod: Italy, 1995.

27. Chrastil, J. Improved colorimetric determination of amylose in starches or flours. *Carbohydr. Res.* **1987**, *159*, 154–158. [CrossRef]

28. Laddomada, B.; Durante, M.; Mangini, G.; D'Amico, L.; Lenucci, M.S.; Simeone, R.; Piarulli, L.; Mita, G.; Blanco, A. Genetic variation for phenolic acids concentration and composition in a tetraploid wheat (*Triticum turgidum* L.) collection. *Genet. Resour. Crop. Evol.* **2017**, *64*, 587–597. [CrossRef]

29. Durante, M.; Milano, F.; Caroli, M.; Giotta, L.; Piro, G.; Mita, G.; Frigione, M.; Lenucci, M.S. Tomato Oil Encapsulation by α-, β-, and γ-Cyclodextrins: A Comparative Study on the Formation of Supramolecular Structures, Antioxidant Activity, and Carotenoid Stability. *Foods* **2020**, *9*, 1553. [CrossRef] [PubMed]

30. Lupi, R.; Denery-Papini, S.; Rogniaux, H.; Lafiandra, D.; Rizzi, C.; De Carli, M.; Moneret-Vautrin, D.A.; Masci, S.; Larré, C. How much does transgenesis affect wheat allergenicity?: Assessment in two GM lines over-expressing endogenous genes. *J. Proteom.* **2013**, *80*, 281–291. [CrossRef] [PubMed]

31. Hammer, O.; Harper, D.A.T.; Ryan, P.D. PAST: Paleontological Statistic Software Package for Education and Data Analysis. *Paleontol. Electron.* **2001**, *4*, 1–9. Available online: http://palaeo-electronica.org/2001_1/past/issue1_01.htm (accessed on 14 November 2021).

32. UNI 10709:1998. *Durum Wheat Grains. Qualitative Requirements, Classification and Test Methods*; Italian National Unification Body: Italy, 1998.

33. Sabanci, K.; Aydin, N.; Sayaslan, A.; Sonmez, M.E.; Aslan, M.F.; Demir, L.; Sermet, C. Wheat Flour Milling Yield Estimation Based on Wheat Kernel Physical Properties Using Artificial Neural Networks. *Int. J. Intell. Syst. Appl. Eng.* **2020**, *8*, 78–83. [CrossRef]

34. Marshall, D.R.; Mares, D.J.; Moss, H.I.; Ellison, F.W. Effects of grain shape and size on milling yields in wheat. II. Experimental studies. *Aust. J. Agric. Res.* **1986**, *37*, 331–342. [CrossRef]

35. Alzuwaid, N.T.; Fellows, C.M.; Laddomada, B.; Sissons, M. Impact of wheat bran particle size on the technological and phytochemical properties of durum wheat pasta. *J. Cereal Sci.* **2020**, *95*, 103033. [CrossRef]

36. Sacchetti, G.; Cocco, G.; Cocco, D.; Neri, L.; Mastrocola, D. Effect of semolina particle size on the cooking kinetics and quality of spaghetti. *Procedia Food Sci.* **2011**, *1*, 1740–1745. [CrossRef]

37. Padalino, L.; Mastromatteo, M.; Lecce, L.; Spinelli, S.; Conte, A.; Del Nobile, M.A. Effect of raw material on cooking quality and nutritional composition of durum wheat spaghetti. *Int. J. Food Sci. Nutr.* **2015**, *66*, 266–274. [CrossRef] [PubMed]

38. Pasqualone, A.; Laddomada, B.; Centomani, I.; Paradiso, V.M.; Minervini, D.; Caponio, F.; Summo, C. Bread making aptitude of mixtures of re-milled semolina and selected durum wheat milling by-products. *LWT-Food Sci. Technol.* **2017**, *78*, 151–159. [CrossRef]

39. Martini, D.; Ciccoritti, R.; Nicoletti, I.; Nocente, F.; Corradini, D.; D'Egidio, M.G.; Taddei, F. From seed to cooked pasta: Influence of traditional and non-conventional transformation processes on total antioxidant capacity and phenolic acid content. *Int. J. Food Sci. Nutr.* **2017**, *69*, 24–32. [CrossRef] [PubMed]

40. Manach, C.; Scalbert, A.; Morand, C.; Rémésy, C.; Jiménez, L. Polyphenols: Food sources and bioavailability. *Am. J. Clin. Nutr.* **2004**, *79*, 727–747. [CrossRef]
41. Cardona, F.; Andres-Lacueva, C.; Tulipani, S.; Tinahones, F.J.; Queipo-Ortuno, I.M. Benefits of polyphenols on gut microbiota and implications in human health. *J. Nutr. Biochem.* **2013**, *24*, 1415–1422. [CrossRef]
42. Fares, C.; Platani, C.; Baiano, A.; Menga, V. Effect of processing and cooking on phenolic acid profile and antioxidant capacity of durum wheat pasta enriched with debranning fractions of wheat. *Food Chem.* **2010**, *119*, 1023–1029. [CrossRef]
43. dos Santos Guilherme, M.; Zevallos, V.F.; Pesi, A.; Stoye, N.M.; Nguyen, V.T.T.; Radyushkin, K.; Schwiertz, A.; Schmitt, U.; Schuppan, D.; Endres, K. Dietary Wheat Amylase Trypsin Inhibitors Impact Alzheimer's Disease Pathology in 5xFAD Model Mice. *Int. J. Mol. Sci.* **2020**, *21*, 6288. [CrossRef]
44. Kalunke, R.M.; Tundo, S.; Sestili, F.; Camerlengo, F.; Lafiandra, D.; Lupi, R.; Larrè, C.; Denery-Papini, S.; Islam, S.; Ma, W.; et al. Reduction of allergenic potential in bread wheat RNAi transgenic lines silenced for CM3, CM16 and 0.28 ATI genes. *Int. J. Mol. Sci.* **2020**, *21*, 5817. [CrossRef]
45. Camerlengo, F.; Frittelli, A.; Sparks, C.; Doherty, A.; Martignago, D.; Larré, C.; Lupi, R.; Sestili, F.; Masci, S. CRISPR-Cas9 multiplex editing of the α-amylase/trypsin inhibitor genes to reduce allergen proteins in durum wheat. *Front Sustain. Food Syst.* **2020**, *4*, 104. [CrossRef]

Communication

Bran-Enriched Milled Durum Wheat Fractions Obtained Using Innovative Micronization and Air-Classification Pilot Plants

Alessandro Cammerata [1], Francesco Sestili [2], Barbara Laddomada [3,*] and Gabriella Aureli [1,*]

1   Council for Agricultural Research and Economics, Research Centre for Engineering and Agro-Food Processing, Via Manziana 30, 00189 Rome, Italy; alessandro.cammerata@crea.gov.it
2   Department of Agriculture and Forest Sciences (DAFNE), University of Tuscia, Via San Camillo de Lellis snc, 01100 Viterbo, Italy; francescosestili@unitus.it
3   Institute of Sciences of Food Production (ISPA), National Research Council (CNR), Via Monteroni, 73100 Lecce, Italy
*   Correspondence: barbara.laddomada@ispa.cnr.it (B.L.); gabriellaureli@gmail.com (G.A.); Tel.: +39-0832-422613 (B.L.); +39-0632-95705 (G.A.)

Abstract: Dietary guidelines recommend the consumption of unprocessed, or minimally processed, wheat foods because they are richer in health-promoting components (i.e., minerals, vitamins, lignans, phytoestrogens, and phenolic compounds) compared to traditionally refined products. The design and implementation of technological solutions applied to the milling process are becoming a key requirement to obtain less refined mill products characterized by healthier nutritional profiles. This study presents the development of an upgraded micronization plant and of a modified air-classification plant to produce several novel types of durum wheat milling fractions, each enriched in bran particles of different sizes (from 425 µm > Ø to Ø < 180 µm) and percentage ratios. A preliminary quality assessment of the milling fractions was carried out by measuring yield percentages and ash content, the latter being related to detect the presence of bran particles. A wide array of milling fractions with different original particle size compositions was provided through the study of the process. Results indicate the ability of the novel pilot plants to produce several types of less refined milling fractions of potential interest for manufacturing durum wheat end-products beneficial for human health.

Keywords: durum wheat; milling fractions; air-classification plant; micronization plant

Citation: Cammerata, A.; Sestili, F.; Laddomada, B.; Aureli, G. Bran-Enriched Milled Durum Wheat Fractions Obtained Using Innovative Micronization and Air-Classification Pilot Plants. *Foods* **2021**, *10*, 1796. https://doi.org/10.3390/foods10081796

Academic Editor: Elena Peñas Pozo

Received: 17 June 2021
Accepted: 2 August 2021
Published: 3 August 2021

Publisher's Note: MDPI stays neutral with regard to jurisdictional claims in published maps and institutional affiliations.

1. Introduction

Durum wheat (*Triticum turgidum* L. ssp. *durum*) is the most important cereal staple food in Mediterranean climates and regions. Traditionally, it is the primary raw material used in the production of pasta, couscous, bulgur, and different types of leavened and unleavened breads [1]. Overall, these products provide a significant portion of calories and proteins to human diets; also, they are an important source of bioactive compounds that may contribute to a healthy diet [2]. Among others, the most common bioactive compounds of wheat include dietary fiber, vitamins, micronutrients, and phytochemicals, which are mainly located in the outer layers of the kernel, typically in the bran, aleurone, and germ tissues [3]. A high number of in vitro, in vivo, and epidemiological studies have shown the significant health-related benefits associated with the consumption of bran-rich or whole-wheat foods [4], together with a decreased risk of non-communicable diseases, such as type 2 diabetes mellitus, cardiovascular disorders, and colorectal cancer [5]. Despite the known positive effects, the consumption of whole wheat food products is still limited in many countries. Such a hindrance may depend on consumer knowledge and behaviors [6], but also on the negative impact of wheat bran on the sensorial quality of end products [7]. Several studies have evaluated the effect of bran particle size, bran pre-treatment, and cooking method on the sensorial quality of final products [8]. The interfering effect of bran on both protein hydration and dilution of gluten network in the dough is due to the occurrence of hydroxyl groups in the bran structure reacting with water through hydrogen bonds, resulting in an increase in water absorption. In fact, arabinoxylans reduce the amount of water available for the gluten network, thus affecting dough stability and development time [9]. New technological solutions in wheat milling processing have many advantages as compared to traditional milling, which, by removing germ and bran, cause the loss of several beneficial compounds that are located in the bran fraction. Industrial pretreatments, such as debranning, micronization, and air-classification processes, can reduce the negative effects of wheat bran [10–13]. Debranning is a dry separation technology based on consecutive abrasions of cereal kernels; the progressive bran removal through the detachment of the outer, intermediate, and inner layers of pericarp leads to different byproduct classes, which can be removed by using pressurized air flowing through the screens and outlets of the debranner [7].

The reduction in particle size obtained through micronization can be followed by an air-classifier treatment in order to obtain two or more fractions that are collected separately. Micronization consists of a milling process able to reduce the starting matrix such as cereal grains in a fine particle product through the use of different technologies (e.g., hammer mill, knife mill). The range of the particle size of the milled product depends on the sieve diameter employed [11]. The general functioning criteria of the air-classification system is based on the pneumatic transportation of the milled particles inside the plant, which operates in depression, thus promoting a pneumatic flow along circular orbits. Inside the orbits, a series of ascending airflows was controlled by setting the airflow inlet valve to different opening conditions. Due to the combined action of different forces, the heavier particles fall in the first "G" housing (heavier gross particles), whereas the lighter particles fail to be collected in the first step but are gathered in the second "F" housing (fine particles). Bran fractions derived from both debranning and micronization were used to enrich semolina, obtaining pasta products with enhanced health properties and minimal impacts on sensory quality [14]. The above technologies are also useful to better manage the natural or chemical contaminants (e.g., mycotoxins, heavy metals, and pesticides) that are typically concentrated in the outermost layers of the wheat kernel [15–17].

The aim of this study was to further improve both the micronization and air-classification of plants by introducing ad hoc modifications into the process in order to obtain a larger choice of less-refined milling fractions. More in detail, the micronizer pilot plant improvement was carried out through the addition of a grinding chamber, equipped with a hammer crusher impeller and a decanting collector, whereas the air-classifier pilot plant was added with both a programmable logic controller (PLC) for the pneumatic flow management and a chamber specially designed with a decreasing section. These latest improvements of pilot plants already supplied in our laboratories allowed us to produce innovative mixtures of milling products. Particular focus was put on studying the process to develop diverse milling fractions, each characterized by a peculiar content and composition in bran particles that could be used to make durum wheat end-products characterized by a higher bran content and healthier nutritional content.

2. Materials and Methods

2.1. Durum Wheat Cultivars

Three Italian durum wheat cultivars, namely Saragolla, Maestà, and Iride, were grown under conventional farming in the experimental field of CREA, Foggia (Italy) in the 2018–2019 growing season. Single grain samples were evaluated for test weight, moisture, protein content, gluten content, and yellow color through near-infrared analysis in transmission mode (NIT) by using Infratec™ mod 1241 (FOSS, Hillerød, Denmark) and the results were expressed as mean value of replicates (n = 10). Figure 1 shows the flow chart of the experimental plan.

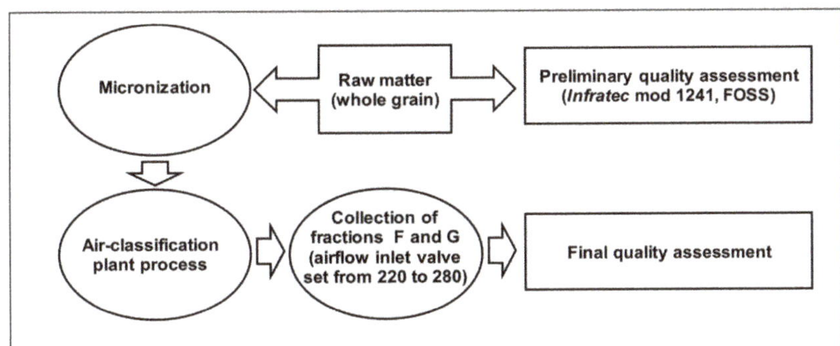

Figure 1. Flow chart of the experimental plan.

2.2. Micronization of Grain Samples and Air Classification of Milling Fractions

Durum wheat grain samples (11 kg for each cultivar) were micronized using a micronizer pilot plan (mod. 32300, KMXi-300-7,5; Separ Microsystem S.a.s, Brescia, Italy). The micronization step did not require preventive conditioning of the grains. An improvement in the micronizing pilot plant was achieved by adding a hammer crusher impeller with a reduced cross-section inside the grinding chamber compared to the normal type. This latter was equipped with a sieving grid ($\varnothing$ = 0.7 mm) suitable for obtaining a more homogeneous product. In addition, the grinding chamber was connected to a decanting chamber suitable for collecting the ground product. Moreover, with the aim to better detach the milled product from the plant walls, an electric vibrator was placed on the top side of the same chamber (Figure 2).

Figure 2. Micronizer pilot plant improved through the addition of a grinding chamber including a hammer crusher impeller (**A**) and a decanting collector (**B**).

Afterwards, 10.5 kg of micronized sample was submitted to an air-classifier pilot plant (turbo-separator unit, model SX-LAB; Separ Microsystem S.a.s, Brescia, Italy) suitable for particle sizes up to a set limit ($\varnothing \leq 1.5$ mm).

The plant was improved through the insertion of a programmable logic control (PLC) able to manage both the frequency and the action time of the airflow inside the separation chamber. Additionally, the chamber was further improved by the modification of the internal orbits that had been made according to a progressive decrease in the internal section. The latter modification allowed us to enhance the separation between the fine fractions and the coarse ones (Figure 3).

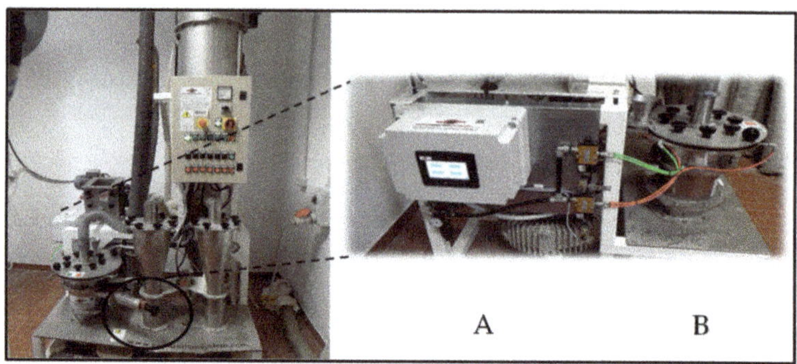

Figure 3. Air-classifier pilot plant augmented through the addition of a programmable logic controller (PLC) for the pneumatic flow management (**A**) and a chamber designed with a decreasing section (**B**). The circle indicates the setting airflow inlet valve. The F and G housings are placed below the steel plan located under the inlet valve.

Micronized aliquots of 1.5 kg were air-classified for one cycle at a time by setting the airflow inlet valve to 220, 230, 240, 250, 260, 270, or 280. At the end of each cycle the fractions of type G and F were collected and submitted for analysis.

2.3. Quality Analysis

Mean yield percentages of samples were evaluated based on the weight percentage of each sample in relation to the starting weight of the same sample. Ash content was also determined by following the official method [18]. The particle sizes of all fractions were measured by using certified test sieves (Giuliani Tecnologie S.r.l., Turin, Italy).

2.4. Statistical Analysis

The analysis of variance was performed on transformed data applying ANOVA (post hoc: Tukey test and Bonferroni correction) using the statistical software PAST 2.12 [19]. The data transformation (log or root square) was needed to achieve a normal distribution of the same data suitable for statistical analysis. The Bonferroni correction was used to counteract the incorrect rejection of a null hypothesis.

3. Results

The qualitative characterization of Saragolla, Maestà, and Iride grains are summarized in Table 1. Test weights among the three cultivars ranged from 81.0 to 84.4 kg/hL, namely in the first class group according to the official standards [20], revealing the absence of significant quantities of shrivelled kernels. The moisture content of the grains varied from 10.2% to 10.9%, below the maximum limit of 14.0%.

Table 1. Durum wheat cultivars, origin (region of Italy), and proximate quality parameters measured by NIT. Mean values of replicated analyses (n = 10).

Cultivar	Origin	Test Weight (kg/hL)	Moisture (%)	Protein Content (% db)	Gluten Content (% db)	Yellow Color
Saragolla	Puglia	81.6	10.9	12.9	8.9	13.9
Maestà	Puglia	81.0	10.2	15.6	11.0	15.4
Iride	Puglia	84.4	10.7	12.6	9.1	14.2

Except for yellow color, which was low for each of the three cultivars, protein and gluten percentages were satisfactory in all samples varying from 12.6 to 15.6, and from

8.9 to 11.0, respectively, suggesting good potential nutritional and technological quality of the cultivars.

Milling yield percentages were determined to test the efficiency of the milling process. The mean recovery of micronized samples was equal to 98.8%, showing a negligible loss of starting sample.

As expected, the air-classified fractions showed marked differences between the F and G fractions (Figure 4). In detail, the F fraction showed yield values with an ascending trend from 23% (registered for setting 220) to 93% (for setting 280). Conversely, a descending trend was observed for the G fractions, varying from 77% (setting 220) to 6% (setting 280). Yield recovery of the F fractions clearly exceeded 60% only for setting conditions ranging from 250 to 280. In the G fractions, this value exceeded only at 220 and 230 settings.

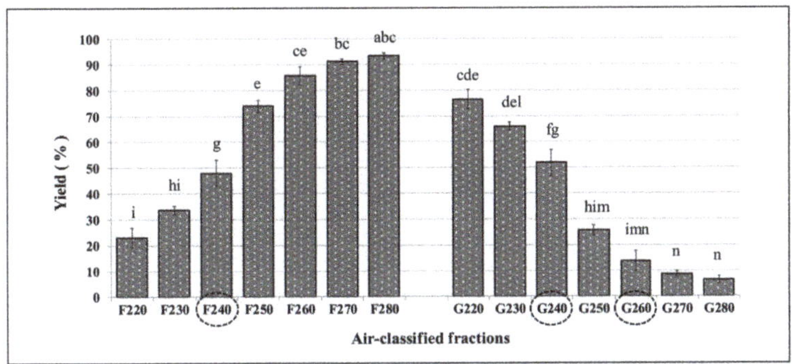

Figure 4. Mean yield percentage (%) of the air-classified fractions of samples belonging to the cultivars Saragolla, Maestà, and Iride. Different letters indicate statistically significant difference ($p < 0.05$, n = 3). Circles highlight the fractions with significant differences ($p < 0.05$) among cultivars (n = 2).

We did not observe significant differences ($p > 0.05$) among the three cultivars at any of the operating airflow conditions, with the exception of F240, G240, and G260 (Figure 4). Results obtained for each cultivar are presented in Table S1.

Further investigations are in progress to test the reliability of the process at those conditions.

The F and G type fractions obtained through the micronization and air-classification of whole durum grains contained a different percentage distribution of bran particles, due to the applied opening rate of the inlet valve. Ash content of F fractions was higher compared to that of G fractions, due to the higher content of bran particles in F fractions, while the G were richer in semolina (Figure 5). Considering the ash percentages across the F fractions, a descending trend was observed, with F220 (3.16%) and F230 (2.74%) showing the highest levels. Conversely, no significant differences ($p > 0.05$) were pointed out across the G fractions, observing a range from 1.67% (G220) to 1.42% (G280). Ash percentage of the air-classified fractions for each cultivar showed a trend that was similar to that of cultivar mean values with no significant differences ($p > 0.05$) from G220 to G280 (Table S2).

Results concerning the particle size distributions ($\varnothing > 425\ \mu m$, $425\ \mu m > \varnothing > 180\ \mu m$, $\varnothing < 180\ \mu m$) within the air-classified fractions are shown in Figure 6. As expected, heavier particles ($\varnothing > 425\ \mu m$) were found mainly in the G fractions, because these contained more semolina than bran residues as compared to the F fractions. The percentage of the heavier particles in G fractions ranged from 11.4% (G220) to 53.4% (G280). Conversely, heavy particles were scarcely present in the F millings, varying from 4.3% (F250) to 19.9% (F220). Intermediate ($425\ \mu m > \varnothing > 180\ \mu m$) and fine ($\varnothing < 180\ \mu m$) particles were predominant in all F fractions.

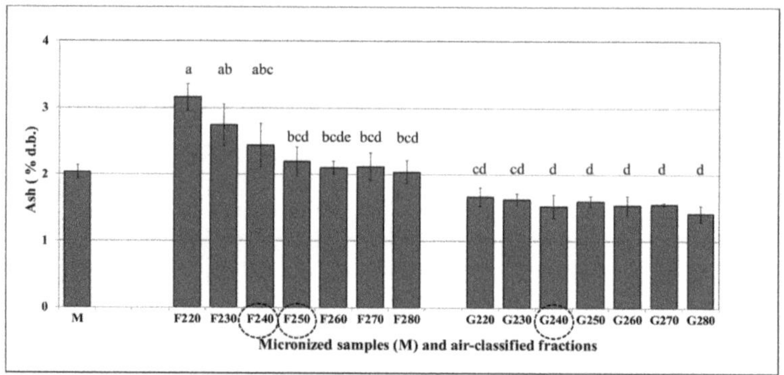

Figure 5. Mean ash percentages (% dry basis) of three durum cultivars: micronized wholemeals (M); air-classified fractions (F and G types). Different letters indicate statistically significant difference ($p < 0.05$); circles highlight the fractions with significant differences ($p < 0.05$) among cultivars (n = 2).

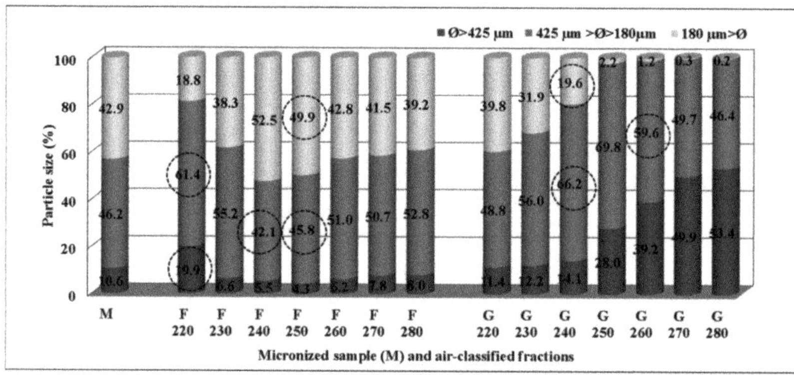

Figure 6. Mean percentages of particle size in micronized samples (M) and air-classified fractions (F and G types) of the mean of three cultivars. Circles highlight the fractions with significant differences ($p < 0.05$) among cultivars (n = 2).

Together, the intermediate and fine particles were prevalent in all G fractions, with the exception of G280 where their amount was 46.6% of the total, which could be explained by the less open inlet valve and the consequent different dynamics of the incoming airflow. Notably, a low percentage of finer particles was detected in F220 and F230 (18.8% and 38.3%, respectively). Detailed data on each cultivar are presented in Table S3, showing a marked trend towards no significant differences ($p > 0.05$) of percentage values from G250 to G280 fractions of all particle sizes assayed. The statistical analysis highlighted the influence of the two sources of variation (cultivar and setting valve point) related to the afore mentioned three particle sizes assayed (heavier, intermediate, and finer). The results showed that each source of variation had a significant ($p < 0.05$) influence on heavier size particles ($\varnothing > 425$ μm) in the fractions F, whereas their interaction was significant ($p < 0.05$) in the case of finer size particles both in F and in G. In the case of the fractions G, only the setting valve point factor was significant ($p < 0.05$) for heavier size particles ($\varnothing > 425$ μm).

4. Discussion

The measure of yield percentages allowed us to evaluate the air-classified fractions from the point of view of milling quality, this latter intended as an important aspect for the choice of more suitable fractions also for large scale production.

The ash content is an important quality parameter influenced both by genetic and environmental factors and associated with bran content [21]. In this study, the assay of the ash percentage proved useful to better characterize each air-classified fraction in any case containing both bran and semolina particles in different ratios.

The evaluation of particle size distribution within each fraction type (F and G) revealed the key role played by the involved physical factors (i.e., weight, air-pressure inside the circuits, turbulence, etc.). The resultant force obtained by the air-classification process, intended as a whole system, determined the distribution of the particles in F or G collectors. In any case, all air-classified millings were less refined compared to traditional semolina and were characterized by a specific percentage of bran residues. In studying the processing plants performance, the use of more than one cultivar with different grain characteristics was considered to test if they could introduce possible "critical" issues influencing the reliability of the system. Although the results need to be confirmed on a higher number of cultivars grown in different years and environments, the preliminary data pointed out that the F220, F250, and G240 fractions showed significant differences ($p < 0.05$) among the cultivars. Nevertheless, the use of the innovative plants allowed the production of several different types of air-classified fractions, allowing us to choose the more suitable for making less refined and more attractive end-products with innovative properties from among them due to the different combinations of heavier, intermediate and finer bran particles. Previous studies already highlighted that the addition of durum wheat bran with a particle size range of 150–500 μm, at levels of 10%, 20%, and 30%, had a negative impact on pasta sensory and technological properties [22]. However, through the addition of 30% of the same size range of bran particles, a comparable texture to that of commercial whole milled wheat pasta was achieved. Conversely, the addition of 10% had similar sensory texture scores to regular durum wheat pasta not supplemented with bran. The use of milling fractions enriched with more fine bran particles is expected to be more suitable for the release and bioavailability of bioactive compounds, due to both the major exposition of bran particle surfaces and aleurone cells to disruption, resulting in a better release of the intra cellular contents [23]. Therefore, the different air-classified fractions here described offer a suitable tool to produce innovative food-products characterized by new technological properties and improved nutritional value that could make these products more attractive to consumers. The inclusion of bran particles in wheat millings leads to technological disadvantages and worsens end-products quality compared to refined semolina or flour-based processes and products, causing, among other effects, a decrease in bread loaf volume, textural changes, and color changes [24]. These effects are more severe, especially at 20% incorporated bran, reducing pasta quality; yet, a reduced impact occurs at the same percentage of incorporation using finer bran [21]. In general, bran supplementation also has a positive effect due to the increase in polyphenols and phytosterols in end-products. These parameters are not influenced by bran particle size above 10% incorporation, except for phenolic acids, which increase at a higher rate of finer bran particles [21]. Therefore, the range size between 180 μm and 425 μm, which is included in the fractions here assayed, has a key role due to the presence of a part of the fine bran fractions (<180 μm, 180 μm, and 250 μm) suitable for better nutritional quality of pasta due to the presence of health-promoting compounds (from the outer kernel layers) compared to traditional pasta produced with refined semolina. The influence of the bran particle size on the water absorption of cooked pasta has been deeply studied. Finer bran particles produce a lower and significant degree of water absorption compared to particles with a larger diameter, regardless of the derivation of the particles (e.g., bran or middlings) [25]. On the other hand, the positive effects of medium coarse and high coarse particle sizes of semolina on the end product (pasta) quality has already been described [26]. Therefore, the particle size composition of less-refined fractions constitutes a very important aspect to be considered, especially in relation to technological aspects and the intended use of the end-products.

With regard to the detailed results concerning the single cultivars, it should be underlined that the choice to use them was aimed only at increasing the variability of samples tested. In any case the results are intended as a study on the cultivar behavior in the process, which would require a different experimental design (greater number of cultivars grown under different and controlled environmental conditions over multiple seasons). Indeed, our study was aimed at assessing the application of new technological solutions to current micronization and air-classification plants in order to improve the quality of the overall milling performances.

5. Conclusions

The improvement of the micronization and air-classified plants yielded a wide range of less-refined milling fractions that could be suitable to make durum wheat end-products characterized by enhanced health-promoting qualities. The technological modifications of the pilot plants already in use in our laboratories regarding both the micronizer (through the addition of a grinding chamber including a hammer crusher impeller and a decanting collector) and the air-classifier (through the inclusion of PLC and a suitable chamber) allowed us to deepen some quality aspects linked to the new mixtures that were developed. Indeed, milling fractions (F and G) have been characterized by new particle size content and composition as compared to those obtained before the update of milling plants. A preliminary assessment of the quality of the milling fractions was based on yield percentages, ash content, and particle size composition. The results revealed that the obtained air-classified fractions offer diverse choices to make a good compromise between high yield percentage, particle size composition, and the health benefits associated with a higher content of bran (fiber, antioxidants, minerals, etc.). Further investigations on technological and qualitative features of the air classified fractions are ongoing and will be the subject of upcoming papers.

Supplementary Materials: The following are available online at https://www.mdpi.com/2304-8158/10/8/1796/s1, Table S1: Mean yield percentage (%± S.D.) of micronized samples and air-classified fractions (cv. Saragolla, Maestà and Iride). Same letter: no significant difference ($p > 0.05$) within all fractions (F and G types) for each cultivar (n = 2), Table S2: Mean ash content (% dry basis ± S.D.) of micronized samples and air-classified fractions (cv. Saragolla, Maestà and Iride). Same letter: no significant difference ($p > 0.05$) within all fractions (F and G types) for each cultivar (n = 2), Table S3: Mean particle size content (% ± Standard Deviation) of micronized samples and air-classified fractions (cv. Saragolla, Maestà and Iride). Same letter: no significant difference ($p > 0.05$) within all fractions (F and G types) for each cultivar (n = 2).

Author Contributions: Conceptualization, G.A. and A.C.; methodology, G.A. and A.C.; validation, A.C. and G.A.; formal analysis, G.A.; data curation, G.A., F.S. and B.L.; writing—original draft preparation, G.A. and B.L.; writing—review and editing, G.A., B.L., and F.S.; supervision, G.A. All authors have read and agreed to the published version of the manuscript.

Funding: This research was carried out within, and was financially supported by, the "AsFRUM" project (Uff. Boll. Reg. Lazio n. 94, suppl.1, 24 November 2016).

Institutional Review Board Statement: Not applicable.

Informed Consent Statement: Not applicable.

Data Availability Statement: The data are contained within the article.

Acknowledgments: The authors wish to thank Alessandra Arcangeli and Roberto Mortaro for their precious technical support and Ernesto Bastoni for the upgrade of the plants.

Conflicts of Interest: The authors declare no conflict of interest.

References

1. Hammami, R.; Sissons, M. Durum wheat Products: Couscous. In *Wheat Quality for Improving Processing and Human Health*; Igregas, G., Ikeda, T., Guzman, C., Eds.; Springer: Cham, Switzerland, 2020; pp. 345–366.
2. Laddomada, B.; Durante, M.; Minervini, F.; Garbetta, A.; Cardinali, A.; D'Antuono, I.; Caretto, S.; Blanco, A.; Mita, G. Phytochemical characterization and anti-inflammatory activity of extracts from the whole-meal flour of Italian durum wheat cultivars. *Int. J. Mol. Sci.* **2015**, *16*, 3512–3527. [CrossRef]
3. Shewry, P.R.; Hassall, K.L.; Grausgruber, H.; Andersson, A.A.M.; Lampi, A.M.; Piironen, V.; Rakszegi, M.; Ward, J.L.; Lovegrove, A. Do modern types of wheat have lower quality for human health? *Nutr. Bull.* **2020**, *45*, 362–373. [CrossRef] [PubMed]
4. Calabriso, N.; Massaro, M.; Scoditti, E.; Pasqualone, A.; Laddomada, B.; Carluccio, M.A. Phenolic extracts from whole wheat biofortified bread dampen overwhelming 1 inflammatory response in human endothelial cells and monocytes: Major role of VCAM-1 and CXCL-10. *Eur. J. Nutr.* **2020**, *59*, 2603–2615. [CrossRef]
5. Budreviciute, A.; Damiati, S.; Sabir, D.K.; Onder, K.; Schuller-Goetzburg, P.; Plakys, G.; Katileviciute, A.; Khoja, S.; Kodzius, R. Management and Prevention Strategies for Non-communicable Diseases (NCDs) and Their Risk Factors. *Front. Public Health* **2020**, *8*, 788. [CrossRef] [PubMed]
6. Foster, S.; Beck, E.; Hughes, J.; Grafenauer, S. Whole Grains and Consumer Understanding: Investigating Consumers' Identification, Knowledge and Attitudes to Whole Grains. *Nutrients* **2020**, *12*, 2170. [CrossRef] [PubMed]
7. Pasqualone, A.; Laddomada, B.; Centomani, I.; Paradiso, V.M.; Minervini, D.; Caponio, F.; Summo, C. Bread making aptitude of mixtures of re-milled semolina and selected durum wheat milling by-products. *LWT Food Sci. Technol.* **2017**, *78*, 151–159. [CrossRef]
8. Onipe, O.O.; Jideani, A.I.O.; Beswa, D. Composition and functionality of wheat bran and its application in some cereal food products. *Int. J. Food Sci. Technol.* **2015**, *50*, 2509–2518. [CrossRef]
9. Coda, R.; Kärki, I.; Nordlund, E.; Heiniö, R.-L.; Poutanen, K.; Katina, K. Influence of particle size on bioprocess induced changes on technological functionality of wheat bran. *Food Microbiol.* **2014**, *37*, 69–77. [CrossRef]
10. Fares, C.; Platani, C.; Baiano, A.; Menga, V. Effects of processing and cooking on phenolic acid profile and antioxidant capacity of durum wheat pasta enriched with debranning fractions of wheat. *Food Chem.* **2010**, *119*, 1023–1029. [CrossRef]
11. Cammerata, A.; Marabottini RAllevato, E.; Aureli, G. Content of minerals and deoxynivalenol in the air-classified fractions of durum wheat. *Cereal Chem.* **2021**. [CrossRef]
12. Taddei, F.; Galassi, E.; Nocente, F.; Gazza, L. Innovative milling processes to Improve the technological and nutritional quality of parboiled brown rice pasta from contrasting amylose content cultivars. *Foods* **2021**, *10*, 1316. [CrossRef] [PubMed]
13. Martín-García, B.; Verardo, V.; de Cerio, E.D.; del Carmen Razola-Díaz, M.; Messia, M.C.; Marconi, E.; Gómez-Caravaca, A.M. Air classification as a useful technology to obtain phenolics-enriched buckwheat flour fractions. *Food Sci. Technol.* **2021**, *150*, 111893. [CrossRef]
14. Ciccoritti, R.; Taddei, F.; Nicoletti, I.; Gazza, L.; Corradini, D.; D'Egidio, M.G.; Martini, D. Use of bran fractions and debranned kernels for the development of pasta with high nutritional and healthy potential. *Food Chem.* **2017**, *225*, 77–86. [CrossRef] [PubMed]
15. Ficco, D.B.M.; Borrelli, G.M.; Miedico, O.; Giovanniello, V.; Tarallo, M.; Pompa, C.; De Vita, P.; Chiaravalle, A.E. Effects of grain debranning on bioactive compounds, antioxidant capacity and essential and toxinc trace elements in purple durum wheats. *LWT Food Sci. Technol.* **2020**, *118*, 108734. [CrossRef]
16. Sovrani, V.; Blandino, M.; Scarpino, V.; Redyneri, A.; Coïsson, J.D.; Travaglia, F.; Locatelli, M.; Bordiga, M.; Montella, R.; Arlorio, M. Bioactive compound content, antioxidant activity, deoxynivalenol and heavy metal contamination of pearled wheat fractions. *Food Chem.* **2012**, *135*, 39–46. [CrossRef]
17. Laddomada, B.; Del Coco, L.; Durante, M.; Presicce, D.S.; Siciliano, P.A.; Fanizzi, F.P.; Logrieco, A.F. Volatile metabolite profiling of durum wheat kernels contaminated by Fusarium poae. *Metabolites* **2014**, *4*, 932–945. [CrossRef]
18. EN ISO 2171:2010. *Cereals, Pulses and By-Products—Determination of Ash Yield by Incineration (ISO 2171:2007)*; ISO: Brussels, Belgium, 2007.
19. Hammer, O.; Harper, D.A.T.; Ryan, P.D. PAST: Paleontological Statistic software package for education and data analysis. *Paleontol. Electron.* **2001**, *4*, 1–9. Available online: http://palaeo-electronica.org/2001_1/past/issue1_01.htm. (accessed on 22 June 2001).
20. UNI 10709:1998. *Durum Wheat Grains. Qualitative Requirements, Classification and Test Methods*; UNI: Rome, Italy, 1998.
21. Alzuwaid, N.T.; Fellows, C.M.; Laddomada, B.; Sissons, M. Impact of wheat bran particle size on the technological and phytochemical properties of durum wheat pasta. *J. Cereal Sci.* **2020**, *95*, 103033. [CrossRef]
22. Aravind, N.; Sissons, M.; Egan, N.; Fellows, C. Effect of insoluble dietary fibre addition on technological, sensory, and structural properties of durum wheat spaghetti. *Food Chem.* **2012**, *130*, 299–309. [CrossRef]
23. Hemery, Y.M.; Anson, N.M.; Havenaar, R.; Haenen, G.R.M.M.; Noort, M.W.J.; Rouau, X. Dry fractionation of wheat bran increases the bioaccessibility of phenolics acids. *Food Res. Int.* **2010**, *43*, 1429–1438. [CrossRef]
24. Hemdane, S.; Jacobs, P.J.; Dornez, E.; Verspreet, J.; Delcour, J.A.; Courtin, C.M. Wheat (*Triticum aestivum* L.) Bran in Bread Making: A Critical Review. *Compr. Rev. Food Sci. food Saf.* **2016**, *15*, 28–42. [CrossRef] [PubMed]

25. Sandberg, E. *The Effect of Durum Wheat Bran Particle Size on the Quality of Bran Enriched Pasta*; Swedish University of Agricultural Sciences, Department of Food Science: Uppsala, Sweden, 2015; Series no: 405. Available online: http://stud.epsilon.slu.se (accessed on 8 July 2015).
26. Sacchetti, G.; Cocco, G.; Cocco, D.; Neri, L.; Mastrocola, D. Effect of semolina particle size on the cooking kinetics and quality of spaghetti. *Procedia Food Sci.* **2011**, *1*, 1740–1745. [CrossRef]

foods

MDPI

Article

Use of Air-Classification Technology to Manage Mycotoxin and Arsenic Contaminations in Durum Wheat-Derived Products

Alessandro Cammerata [1], Rosita Marabottini [2], Viviana Del Frate [1], Enrica Allevato [3], Samuela Palombieri [4], Francesco Sestili [4,*] and Silvia Rita Stazi [5,*]

[1] Council for Agricultural Research and Economics, Research Centre for Engineering and Agro-Food Processing, Via Manziana 30, 00189 Rome, Italy; alessandro.cammerata@crea.gov.it (A.C.); viviana.delfrate@crea.gov.it (V.D.F.)

[2] Department for Innovation in Biological, Agro-Food and Forest Systems (DIBAF), University of Tuscia, Via San Camillo de Lellis snc, 01100 Viterbo, Italy; marabottini@unitus.it

[3] Department of Environmental and Prevention Sciences (DiSAP), University of Ferrara, Via Luigi Borsari 46, 44121 Ferrara, Italy; enrica.allevato@unife.it

[4] Department of Agriculture and Forest Sciences (DAFNE), University of Tuscia, Via San Camillo de Lellis snc, 01100 Viterbo, Italy; palombieri@unitus.it

[5] Department of Chemical, Pharmaceutical and Agricultural Sciences (DOCPAS), University of Ferrara, Via Luigi Borsari 46, 44121 Ferrara, Italy

* Correspondence: francescosestili@unitus.it (F.S.); silviarita.stazi@unife.it (S.R.S.); Tel.: +39-328-816-6276 (F.S.); +39-393-007-1671 (S.R.S.)

Citation: Cammerata, A.; Marabottini, R.; Del Frate, V.; Allevato, E.; Palombieri, S.; Sestili, F.; Stazi, S.R. Use of Air-Classification Technology to Manage Mycotoxin and Arsenic Contaminations in Durum Wheat-Derived Products. *Foods* **2022**, *11*, 304. https://doi.org/10.3390/foods11030304

Academic Editors: Giuseppa Di Bella and Luís Abrunhosa

Received: 21 December 2021
Accepted: 20 January 2022
Published: 24 January 2022

Publisher's Note: MDPI stays neutral with regard to jurisdictional claims in published maps and institutional affiliations.

Abstract: Mycotoxins are the most common natural contaminants and include different types of organic compounds, such as deoxynivalenol (DON) and T-2 and HT-2 toxins. The major toxic inorganic elements include those commonly known as heavy metals, such as cadmium, nickel, and lead, and other minerals such as arsenic. In this study, micronisation and air classification technologies were applied to durum wheat (*Triticum turgidum* ssp. *durum* L.) samples to mitigate inorganic (arsenic) and organic contaminants in unrefined milling fractions and final products (pasta). The results showed the suitability of milling plants, providing less refined milling products for lowering amounts of mycotoxins (DON and the sum of T-2 and HT-2 toxins) and toxic inorganic elements (As, Cd, Ni, and Pb). The results showed an As content (in end products) similar to that obtained using semolina as raw material. In samples showing high organic contamination, the contamination rate detected in the more bran-enriched fractions ranged from 74% to 150% (DON) and from 119% to 151% (sum of T2 and HT-2 toxins) as compared to the micronised samples. Therefore, this technology may be useful for manufacturing unrefined products with reduced levels of organic and inorganic contaminants, minimising the health risk to consumers.

Keywords: durum wheat; air classification; inorganic contaminants; organic contaminants; arsenic; mycotoxins

1. Introduction

Currently, cereal grains are the primary source of carbohydrates in human nutrition worldwide. The Food and Agriculture Organization (FAO) forecasts that world grain utilisation in 2021/2022 will reach a record level of 2809 million tons. The FAO's forecast for total wheat utilisation is 777 million tons, which is 2.4% (18.5 million tons) higher than that in 2020/2021 [1]. Durum wheat is a crop of primary importance, as semolina (obtained by milling durum wheat grain) is the basic raw material for the realisation of several highly consumed end products (pasta, couscous, bulgur, local breads) and is mainly used as a traditional food for Mediterranean populations, and it is highly appreciated worldwide.

Cereals such as wheat provide most of the energy in diets, and their consumption in the form of unrefined end products provides health advantages because they are rich in bioactive compounds. Among these, antioxidant molecules are of great interest because they are involved in lowering the risk of cardiovascular diseases, cancers, gastrointestinal disorders, and type 2 diabetes [2–6].

In general, wholemeal products are realised by mixing suitable amounts of bran, germ, and endosperm fractions so that they are as similar as possible to the natural grains [7]. However, the consumption of bran-enriched products could increase the risk of consumers' exposure to food-borne organic and inorganic contaminants, the main ones being mycotoxins, toxic elements such as arsenic, and heavy metals. All of these compounds can persist from the raw matter to the end products, in addition to so-called processing compounds, such as acrylamide [8]. The environment plays a key role as the main source of contamination in grains, such as wheat, starting from the cultivation field and ending in storage. Even though minerals are ubiquitous in the Earth's crust, anthropogenic activities are increasingly resulting in the contamination of water and soil with toxic metals and metalloids [9–11]. Among toxic minerals, arsenic (As), cadmium (Cd), and lead (Pb) are ranked as the most hazardous substances owing to their toxicity, prevalence, and potential for human exposure [12]. Among the metals of great concern, cadmium (Cd) is uniformly distributed within the endosperm, whereas lead (Pb) and nickel (Ni) are mainly located in the outer teguments [13–15]. European legislation has established maximum tolerable levels of several metals and metalloids in foodstuffs. However, the maximum levels of Pb, Cd, and tin (Sn) are currently established and monitored in cereals, whereas the maximum arsenic content has been established only for rice (0.200 µg/g) [16,17].

Mycotoxins, which are produced by several fungi growing on cereal grains, are of great concern, as they add organic contaminants to a wide range of food commodities [17,18]. Deoxynivalenol (DON) and T-2/HT-2 toxins belong to type B and A, respectively, groups of chemically related compounds named trichothecenes. They are the most widespread *Fusarium* toxins in small grain cereals, including durum wheat (*Triticum turgidum* ssp. *durum* L.), which constitutes the main source through which humans and animals are exposed to these types of mycotoxins [19,20]. These mycotoxins are produced by fungi that colonise the kernel, starting in its outer layers, where they mostly accumulate. Since the coating structures end up in the bran milling fraction, bran-enriched foods represent an increased risk of human exposure to mycotoxins [21–25].

The maximum tolerable limits for DON in food products have been set by the European Commission [17]. However, to date, there are no legal limits for T-2 and HT-2 toxins in food and feed. Nevertheless, the European Commission recommendation on the content of T-2 and HT-2 in cereals and cereal indicates 100 µg/kg for the sum of T-2 and HT-2 as the maximum level in cereal grains [26].

The use of new technologies, such as debranning, micronisation, and air classification, at the starting phase of wheat processing is effective in reducing the presence of undesirable compounds within the milling products [27–29]. In fact, micronisation combined with air classification provides suitable solutions to obtain less refined products with improved safety and nutritional aspects, ensuring optimal technological quality. Recently, a study of the process carried out using updated pilot plants allowed the selection of several unrefined milling fractions obtained from micronised and air-classified durum wheat samples. Therefore, three types of fractions were selected based on the results obtained in previous assays [30,31]. These fractions proved to be of interest owing to the yield percentage, particle size composition, and health benefits associated with a higher content of bran (fibre, antioxidants, minerals, etc.).

The goal of this study was to assess the reliability of recently updated milling plants (microniser and air classifier) to minimise the presence of DON, T-2 and HT-2 toxins, and toxic elements (mainly arsenic) along the production chain.

2. Materials and Methods

2.1. Durum Wheat Samples

A process study was carried out on three durum wheat grain samples (group I) of cultivars Saragolla, Maestà, and Iride. All of them were grown in conventional farming fields in Apulia (Southern Italy) during the 2018–2019 season (Figure 1). The use of these three samples allowed the identification of air-classifying conditions that were suitable for the aim of this study. In particular, group I samples were selected from those collected within the "AsFRUM" research project. The criteria adopted in the choice of samples concerned both the type and the rate of contamination. Moreover, the use of three different cultivars only served to achieve greater variability in the grain characteristics without accounting for the cultivar as a factor. On the basis of the results obtained in the process study, three additional durum wheat grain samples (group II) were used only under the selected air-classifying conditions previously identified in the process study. These group II samples were collected within the aforesaid research project during the same cropping year (2018–2019) in conventional growing fields located in three regions of the centre-south of Italy. More specifically, one durum wheat sample (Saragolla) was grown in the Lazio region (Central Italy), and the other two samples from the cv. Antalis were grown in the Marche region (Central Italy) and in the Basilicata region (Southern Italy) (Figure 1). In this case (group II), the selection criteria adopted for grain samples also concerned both the type and the rate of contamination. Three analytical replicates were carried out for all of the materials tested, as detailed below.

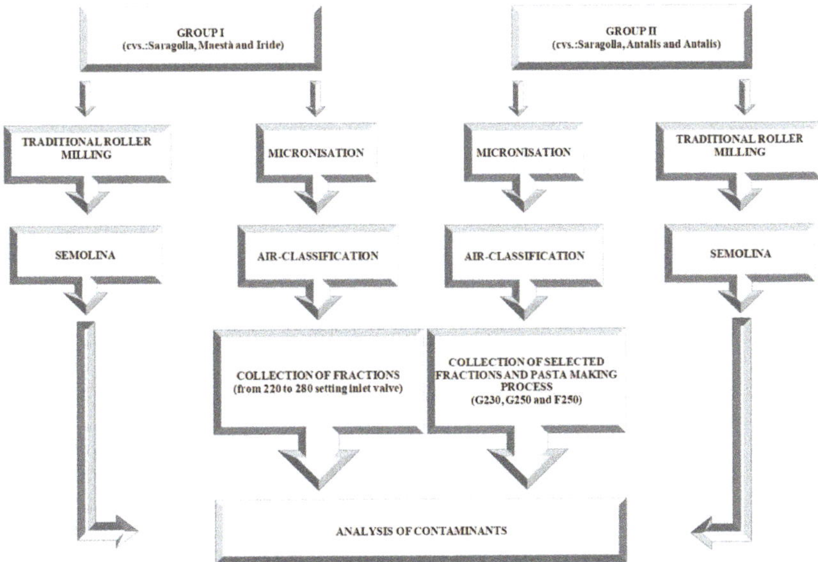

Figure 1. Flow chart of the experimental plan.

2.2. Micronisation and Air Classification Processes

The milling plants (microniser and air classifier) used in this study were already supplied to our laboratories and recently updated [30].

Briefly, innovative advances in the microniser pilot plant (model 32,300, KMXi-300-7,5; Separ Microsystem S.a.s, Brescia, Italy) included the introduction of a hammer crusher impeller, which reduced the internal cross-section and a sieving grid ($\varnothing$ = 0.7 mm), which produced a more homogeneous milled product than the traditional type. The air-classifier pilot plant (model SX-LAB; Separ Microsystem S.a.s, Brescia, Italy) designed for a sample particle size up to a set limit ($\varnothing \leq 1.5$ mm) was improved through the insertion of a programmable logic controller (PLC) capable of managing the airflow inside the separation chamber. Moreover, the orbits located in the inner part of the chamber were improved by designing a progressively decreasing inside diameter. Each micronised sample (groups I and II) was split into 1.5 kg aliquots. The micronisation step did not require preventive conditioning of the grains. The samples included in group I were subjected to the air classification process by setting the airflow inlet valve from 220 to 280 as opening conditions for each cycle at a time. At the end of each cycle, fractions of types G (heavier coarse particles) and F (fine particles) were collected. The samples of group II were only air-classified at the 230 and 250 settings based on a previous evaluation (see Figure 1). Therefore, the three air-classified fractions (F250, G250, and G230) were collected and used in the subsequent steps.

2.3. Traditional Roller Milling Process

The traditional roller milling process was carried out as a reference process. Grain samples were conditioned by adding tap water until a moisture value of 17% was achieved, and the samples were left to rest for 24 h. This treatment facilitated the processes of both stripping the kernels and softening the endosperm, making grain samples suitable for processing in the traditional roller milling plant (Bühler, model MLU 202, Uzwil, Switzerland). The main milling products were collected: semolina refined through a sieving treatment (sieve types: 38GG, 40GG, and 44GG) using a suitable pilot plant sieving system (NAMAD Impianti, Rome, Italy), bran fractions (coarse and refined types), and fine middlings.

2.4. Pasta-Making Process

A pilot press plant (NAMAD Impianti, Rome, Italy) was used, and the "spaghetti" format ($\emptyset$ = 1.6 mm) was adopted in the pasta-making process. The pasta samples were dried for 18 h at low temperature (50 °C) using a suitable dryer plan (AFREM, Lyon, France).

2.5. Mycotoxin Analysis

Sampling was carried out based on representative criteria, and all milled samples (from micronisation, air classification, and traditional roller milling) were stored at 2–8 °C until analysis [32]. DON content was assessed using the Ridascreen® DON kit (Ridascreen® DON method, R-Biopharm AG, Darmstadt, Germany) following the manufacturer's instructions. Specifically, extraction at a ratio of 1 part sample to 5 parts deionised water was performed, and then the sample was shaken vigorously for 3 min. The extract was filtered through a Whatman no. 1 filter, and the filtrate was collected as a sample. The limit of detection (LOD) of the enzyme-linked immunosorbent assay (ELISA) was 18.5 µg/kg. The recovery range obtained with this method was 85–110% of the sample. Deionised water was obtained from the Water Purification System Zeener Power I (Human Corporation, Seoul, Korea). The basic robotic immunoassay operator (BRIO, SEAC, Radim Group, Florence, Italy) was employed, and the absorbance data were acquired using a Sirio-S microplate reader (SEAC, Radim Group, Florence, Italy). The ELISA method for the DON content in durum wheat samples had already been validated by comparison with chromatographic (HPLC) analysis [33]. The sum of T-2 and HT-2 toxins was measured by ELISA analysis (Veratox® T2/HT2 toxin, Neogen, Lansing, MI, USA) according to the manufacturer's instructions. Extraction at a ratio of 1 part sample to 5 parts 70% methanol was carried out, and then the sample was shaken vigorously for 3 min. The extract was filtered through a Whatman no. 1 filter, and the filtrate was collected as a sample. The limit of quantification (LOQ) was 25 µg/kg. The ELISA method for measuring the sum of T-2 and HT-2 toxin content in durum wheat samples was assessed by comparison with ultra-performance liquid chromatography (UPLC) analysis [34]. Absorbance values were read using the Sirio-S microplate reader (SEAC, Radim Group, Florence, Italy), and the RIDA® Soft Win software (R-Biopharm AG, Darmstadt, Germany) was used for the quantification of mycotoxins in samples.

2.6. Toxic Elements

Micronised samples and all air-classified G and F fractions were processed to determine the concentrations of As, Cd, Pb, and Ni. Each sample underwent an initial step of acid digestion using a laboratory high-pressure microwave oven (Mars plus CEM, Cologno al Serio (BG), Italy) with a power of 1.200 W. Approximately 200 mg of the dry sample was placed in contact with 7.5 mL of 67% v/v HNO_3 for one hour, and then 0.5 mL of HCl was added and left to digest for one hour; finally, 2 mL of H_2O_2 was added and left to digest for 30 min. At the end of the predigestion process, the obtained acid solution was mineralised inside a microwave, with a heating program in which a temperature of 180 °C was reached in 37 min and incubation at 180 °C continued for 15 min. At the end of the digestion procedure, the samples were appropriately diluted with high-purity water (18 MΩ/cm) obtained from a Milli-Q water purification system (Millipore, Bedford, MA, USA) and then filtered (DISMIC 25HP PTFE syringe filter, pore size = 0.45 mm, Toyo Roshi Kaisha, Ltd., Tokyo, Japan) and stored in a plastic screw-cap tube (Nalgene, New York, NY, USA). Each experiment was performed in triplicate.

Elemental quantification was performed using an inductively coupled plasma optical emission spectrometer (ICP-OES) with an axial configuration (8000 DV, Perkin Elmer Inc. Waltham, MA, USA) equipped with an ultrasonic nebuliser (Teledyne Cetac Tecnologies, Omaha, USA). To assess the concentration levels of trace elements, calibration standards were prepared and treated in the same way as the samples before dilution (multi-element standard solution, CaPurAn, CPAchem, Stara Zagora, Bulgaria). The frequencies used for the determinations were: Cd 228.8 nm, Fe 259.9 nm, Cu 324.7 nm, Pb 220.3, Ni 231.6 nm, and As 197.69 nm.

The European Reference Material ERM-BC21 was used as the standard material to assess the accuracy of the measurements.

The super-pure-grade reagents used for microwave-assisted digestion were as follows: hydrochloric acid (36% HCl), nitric acid (69% HNO_3), and hydrogen peroxide (30% H_2O_2); highly pure water (18 MΩ/cm) was used to dilute the standards and prepare the samples throughout the chemical procedure.

2.7. Statistical Analysis

The results for the mycotoxin content were subjected to analysis of variance (ANOVA and Tukey test) using the statistical software PAST v. 2.12, Oslo, Norway [35].

3. Results

3.1. Determination of Mycotoxin Concentration

All air-classified G and F fractions (220–280) collected from group I samples were analysed to show the distribution of DON and T-2 and HT-2 toxins (sum of the two types) within the same samples. The results clearly highlight the influence of the process conditions on the DON content in the milling fractions collected (Figure 2).

The average levels of DON In micronised samples ranged between 28 µg/kg (cv. Iride) and 156 µg/kg (cv. Maestà). The highest mycotoxin content was detected in the F fractions, with a clear decreasing trend from fraction F220 to fraction F280 in all samples tested. From F220 to F240, which had the highest degree of contamination, the differences were statistically significant ($p < 0.05$) as compared with the corresponding micronised samples. A marked reduction in mycotoxin content was also found in all G fractions, with no significant differences ($p > 0.05$) between the most contaminated cultivar samples (Saragolla and Maestà). In this regard, all G fractions were significantly different ($p < 0.05$) from the F fractions.

The aforementioned general trend was even more marked in the case of the residual content, expressed as a percentage, compared to the micronised samples, all of which were set equal to 100%. In fact, the contamination rate varied from a minimum of 155% (cv. Maestà) to a maximum of 215% (cv. Iride) in the F220 fractions, whereas it decreased in the F280 fractions to values even lower than those of the micronised samples (cv. Maestà = 86% and cv. Iride = 94%, both in F280). Among the G fractions, the maximum rate achieved was 66% (G250) as compared to the micronised sample (cv. Iride).

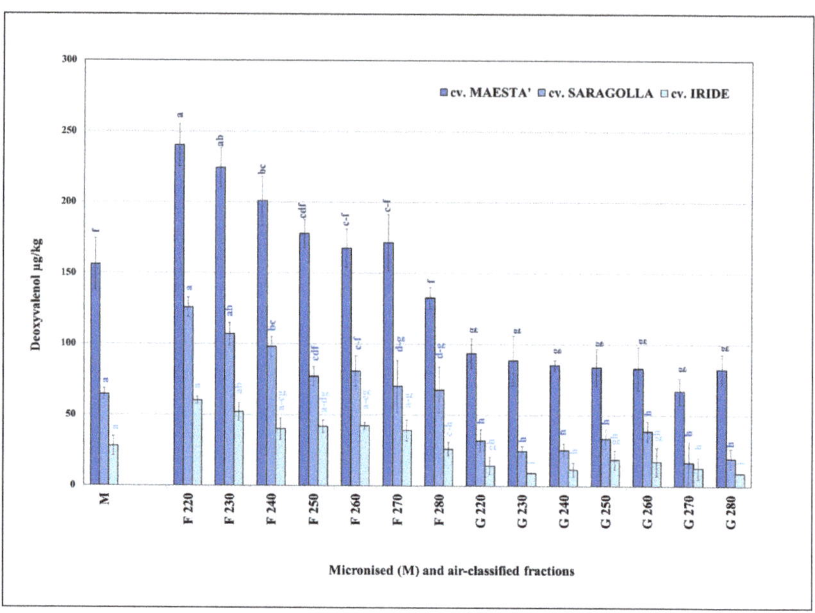

Figure 2. Average content of DON in the micronised (M) and air-classified fractions (F and G) of the durum wheat samples belonging to group I; different letters within each cultivar indicate significant differences ($p < 0.05$), $n = 3$.

A trend similar to DON was also found for the sum of T-2 and HT-2 toxin content, though to a greater extent (Figure 3). In fact, the F fractions F220, F230, and F240 showed significantly higher levels of contamination ($p < 0.05$) with a clear further reduction in the remaining F fractions as compared to the average values of the corresponding micronised samples, ranging from 58 µg/kg (cv. Iride) to 256 µg/kg (cv. Saragolla). A marked reduction in mycotoxin content was detected in all G fractions without significant differences ($p > 0.05$) in any samples between fractions G240 and G280, but they were significantly different ($p < 0.05$) from the F fractions. Moreover, similarly to DON, a marked effect of the air classification process was evident in the percentage evaluation as compared to micronised samples (set equal to 100%) In fact, within all F220 fractions, the percentage rates varied from a minimum of 164% (cv. Maestà) to a maximum of 225% (cv. Iride), with a tendency to decrease as the opening of the airflow inlet valve increased until reaching levels close to those of the micronised samples (Maestà = 105% in F270). A further decreasing trend was observed in G fractions, which did not exceed the percentage value of 46% for G230 (cv. Maestà), except for G220 in all samples (maximum value = 67% for cv. Iride).

The results for the DON contents only in the selected fractions (group II) showed a trend similar to that of group I (Figure 4). In fact, in the case of the more contaminated sample (cv. Antalis, Marche region) showing the highest DON content (1350 µg/kg) in the micronised sample, there was no significant difference ($p > 0.05$) between micronised and F250 samples, and significant differences ($p < 0.05$) between micronised and G fractions were confirmed. The pasta-making process effect tended to limit the concentration of DON in the F fractions as compared to the micronised sample, and only the F250 fraction (1043 µg/kg) was significantly different ($p < 0.05$) from semolina in Antalis (Marche).

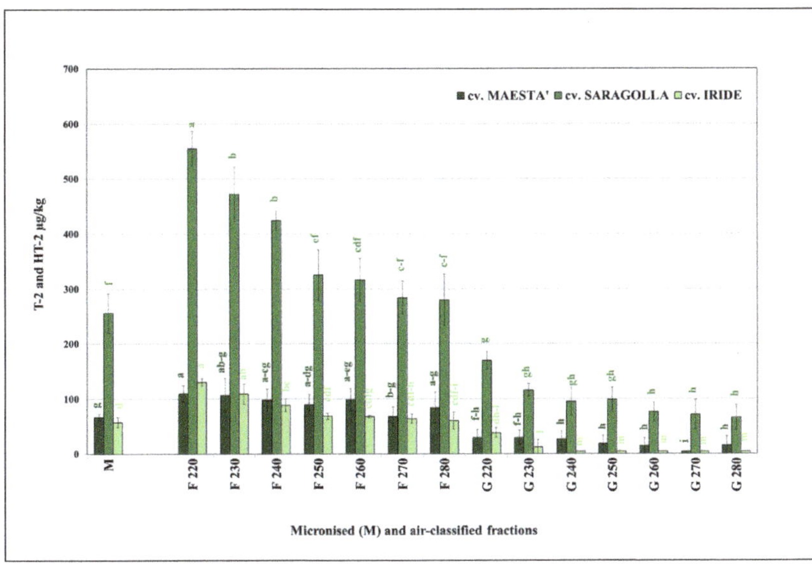

Figure 3. Average content of the sum of T-2 and HT-2 toxins in the micronised (M) and air-classified fractions (F and G) of the durum wheat samples belonging to group I; different letters within each cultivar indicate significant differences ($p < 0.05$), $n = 3$.

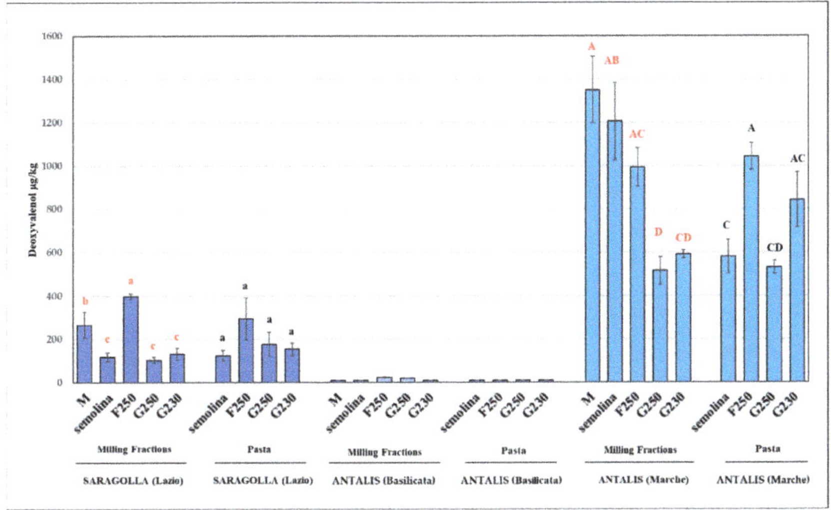

Figure 4. Group II: average content of DON in the micronised (M) and air-classified fractions (F and G) of the durum wheat samples that tested positive; different letters within each group indicate significant difference ($p < 0.05$). Different colored letters are referred to different sample types (red for milling fractions and black for pasta), $n = 3$.

As regards contamination by T-2 and HT-2 toxins, which is defined as the sum of the two, in the most contaminated sample (cv. Antalis, Basilicata), there was a trend showing the highest contamination rate in the bran-enriched F250 fraction (97 µg/kg), but the difference was not significant ($p > 0.05$) compared to the micronised sample (64 µg/kg) (Figure 5). Conversely, a marked decrease from 34 µg/kg in the G230 fraction to undetectable levels (less than the limit of quantification (<LOQ)) in the G250 fraction was observed. Moreover, in micronised samples that did not show detectable levels of the sum of T-2 and HT-2 toxins (cv. Saragolla, Lazio region, and cv. Antalis, Marche region), the "latent" presence of toxins, just below the LOQ (25 µg/kg), led to their detection only in the F250 fraction (maximum value: 37 µg/kg). The absence of contamination was confirmed in all G250 fractions analysed, whereas a marked effect of the pasta-making process on the most contaminated fractions (F250) was observed.

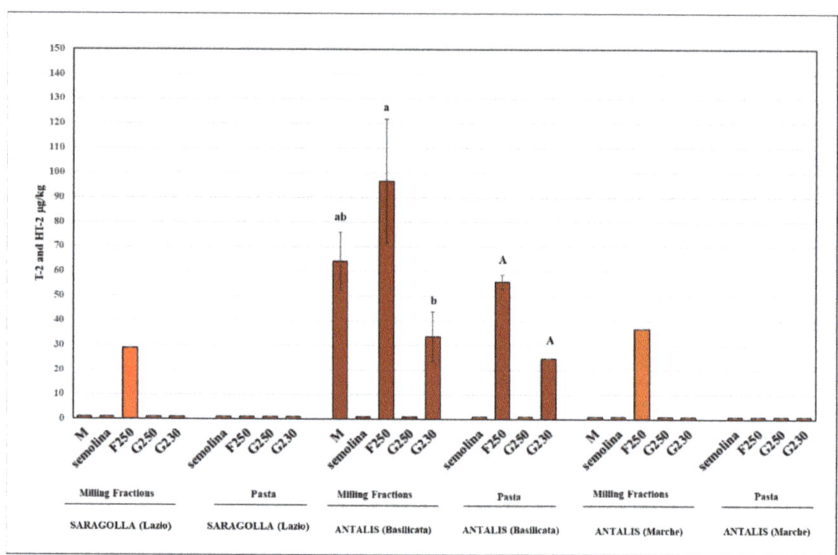

Figure 5. Group II: average values of the sum of the T-2 and HT-2 toxins in the micronised (M) and in the air-classified fractions (F and G) of the durum wheat samples that tested positive; different letters within each group indicate significant difference ($p < 0.05$), $n = 3$; data without a letter indicate single values (F250).

3.2. Determination of Toxic Elements

All air-classified G and F fractions (220–280) collected from the group I samples were analysed to determine the distribution of As, Cd, Ni, and Pb (Tables 1 and 2).

Table 1. Group I: average values ± standard error (SE) of arsenic (As), cadmium (Cd), nickel (Ni), and lead (Pb) contents in the micronised samples and F-type air-classified fractions; d.m.: dry matter; LOD: limit of detection. Different letters indicate significant difference ($p < 0.05$), $n = 3$.

		cv. Maestà (Apulia Region)			
		As (µg/g d.m.) ± SE	Cd (µg/g d.m.) ± SE	Ni (µg/g d.m.) ± SE	Pb (µg/g d.m.) ± SE
Micronised sample		0.478 ± 0.037 [a]	0.002 ± 0.0004 [a]	0.043 ± 0.001 [b]	0.157 ± 0.019 [a]
	F 220	0.143 ± 0.004 [d]	0.001 ± 0.0002 [a]	0.031 ± 0.001 [b]	0.059 ± 0.020 [bc]
	F 230	0.190 ± 0.008 [cd]	<LOD	0.027 ± 0.003 [b]	0.030 ± 0.005 [c]
	F 240	0.274 ± 0.012 [bc]	0.001 ± 0.0002 [a]	0.034 ± 0.002 [b]	0.039 ± 0.004 [c]
Air-classified fractions	F 250	0.359 ± 0.022 [b]	<LOD	<LOD	0.005 ± 0.001 [c]
	F 260	0.522 ± 0.016 [a]	<LOD	0.355 ± 0.012 [a]	0.038 ± 0.005 [bc]
	F 270	0.511 ± 0.018 [a]	0.074 ± 0.025 [a]	0.058 ± 0.012 [b]	0.103 ± 0.004 [ab]
	F 280	0.513 ± 0.018 [a]	0.001 ± 0.0003 [a]	0.041 ± 0.013 [b]	0.025 ± 0.006 [c]
		cv. Saragolla (Apulia Region)			

Table 1. *Cont.*

		cv. Maestà (Apulia Region)			
		As (µg/g d.m.) ± SE	Cd (µg/g d.m.) ± SE	Ni (µg/g d.m.) ± SE	Pb (µg/g d.m.) ± SE
Micronised sample		0.518 ± 0.069 [a]	0.008 ± 0.0013 [abc]	0.143 ± 0.004 [b]	0.152 ± 0.007 [ab]
	F 220	0.073 ± 0.012 [c]	0.001 ± 0.002 [c]	0.014 ± 0.001 [e]	0.074 ± 0.002 [c]
	F 230	0.207 ± 0.009 [bc]	0.005 ± 0.0004 [bc]	0.039 ± 0.004 [d]	0.077 ± 0.004 [c]
	F 240	0.286 ± 0.014 [b]	0.011 ± 0.003 [a]	0.051 ± 0.002 [cd]	0.081 ± 0.015 [bc]
Air-classified fractions	F 250	0.486 ± 0.023 [a]	0.005 ± 0.001 [bc]	0.055 ± 0.002 [c]	0.086 ± 0.011 [bc]
	F 260	0.555 ± 0.029 [a]	0.013 ± 0.003 [ab]	0.057 ± 0.001 [c]	0.163 ± 0.031 [a]
	F 270	0.590 ± 0.038 [a]	0.007 ± 0.002 [abc]	0.126 ± 0.001 [b]	0.118 ± 0.011 [abc]
	F 280	0.297 ± 0.061 [b]	<LOD	0.171 ± 0.003 [a]	0.101 ± 0.026 [abc]
		cv. Iride (Apulia Region)			
Micronised sample		0.582 ± 0.013 [a]	<LOD	<LOD	0.022 ± 0.005 [b]
	F 220	0.110 ± 0.008 [f]	0.001 ± 0.0002 [b]	0.014 ± 0.001 [d]	0.074 ± 0.002 [d]
	F 230	0.246 ± 0.008 [e]	<LOD	0.007 ± 0.002 [d]	0.016 ± 0.001 [d]
	F 240	0.342 ± 0.013 [d]	<L.R.	0.009 ± 0.001 [bcd]	0.038 ± 0.002 [cd]
Air-classified fractions	F 250	0.469 ± 0.022 [c]	<LOD	0.009 ± 0.001 [cd]	0.038 ± 0.002 [bc]
	F 260	0.502 ± 0.033 [bc]	0.005 ± 0.0007 [b]	0.031 ± 0.003 [a]	0.055 ± 0.009 [ab]
	F 270	0.618 ± 0.021 [a]	<LOD	0.019 ± 0.003 [bc]	0.044 ± 0.005 [ab]
	F 280	0.638 ± 0.021 [a]	0.028 ± 0.001 [a]	0.021 ± 0.001 [ab]	0.063 ± 0.006 [a]

Table 2. Group I: average values ± standard error (SE) of arsenic (As), cadmium (Cd), nickel (Ni), and lead (Pb) contents in the micronised samples and G-type air-classified fractions; d.m.: dry matter; LOD: limit of detection , different letters indicate significant difference ($p < 0.05$), $n = 3$.

		cv. Maestà (Apulia Region)			
		As (µg/g d.m.) ± SE	Cd (µg/g d.m.) ± SE	Ni (µg/g d.m.) ± SE	Pb (µg/g d.m.) ± SE
Micronised sample		0.478 ± 0.037 [a]	0.002 ± 0.000 [a]	0.043 ± 0.001 [b]	0.157 ± 0.019 [a]
	G 220	0.325 ± 0.034 [b]	0.001 ± 0.000 [ab]	0.102 ± 0.024 [a]	0.109 ± 0.034 [ab]
	G 230	0.313± 0.009 [b]	0.002 ± 0.000 [ab]	0.019 ± 0.001 [c]	0.049 ± 0.0059 [bc]
	G 240	0.372 ± 0.003 [b]	<LOD	0.019 ± 0.002 [c]	0.026 ± 0.003 [c]
Air-classified fractions	G 250	0.145 ± 0.002 [c]	0.001 ± 0.000 [b]	0.001 ± 0.000 [d]	0.006 ± 0.002 [c]
	G 260	0.055 ± 0.000 [d]	<LOD	0.003 ± 0.001 [d]	0.003 ± 0.000 [c]
	G 270	0.037 ± 0.001 [d]	0.001 ± 0.000 [ab]	0.002 ± 0.001 [d]	0.007 ± 0.001 [c]
	G 280	0.025 ± 0.001 [d]	0.0001 ± 0.000 [b]	0.003 ± 0.000 [d]	0.007 ± 0.001 [c]
		cv. Saragolla (Apulia Region)			
Micronised sample		0.518 ± 0.069 [a]	0.008 ± 0.0013 [b]	0.143 ± 0.004 [a]	0.152 ± 0.007 [a]
	G 220	0.420 ± 0.044 [ab]	0.021 ± 0.0056 [a]	0.014 ± 0.001 [bc]	0.213 ± 0.044 [a]
	G 230	0.363 ± 0.001 [bc]	<LOD	0.058 ± 0.0017 [b]	0.061 ± 0.001 [b]
	G 240	0.243 ± 0.005 [cd]	0.001 ± 0.000 [c]	0.008 ± 0.001 [c]	0.036 ± 0.005 [b]
Air-classified fractions	G 250	0.119 ± 0.002 [de]	<LOD	0.007 ± 0.001 [c]	0.015 ± 0.002 [b]
	G 260	0.100 ± 0.002 [de]	0.0007 ± 0.0001 [c]	0.057 ± 0.001 [bc]	0.020 ± 0.002 [b]
	G 270	0.045 ± 0.001 [e]	0.0005 ± 0.0001 [c]	0.009 ± 0.003 [c]	0.019 ± 0.001 [b]
	G 280	0.034 ± 0.001 [e]	0.0002 ± 0.000 [c]	0.005 ± 0.001 [c]	0.010 ± 0.001 [b]
		cv. Iride (Apulia Region)			
Micronised sample		0.582 ± 0.013 [a]	<LOD	<LOD	0.022 ± 0.005 [b]
	G 220	0.455 ± 0.038 [b]	<LOD	<LOD	0.044 ± 0.003 [a]
	G 230	0.412± 0.002 [b]	<LOD	<LOD	0.023 ± 0.0012 [b]
	G 240	0.322 ± 0.002 [c]	<LOD	<LOD	0.022 ± 0.002 [b]
air-classified fractions	G 250	0.160 ± 0.001 [d]	0.008 ± 0.000 [a]	0.035 ± 0.010 [a]	0.015 ± 0.001 [bc]
	G 260	0.088 ± 0.002 [de]	<LOD	0.004 ± 0.001 [b]	0.015 ± 0.002 [bc]
	G 270	0.065 ± 0.001 [e]	<LOD	<LOD	0.015 ± 0.001 [bc]
	G 280	0.050 ± 0.001 [e]	0.0003 ± 0.000 [b]	<LOD	0.008 ± 0.001 [c]

As in the case of As and Cd, a more marked reduction in the Ni content of the G fractions than that in the F fractions was observed. In particular, F220 and F230 (cv. Maestà and cv. Saragolla, respectively) showed a trend toward lower values of Ni content than the corresponding micronised samples, but the difference was significant ($p < 0.05$) only for the cv. Saragolla samples. This trend was not confirmed in cv. Iride samples because all F fractions showed values higher than the micronised fractions, with relevance for F260 (0.031 µg/g). In addition, this fraction type tended toward higher levels of Cd in all cultivar samples tested. The G fractions had a lower rate of contamination than the corresponding micronised samples, especially with regard to G280, which reached a maximum value of 0.005 µg/g (cv. Saragolla). Regarding the cv. Iride G fractions, only G250 (0.035 µg/g) and G260 (0.004 µg/g) had detectable Ni content, which was undetectable for the micronised and the remaining G samples.

Higher Pb content in micronised samples from cv. Maestà (0.157 µg/g) and cv. Saragolla (0.152 µg/g) compared to cv. Iride (0.022 µg/g) was detected. An appreciable trend toward a reducing rate due to the technological process of milling fractions was confirmed in all cultivar samples, especially for G fractions from G250 to G280. Although not always significant, a general tendency toward values lower than those of the micronised samples was observed from F230 to F250 in all cultivar samples. However, in the case of cv. Iride, only the F230 value (0.016 µg/g) was significantly lower than that of the micronised sample. Table 3 shows the results for As, Cd, Ni, and Pb for group II. The As contamination was significantly decreased in all fractions as compared to the corresponding micronised samples. More specifically, the F250 and G250 fractions maintained the lowest rate of contamination (maximum percentage residue: 52% in F250, cv. Saragolla), which is even lower than that of semolina.

Table 3. Group II: average values ± standard error (SE) of arsenic (As), cadmium (Cd), nickel (Ni), and lead (Pb) contents in the micronised, semolina, and selected air-classified fractions (F250, G250, and G230); d.m.: dry matter; LOD: limit of detection, different letters indicate significant difference ($p < 0.05$), $n = 3$.

		As (µg/g d.m.) ± SE	Cd (µg/g d.m.) ± SE	Ni (µg/g d.m.) ± SE	Pb (µg/g d.m.) ± SE
cv. Saragolla (Lazio Region)	Micronised sample	0.100 ± 0.002 [a]	0.01 ± 0.002 [b]	<LOD	<LOD
	Semolina	0.053 ± 0.002 [c]	0.002 ± 0.000 [b]	<LOD	0.013 ± 0.001 [a]
	F250	0.052 ± 0.004 [c]	<LOD	<LOD	<LOD
	G250	0.025 ± 0.001 [d]	0.01 ± 0.000 [a]	<LOD	<LOD
	G230	0.063 ± 0.005 [b]	0.002 ± 0.001 [b]	<LOD	0.008 ± 0.003 [a]
cv. Antalis (Basilicata Region)	Micronised sample	0.111 ± 0.05 [a]	0.01 ± 0.001 [a]	<LOD	0.138 ± 0.006 [a]
	Semolina	0.066 ± 0.005 [b]	0.001 ± 0.000 [a]	<LOD	0.144 ± 0.026 [a]
	F250	0.039 ± 0.001 [c]	0.001 ± 0.000 [a]	<LOD	0.042 ± 0.009 [b]
	G250	0.015 ± 0.001 [d]	0.001 ± 0.000 [a]	<LOD	0.016 ± 0.002 [c]
	G230	0.062 ± 0.007 [b]	0.002 ± 0.000 [a]	<LOD	0.038 ± 0.004 [b]
cv. Antalis (Marche Region)	Micronised sample	0.164 ± 0.016 [a]	<LOD	<LOD	<LOD
	Semolina	0.089 ± 0.007 [b]	<LOD	<LOD	0.042 ± 0.002
	F250	0.041 ± 0.001 [d]	<LOD	<LOD	<LOD
	G250	0.034 ± 0.002 [e]	<LOD	<LOD	<LOD
	G230	0.054 ± 0.004 [c]	<LOD	<LOD	<LOD

Regarding Cd contamination, the presence of the element was unchanged compared to micronised samples. Undetectable Ni content was observed in all micronised and milled samples, and the same results were obtained for all fraction types tested.

Regarding Pb contamination, the fractions F250 and G250 had undetectable levels of this element (content < LOD) in micronised samples. Moreover, the F250 and G250 fractions showed a significantly ($p < 0.05$) lower rate of contamination than the micronised sample (0.138 µg/g) from cv. Antalis (Basilicata).

Through the pasta-making process, a slight decrease in the content of As in semolina was achieved, which proved to not be significantly ($p > 0.05$) different from the air-classified fractions (Table 4).

Table 4. Pasta: average values ± standard error (SE) of arsenic (As) content in semolina and selected air-classified fractions (F250, G250, and G230); d.m.: dry matter, different letters indicate significant difference ($p < 0.05$), $n = 3$.

| | As (µg/g d.m.) ± SE | | |
	cv. Saragolla (Lazio Region)	cv. Antalis (Basilicata Region)	cv. Antalis (Marche Region)
Semolina	0.050 ± 0.012 [ab]	0.063 ± 0.009 [ab]	0.048 ± 0.011 [a]
F250	0.026 ± 0.004 [b]	0.084 ± 0.019 [a]	0.037 ± 0.004 [a]
G250	0.023 ± 0.004 [b]	0.031 ± 0.006 [ab]	0.044 ± 0.010 [a]
G230	0.053 ± 0.003 [a]	0.029 ± 0.013 [b]	0.039 ± 0.013 [a]

4. Discussion

This study aimed to address the issue of organic and inorganic contamination in durum wheat-derived food products, which has already been preliminarily examined [29]. In the present assay, updated micronising and air classification plants were used as reliable tools to improve the quality characteristics of products. Moreover, the choice of the three fractions assayed in the second series of samples (group II) also took into account the previous assessment of the main quality characteristics (i.e., milling yield, particle size composition, and ash content). These characteristics shown by all air-classified fractions (from the 220 to 280 setting) were previously assayed in a process study [30].

As a whole, the three air-classified fractions assayed in this study were the best compromise between the presence of bran particles and several contaminants in the products.

Moreover, the use of several cultivars showing different grain characteristics was deemed suitable only to introduce more variability within the durum wheat samples tested.

The results obtained show a clear effect of the process on the mycotoxin contamination of both milling and final products while retaining a non-negligible amount of bran particles in the air-classified mixes. Both types of mycotoxins were concentrated in the F250 fraction, with a clear trend toward a strong reduction in the G fractions. However, different behaviours of DON and T-2/HT-2 toxins were observed (Figures 4 and 5). In fact, the only semolina sample that was significantly different ($p < 0.05$) from both micronised and F250 fractions was the sample with slight DON contamination (cv. Saragolla, region Lazio), whereas in the case of high contamination (cv. Antalis, Marche region), no significant difference ($p > 0.05$) was detected for these fractions. These results confirm the high capacity of DON-producing fungi to penetrate the inner parts of the kernel, thus colonising the endosperm structure with negative effects on semolina, as previously shown [36]. Moreover, the pasta-making process did not produce significant differences ($p > 0.05$) among the semolina and G fractions, probably due to the high content of semolina particles in the G fractions [30].

In the case of a higher rate of contamination of the sum of T-2 and HT-2 toxins (cv. Antalis, Basilicata region), a strong mycotoxin accumulation in F250 (151%) and a decreased amount in G230 (53%) were noted as compared to the micronised sample. Mycotoxin contamination persisted in these air-classified fractions (F250 and G230), albeit to a lesser extent (88% and 39%, respectively) in the end product (pasta). Conversely, the presence of toxins was not detected in semolina. This result is consistent with previous studies showing the greater difficulty of managing toxin-producing fungi (i.e., *F. langsethiae*) colonising the inner endosperm of the kernel and thus strongly limiting semolina contamination [21–37]. Interestingly, the toxin content (sum of T-2 and HT-2) tended to concentrate in the F250 fraction, and a higher bran content was clearly demonstrated by the measurement that was close to the limit of quantification of the ELISA test (25 µg/g). This limit must be defined as the "zone" with the greatest uncertainty of quantification. However, this was not observed in the corresponding pasta samples analysed.

As a whole, the results for both types of mycotoxins underline the importance of the starting levels of grain contamination for determining the rate of contamination in both the air-classified fractions and final product (pasta). For this reason, the air classification technology, when intended as a management tool for mycotoxin contamination in products, must be used only for grain samples with absent or slight contamination below the mandatory limits [16–26]. In fact, only in this manner can the residual contamination rate in milling and final products be maintained below the established safe levels.

The data from the detection of inorganic contaminants showed an important effect of the technological process on the content of toxic elements in the different fractions, which is in agreement with previous assays [14,15,38]. In particular, the latter assays highlighted the effect of processing under controlled pilot plant conditions on the levels of As, Cd, Pb, and Ni, mainly by producing a marked trend toward a reduction in As content in the milling fractions. This latter aspect was confirmed in our study for pasta manufactured both with semolina and with the use of each of the three selected fractions (F250, G250, and G230).

The residual content of elements assayed in the fractions decreased as compared to the micronised samples, except for several fractions in which the opposite trend was observed. Moreover, considering the lower starting contamination rates of micronised samples with Cd, Ni, and Pb as compared to As, only the latter element was analysed in the end product (pasta) as the inorganic contaminant of major interest for the purposes of this work.

From the results obtained, we cannot select a fraction where the average content of all toxic elements analysed was consistently significantly lower than that of the micronised sample. However, as shown by the first series tested (group I), a significant trend of the G fractions toward a reduction in inorganic element contents, as compared to the F fractions, was clearly observed. This could be due to the higher content of bran particles in F fractions than in G fractions, which conversely contained higher amounts of semolina residues. Furthermore, the latter effect was evident as the progressive airflow opening increased. In addition, based on the contamination rates observed, the three air-classified fractions (F250, G250, and G230), previously selected based on qualitative evaluation, were confirmed as suitable products for the pasta-making process [30,31].

The results for the second series of samples (group II) proved to be of particular interest because it seemed to confirm the process effect shown in the first series (group I), mainly regarding As contamination. In fact, in all samples, the lowest As content was found in the G250 fraction, with decreasing rates ranging from 75% (cv. Saragolla, Lazio region) to 86% (cv. Antalis, Basilicata region). The presence of Cd and Pb was reduced, and thus, lower levels were present in the chosen fractions than in the semolina samples.

The assessment of As content in the end product was of great interest due to its role as an important source of consumers' direct exposure to food-borne contaminants [11]. The clear trend toward lower rates of As contamination shown by the three selected fractions (F250, G250, and G230) can provide greater safety in manufacturing unrefined end products. In fact, the same three air-classified fractions showed no significant difference as compared to semolina, even in the fraction with a higher content of bran residue (F250). The lower As content as a process effect in pasta manufacturing is in agreement with a previous assay [15].

5. Conclusions

This work addressed the management of several organic (mycotoxins) and inorganic (mainly arsenic) contaminants in the manufacturing of unrefined durum wheat products. The importance of reducing the risk of consumers' exposure to toxic compounds derives from the role of pasta as an appreciated and widely consumed food worldwide. The results obtained confirmed the reliability of the upgraded technological process to provide air-classified milling fractions characterised by the strong containment of several contaminants, and thus, the process is suitable for manufacturing high-quality, unrefined end products. Therefore, from a technological point of view, suitable management protocols and desirable results were achieved for preparing the main product of durum wheat (pasta), with a focus on ensuring consumers' safety.

Author Contributions: Conceptualisation, A.C., S.R.S., R.M. and E.A.; methodology, A.C., S.R.S., R.M. and E.A.; validation, A.C., S.R.S., R.M. and E.A.; formal analysis, A.C., V.D.F., S.P., S.R.S., R.M. and E.A.; data curation, A.C., V.D.F., S.R.S., R.M., E.A. and F.S.; writing—original draft preparation, A.C., S.R.S., R.M., E.A. and F.S.; writing—review and editing, A.C., S.P., S.R.S., R.M., E.A. and F.S.; supervision, S.R.S. and F.S. All authors have read and agreed to the published version of the manuscript.

Funding: This research was carried out within the "AsFRUM" project (Uff. Boll. Reg. Lazio n. 94, suppl.1, 24 November 2016); the project was funded by Region Lazio, Italy. This research was partially funded by the Italian Ministry of Education, University, and Research (MIUR) in the frame of the MIUR initiative "Departments of excellence", Law 232/2016.

Institutional Review Board Statement: Not applicable.

Informed Consent Statement: Not applicable.

Data Availability Statement: All data are contained within the article.

Acknowledgments: The authors wish to thank Gabriella Aureli, Alessandra Arcangeli, and Roberto Mortaro for their valuable support.

Conflicts of Interest: The authors declare that they have no conflict of interest.

References

1. FAO. World Food Situation. 2021. Available online: http://www.fao.org/worldfoodsituation/csdb/en/ (accessed on 13 August 2018).
2. Barrett, E.M.; Foster, S.I.; Beck, E.J. Whole Grain and High-Fibre Grain Foods: How Do Knowledge, Perceptions and Attitudes Affect Food Choice? *Appetite* **2020**, *149*, 104630. [CrossRef] [PubMed]
3. Reynolds, A.; Mann, J.; Cummings, J.; Winter, N.; Mete, E.; Te Morenga, L. Carbohydrate Quality and Human Health: A Series of Systematic Reviews and Meta-Analyses. *Lancet* **2019**, *393*, 434–445. [CrossRef]
4. He, M.; Van Dam, R.M.; Rimm, E.; Hu, F.B.; Qi, L. Response to Letter Regarding Article, Whole-Grain, Cereal Fiber, Bran, and Germ Intake and the Risks of All-Cause and Cardiovascular Disease-Specific Mortality among Women with Type 2 Diabetes Mellitus. *Circulation* **2010**, *121*, 2162–2168. [CrossRef]
5. Ferlay, J.; Soerjomataram, I.; Dikshit, R.; Eser, S.; Mathers, C.; Rebelo, M.; Parkin, D.M.; Forman, D.; Bray, F. Cancer Incidence and Mortality Worldwide: Sources, Methods and Major Patterns in GLOBOCAN 2012. *Int. J. Cancer* **2015**, *136*, E359–E386. [CrossRef]
6. Torre, L.A.; Bray, F.; Siegel, R.L.; Ferlay, J.; Lortet-Tieulent, J.; Jemal, A. Global Cancer Statistics, 2012. *CA Cancer J. Clin.* **2015**, *65*, 87–108. [CrossRef]
7. Barrett, E.M.; Batterham, M.J.; Ray, S.; Beck, E.J. Whole Grain, Bran and Cereal Fibre Consumption and CVD: A Systematic Review. *Br. J. Nutr.* **2019**, *121*, 914–937. [CrossRef] [PubMed]
8. Thielecke, F.; Nugent, A.P. Contaminants in Grain—A Major Risk for Whole Grain Safety? *Nutrients* **2018**, *10*, 1213. [CrossRef]
9. Bowell, R.J.; Alpers, C.N.; Jamieson, H.E.; Nordstrom, D.K.; Majzlan, J. The Environmental Geochemistry of Arsenic—An Overview. *Rev. Mineral Geochem.* **2014**, *79*, 1–16. [CrossRef]
10. Zhu, Y.G.; Yoshinaga, M.; Zhao, F.-J.; Rosen, B.P. Earth Abides Arsenic Biotransformations. *Annu. Rev. Earth Planet. Sci.* **2014**, *42*, 443–467. [CrossRef]
11. Zhao, F.J.; Ma, Y.; Zhu, Y.G.; Tang, Z.; McGrath, S.P. Soil Contamination in China: Current Status and Mitigation Strategies. *Environ. Sci. Technol.* **2015**, *49*, 750–759. [CrossRef] [PubMed]
12. Clemens, S.; Ma, J.F. Toxic Heavy Metal and Metalloid Accumulation in Crop Plants and Foods. *Annu. Rev. Plant Biol.* **2016**, *67*, 489–512. [CrossRef] [PubMed]
13. Conti, M.E.; Cubadda, F.; and Carcea, M. Trace Metals in Soft and Durum Wheat from ITALY. *Food Addit. Contam.* **2000**, *17*, 45–53. [CrossRef] [PubMed]
14. Ficco, D.B.M.; Borrelli, G.M.; Miedico, O.; Giovanniello, V.; Tarallo, M.; Pompa, C.; De Vita, P.; Chiaravalle, A.E. Effects of Grain Debranning on Bioactive Compounds, Antioxidant Capacity and Essential and Toxinc Trace Elements in Purple Durum Wheats. *LWT-Food Sci. Technol.* **2020**, *118*, 108734. [CrossRef]
15. Cubadda, F.; Raggi, A.; Zanasi, F.; Carcea, M. From Durum Wheat to Pasta: Effect of Technological Processing on the Levels of Arsenic, Cadmium, Lead and Nickel-A Pilot Study. *Food Addit. Contam.* **2003**, *20*, 353–360. [CrossRef]
16. European Commission. Commission regulation (EC) No 1881/2006 of 19 December 2006 Setting Maximum Levels for Certain Contaminants in Foodstuffs. *Off. J. Eur. Union* **2006**, *L364*, 5–24.
17. *Commission Regulation (EU) 2015/1006 of 25 June 2015 Amending Regulation (EC) No 1881/2006 as Regards Maximum Levels of Inorganic Arsenic in Foodstuffs*; European Commission: Brussels, Belgium, 2015.
18. Bennett, J.W.; Klich, M. Mycotoxins. *Clin. Microbiol. Rev.* **2003**, *16*, 497–516. [CrossRef] [PubMed]
19. European Food Safety Authority. Scientific Opinion on the Risks for Animal and Public Health Related to the Presence of T-2 and HT-2 Toxin in Food and Feed. *EFSA J.* **2011**, *9*, 2481. [CrossRef]

20. European Food Safety Authority. Scientific Opinion on the Risks to Human and Animal Health Related to the Presence of Deoxynivalenol and Its Acetylated and Modified Forms in Food and Feed. *EFSA J.* **2017**, *15*, 4718.
21. Pascale, M.; Haidukowski, M.; Lattanzio, V.M.T.; Silvestri, M.; Ranieri, R.; Visconti, A. Distribution of T-2 and HT-2 Toxins in Milling Fractions of Durum Wheat. *J. Food Prot.* **2011**, *74*, 1700–1707. [CrossRef] [PubMed]
22. Tibola, C.S.; Guarient, E.M.; Guerra Dias, A.R.; Nicolau, M.; Barboza Devos, R.J.; Dalbosco Teixeira, D. Effect of Debranning Process on Deoxynivalenol Content in Whole-Wheat Flours. *Cereal Chem.* **2019**, *96*, 717–724. [CrossRef]
23. Khaneghah, A.M.; Martins, L.M.; von Hertwig, A.M.; Bertoldo, R.; Sant'Ana, A.S. Deoxynivalenol and Its Masked forms: Characteristics, Incidence, Control and Fate during Wheat and Wheat Based Products Processing—A Review. *Trends Food Sci. Technol.* **2018**, *71*, 13–24. [CrossRef]
24. Freire, L.; Sant'Ana, A.S. Modified Mycotoxins: An Updated Review on Their Formation, Detection, Occurrence, and Toxic Effects. *Food Chem. Toxicol.* **2018**, *111*, 189–205. [CrossRef]
25. Aureli, G.; D'Egidio, M.G. Efficacy of Debranning on Lowering of Deoxynivalenol (DON) Level in Manufacturing Processes of Durum Wheat. *Tech. Molit.* **2007**, *58*, 729–733.
26. Commission Recommendation (2013/165/EU) on the Presence of T-2 and HT-2 Toxin in Cereals and Cereal Products. *Off. J. Eur. Union* **2013**, *L91*, 12.
27. Brera, C.; Gregori, E.; De Santis, B.; Pantanali, A.; Monti, M. Ricerca e Sviluppo di una Farina Innovativa Destinata ai Bambini di Seconda e Terza Infanzia. In Proceedings of the ISTISAN Congressi, Riassunti: Istituto Superiore di Sanità, 2019 19/C4. IV Congresso Nazionale. Micotossine e Tossine Vegetali nella Filiera Agro-Alimentare, Roma, Italy, 10–12 June 2019; p. 61.
28. Wang, J.; Xie, A.; Zhang, C. Feature of Air Classification Product in Wheat Milling: Physicochemical, Rheological Properties of Filter Flour. *J. Cereal Sci.* **2013**, *57*, 537–542. [CrossRef]
29. Cammerata, A.; Marabottini, R.; Allevato, E.; Aureli, G.; Stazi, S.R. Content of Minerals and Deoxynivalenol in the Air-Classified Fractions of Durum Wheat. *Cereal Chem.* **2021**, *98*, 1101–1111. [CrossRef]
30. Cammerata, A.; Sestili, F.; Laddomada, B.; Aureli, G. Bran-Enriched Milled Durum Wheat Fractions Obtained Using Innovative Micronization and Air-Classification Pilot Plants. *Foods* **2021**, *10*, 1796. [CrossRef] [PubMed]
31. Cammerata, A.; Laddomada, B.; Milano, F.; Camerlengo, F.; Bonarrigo, M.; Masci, S.; Sestili, F. Qualitative Characterization of Unrefined Durum Wheat Air-Classified Fractions. *Foods* **2021**, *10*, 2817. [CrossRef] [PubMed]
32. European Commission. Commission Regulation (EC) No. 401/2006 of 23 February 2006 Laying Down the Methods of Sampling and Analysis for the Official Control of the Levels of Mycotoxins in Foodstuffs. *Off. J. Eur. Union* **2006**, *L70*, 12.
33. Brera, C.; Debegnach, F.; De Santis, B.; Pannunzi, E.; Berdini, C.; Prantera, E.; Miraglia, M. Validazione di Metodi Immunoenzimatici per la Determinazione Delle Micotossine in Campioni di Cereali. In *I Georgofili Quaderni 2008*; Polistampa, Ed.; Edizioni Polistampa: Firenze, Italy, 2009; Volume IV, pp. 3–45.
34. Pascale, M.; Aureli, G.; Haidukowski, M.; Melloni, S.; D'Egidio, M.G.; Visconti, A. Comparison between ELISA and UPLC Methods for the Determination of T-2 and HT-2 Toxins in Wheat. In Proceedings of the International Mycotoxin Conference, Beijing, China, 19–23 May 2014; p. 258.
35. Hammer, O.; Harper, D.A.T.; Ryan, P.D. PAST: Paleontological Statistic Software Package for Education and Data Analysis. *Paleontol. Electron.* **2001**, *4*, 1–9.
36. Ríos, G.; Zakhia-Rozis, N.; Chaurand, M.; Richard-Forget, F.; Samson, M.F.; Abecassis, J.; Lullien-Pellerin, V. Impact of Durum Wheat Milling on Deoxynivalenol Distribution in the Outcoming Fractions. *Food Addit. Contam.* **2009**, *26*, 487–495. [CrossRef] [PubMed]
37. Aureli, G.; Melloni, S.; . D'Egidio, M.G.; Quaranta, F.; Haidukowski, M.; Pascale, M.; Lattanzio, V.M.T. Effect of Debranning on T-2 and HT-2 Toxin Content in Durum Wheat Kernels and Milling Fractions. In Proceedings of the 10th AISTEC Conference "Grain for Feeding the World", Milano, Italy, 1–3 July 2015; pp. 156–161.
38. Cheli, F.; Campagnoli, A.; Ventura, V.; Brera, C.; Berdini, C.; Palmaccio, E.; Dell'Orto, V. Effects of Industrial Processing on the Distributions of Deoxynivalenol, Cadmium and Lead in Durum Wheat Milling Fractions. *Food Sci. Technol.* **2010**, *43*, 1050–1057. [CrossRef]

 *foods*

MDPI

Article

Modeling and Optimization of Process Parameters for Nutritional Enhancement in Enzymatic Milled Rice by Multiple Linear Regression (MLR) and Artificial Neural Network (ANN)

Anjineyulu Kothakota [1,*], Ravi Pandiselvam [2], Kaliramesh Siliveru [3,*], Jai Prakash Pandey [4], Nukasani Sagarika [5], Chintada H. Sai Srinivas [6], Anil Kumar [7], Anupama Singh [4] and Shivaprasad D. Prakash [3]

[1] Agro-Processing & Technology Division, CSIR-National Institute for Interdisciplinary Science and Technology, Thiruvananthapuram 695019, Kerala, India

[2] Physiology, Biochemistry and Post-Harvest Technology Division, ICAR-Central Plantation Crops Research Institute, Chowki 671124, Kerala, India; anbupandi1989@yahoo.co.in

[3] Department of Grain Science & Industry, Kansas State University, Manhattan, KS 66502, USA; shivdp1994@ksu.edu

[4] Department of Post-Harvest Process and Food Engineering, College of Technology, G.B. Pant University of Agriculture and Technology, Pantnagar 263145, Uttarakhand, India; jppandey55@gmail.com (J.P.P.); asingh3@gmail.com (A.S.)

[5] Department of Food Process Engineering, College of Food Processing Technology & Bio-Energy, Anand Agricultural University, Anand 388110, Gujarat, India; sagarikanukasani@gmail.com

[6] Foods Business Division, ITC Limited, Visakhapatnam 530001, Andhra Pradesh, India; saisrinivas36cfst@gmail.com

[7] Department of Food Science and Technology, College of Agriculture, G.B. Pant University of Agriculture and Technology, Pantnager 263145, India; anilkumargbpuat@gmail.com

* Correspondence: anjineyuluk@niist.res.in (A.K.); kaliramesh@k-state.edu (K.S.); Tel.: +91-9719427663 (A.K.); +1-(630)210-2462 (K.S.)

Citation: Kothakota, A.; Pandiselvam, R.; Siliveru, K.; Pandey, J.P.; Sagarika, N.; Srinivas, C.H.S.; Kumar, A.; Singh, A.; Prakash, S.D. Modeling and Optimization of Process Parameters for Nutritional Enhancement in Enzymatic Milled Rice by Multiple Linear Regression (MLR) and Artificial Neural Network (ANN). *Foods* **2021**, *10*, 2975. https://doi.org/10.3390/foods10122975

Academic Editors: Weiqun Wang and Barbara Laddomada

Received: 9 September 2021
Accepted: 5 November 2021
Published: 3 December 2021

Publisher's Note: MDPI stays neutral with regard to jurisdictional claims in published maps and institutional affiliations.

Abstract: This study involves information about the concentrations of nutrients (proteins, phenolic compounds, free amino acids, minerals (Ca, P, and Iron), hardness) in milled rice processed with enzymes; xylanase and cellulase produced by *Aspergillus awamori*, MTCC 9166 and *Trichoderma reese*, MTCC164. Brown rice was processed with 60–100% enzyme (40 mL buffer -undiluted) for 30 to 150 min at 30 °C to 50 °C followed by polishing for 20–100 s at a safe moisture level. Multiple linear regression (MLR) and artificial neural network (ANN) models were used for process optimization of enzymes. The MLR correlation coefficient (R^2) varied between 0.87–0.90, and the sum of square (SSE) was placed within 0.008–8.25. While the ANN R^2 (correlation coefficient) varied between 0.97 and 0.9999(1), MSE changed from 0.005 to 6.13 representing that the ANN method has better execution across MLR. The optimized cellulase process parameters (87.2% concentration, 80.1 min process time, 33.95 °C temperature and 21.8 s milling time) and xylanase process parameters (85.7% enzyme crude, 77.1 min process time, 35 °C temperature and 20 s) facilitated the increase of Ca (70%), P (64%), Iron (17%), free amino acids (34%), phenolic compounds (78%) and protein (84%) and decreased hardness (20%) in milled rice. Scanning electron micrographs showed an increased rupture attributing to enzymes action on milled rice.

Keywords: multiple linear regression (MLR); artificial neural network (ANN); milled rice; enzymes

1. Introduction

Rice (*Oryza sativa* L.) exists as an essential commodity all over the world. This must be a prominent subsistence grain for growing nations and a primary energy source containing bioactive compounds, vitamins, minerals, amino acids, and fiber [1]. Brown rice dehusked from paddy consists of an embryo (2–3% of its total weight), endosperm (90%), and bran layer (6–7%) [2]. The bran layer's major constituents are protein, fat, crude fiber, ash, carbohydrates, cellulose, arabinoxylans, mannans, Galatians, pentosans, and uronic acid.

It additionally furnishes a substantial quantity of B1, B2, B3, Zn, and smaller amounts of different trace elements [3].

The commercial value of rice grains is established over these proportions and moisture of crop; these characteristics further support more portion of fracture throughout automatic polishing. However, it may not be suitable for consumer consumption. Polishing the grain to a proper degree may be essential for preserving the quality and characteristics of the grain. Degrading the bran structure in paddy grains is a critical stage for the grain in an individual's ingestion [4]. The loss of essential nutrients in case milling have been shown to occur through over-processing; reducing such losses throughout automatic milling may be conquered via novel pre-treatment: cold plasma [5], high-pressure processing [6], ultrasonication [7], pulsed electric field processing [8], and enzymes treatment [9], before polishing to improve the milling properties. We have attempted to apply enzymes as a pre-treatment to enhance the nutrient content after milling. Pretreatments are evidence to enhance seed functioning by affecting the biochemical and physiological qualities of grains without adversely affecting the atmosphere. Polishing is a primary treatment step for cereals to eliminate their hard-cellulosic cover and bran adhering to the surface. Several pretreatment methods, such as high hydrostatic pressure (HHP), cold plasma, pulse electric filed, UV light, microwave and enzymatic treatment can ensure the nutritional improvement in cereals and millets for shorter processing time at various conditions [9,10].

Some investigations have described how the complete breakdown of bran structure through different enzymes (amylase and glucanase) may improve the textural characteristics of milled rice due to the bran structure interface through polysaccharides (cellulose, hemicelluloses, and amylase) via various bonds; glycosidic, covalent, and hydrolytic bonds may be adhered and hydrolyzed with endoglucanase, cellulase, and xylanase [9,10]. These bonds may perform to limit bran structure decline during the degree of milling materials.

Considering the above factors, both Cellulase and xylanase enzymes, have been selected for the process of brown rice, improving the polishing nature, cooking characteristics, retaining the nutrients in polished rice, and textural characteristics. Few investigations have been carried out, but none have attempted to model and optimize processing conditions by using MLR (multiple linear regression) and ANN (artificial neural networks) to enhance nutritional properties in milled rice.

The modeling and optimization of treatment conditions for nutritional enhancement has had a problematic method, which has been analyzed as far as agriculture based products, food products, beverages, dairy products, and oil extraction industry products have pertained to different design methods to attain reasonable valuable resources [11]. Multiple linear regression (MLR) may be described while an experimental modeling design is employed as formulating, enhancing, and reducing complicated operations [12,13]. This method has the benefit of reducing the number of empirical tests and may be sufficient to provide a significantly acceptable outcome [13]. It is used for modeling and optimizes nutritional increments in food products: Extrusion [13,14], bakery foods [15], meat products [16], oils [17], and enzymes [18] are enzymatically treated during processing [19,20]. ANN sought tools and design for learning a simple process from the nonlinear association between input data to output data in a system compared to MLR. Lately, several researchers have mentioned ANN utilization in favor of optimizing conditions for food processing, including the enzymes production for beverages [11] and enzyme application in food [18]. This study aimed to (1) generate cellulase and xylanase enzymes with *Aspergillus awamori* (MTCC 9166) as well as *Trichoderma reesei* Rut C-30(MTCC16675) for the intent of processing, (2) modeling and optimizing enzyme-treated parameters for rice by applying MLR and ANN, and (3) to evaluate the nutritional and textural properties of enzyme treated milled rice.

2. Materials and Methods

2.1. Rice Samples

The Pant Sugandh Dhan15 (Aromatic, long and slender) rough rice was acquired with a Crop Research Centre, GB Panth University of Agriculture and Technology, Pantnagar. The sample was stored in an airtight container to avert the moisture interchange with the atmosphere. Rough rice was dehusked for further experimentation.

All Reagents were acquired with Hi-media Laboratories Pvt Ltd., Mumbai, and Sigma–Aldrich, New Delhi, India.

2.2. Enzyme Preparation

The Fungal crude cellulase and xylanase were produced from *Trichoderma reesei* Rut C-30 (MTCC16675) and *Aspergillus awamori* (MTCC 9166) submerged fermentation (Figure 1) was used as the enzyme activity. Enzyme activities can be expressed in the enzyme unit (U). 1 U was determined as the quantity by which the transformation is induced of about one micromole of matrix materials per minute through the particular circumstances of the analysis method [9,20]. The produced enzyme were diluted with different ratios: 100% (undiluted), 90% (90 mL crude + 10 mL buffer), 80% (80 mL crude + 20 mL buffer), 70% (70 mL crude + 30 mL buffer), and 60% (60 mL crude + 40 mL buffer) [19,20].

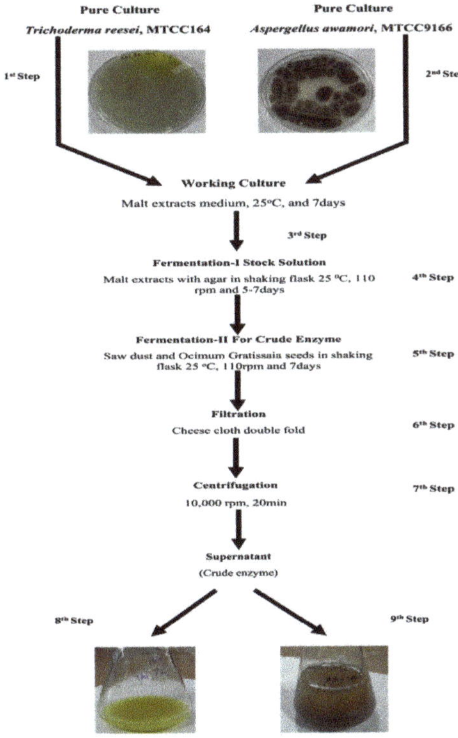

Figure 1. Flow diagram of cellulase (**A**) and xylanase (**B**) enzyme production.

2.3. Experimental Process

The head brown rice (100 g) was soaked in 50 mL water for 24 h, and the water was altered at specific gaps of time to decrease microbial infection. Soaked grains were again treated to an additional one hour in 100 mL sterile water about 5 g of calcium carbonate on 55 °C to create calcium ions enforce as a promoter to the enzyme action. These soaked samples processed cellulase and xylanase at various concentrations in the ratio 100% to

60% appropriately, for process the brown rice at various temperatures 30 °C to 50 °C (with 5 °C variation) in distinct time: 30–150 min (with 30 min variation). The processed samples were polished at various times, 20–100 s (20 s variation) through an abrasive Satake polisher. After de-husking, the polished rice was removed by sieving [19,20].

2.3.1. Estimation of Mineral Content

Determinations of mineral content in rice were measured by atomic absorption spectroscopy (Spectro Ciros C CD, Spectro, and Dusseldorf, Germany) in ppm according to [9].

2.3.2. Total Phenolic Substance Estimation

The phenolic content of milled rice was estimated through the Follin–Ciocalteu reagent method by spectrophotometer. The details were precise for mg gallic acid equivalent (GAE) per 100 g of a crude sample [21].

2.3.3. Total Free Amino Acid Assessment

The free amino acid substance of samples were determined by using Moore and Stein method (Spectrophotometer: ninhydrin solution and n-propanol at 570 nm absorbance) referred by [22].

2.3.4. Grain Hardness Assay

The Hardness of cooked rice used was measured by applying a texture analyzer (TA-XT2, Stable microsystems) with a 5 kg load cell [23]. The cooked sample of one kernel was directly situated on the inner cylindrical compressed probe with 100 mm diameter with a test speed of 0.5 mm/min.

2.3.5. Total Protein Content Determination

Protein was estimated by using the micro-Kjeldahl method and showed by way of total nitrogen $\times$ 5.95 g/100 g [24].

2.3.6. Enzyme Interaction through Scanning Electron Microscopy (SEM)

SEM was utilized to observe the action of interaction enzymes with treated and untreated brown rice (JEOL-JSM 6610 LV, Japan; Plate No.18) using 15 kv electron voltage. The freeze-dried (5%) specimens were situated with two-layered adhesive tape fastened over metallic but sheeted with gold.

2.4. Statistical Analysis

In addition to observational optimization, the enzyme treatment analysis turned out to be performed using multiple linear regression (MLR) and artificial neural network (ANN). The treatment variables (enzyme concentration, treatment time, temperature, and polishing time) influence quality attributes viz. mineral content (Ca, P, Fe). Phenolic content, free amino acid content, protein, and hardness was evaluated through multiple experimental designs. The multiple polynomial regression equations were used for the experimental design and produced to fit the experimental data; the applicable model terms are shown at Equation (1)

$$Y = b0 + b1X_1 + b2X_2 + b3X_3 + b11X_{12} + b22X_{22} + b33\,X_{32} + b12X_1X_2 + b13X_1X_3 + b23X_2X_3 \tag{1}$$

Considering Y states expected inconsistent, b0, b1, b2, and b3 represents linear interval, b11, b22, and b33 portray quadratic gap, b12, b13, and b23 exists interlinkage interval, X_1, X_2, and X_3 indicates explanatory variables for enzyme processed rice. This statistical evaluation and analysis of variance were stated applying Design Expert Version 11.0 (Stat-Ease, Inc., Minneapolis, MN, USA). The significance exists at 0.01%, 1%, and 5% with the linear, cross-product, and square terms.

2.4.1. Multiple Linear Regression (MLR)

The significant variables determining the target variable were chosen over the central composite rotatable design resultant in addition to being applied to create a multivariate analysis (MLR) equation implementing MATLAB's fitlm function. A design elucidation was evaluated using the correlation coefficient (R^2) and the sum of square error (SSE).

2.4.2. Artificial Neural Network

An ANN model consists of uncomplicated treatment components termed neurons that are interlinked with each other in a fuzzy logic configuration. A neuron receives a series of inputs that are filtered by an activation function to generate a primary output signal that serves as the stimulus for the next neuron. Training of the network is carried out by fine-tuning the progressive input neuron signals. MATLAB software R2018a was used for developing and testing the ANN design. The positive reaction neural network with a backpropagation algorithm comprising three strata, viz. an entry t layer, one concealed layer, and an exit layer, was employed as shown in Figure 2 below. The signals coming from the previous layer were processed, followed by transmission of output to the next layer on the basis of convergence criteria [13,25,26]. The variables selected for the input layer were cellulase—X_1 X_2 X_3 X_4 and xylanase—X_1 X_2 X_3 X_4, and the variables in the output layer were cellulase—Y_1 Y_2 Y_3 Y_4 Y_5 Y_6 and xylanase—Y_7 Y_8 Y_9 Y_{10} Y_{11} Y_{12} Y_{13} Y_{14} Y_{15}. The input stratum included 4 neurons. The exit stratum comprised 6 neurons, whereas the number of neurons inside the concealed layer was optimized to be 7. The sigmoid transfer function "transit" was selected for activation of neurons at the hidden layer. For neurons of the exit layer, linear alienate operate "purelin" was utilized as this function is regarded as most suitable for backpropagation networks [25]. The Levenberg–Marquardt training algorithm was selected for training the network as this algorithm has now been calculated as the quickest technique to learn moderate-sized feed-forward neural networks until various hundred weights. Throughout fine-tuning, the actual observational data (30 runs) were been reproduced threefold (90 entries) and ruptured with three portions: 80:10:10 (%) for training, validation, and testing [25].

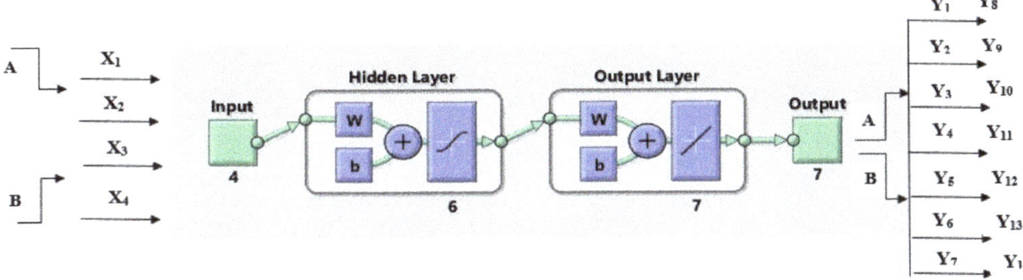

X_1–enzyme concentration (%), X_2–treatment time (min), X_3–polishingtime (s), X_4–treatment temperatuer (^{0}C), Cellulose (A), Y_1- Ca (Mg), Y_2- P(Mg), Y_3-Iron(Mg), Y_4- Phenolic content(μg), Y_5- Hardness(mg), Y_6- Free amino acid (Mg), Y_7- Protein(g) Xylanase (B), Y_8- Ca (Mg), Y_9- P(Mg), Y_{10}-Iron(Mg), Y_{11}- Phenolic content(μg), Y_{12}- Hardness(mg), Y_{13}- Free amino acid (Mg), Y_{14}- Protein(g)

Figure 2. Configuration of multilayer ANN model with four input neurons, six hidden neurons and seven output neurons.

3. Results and Discussion

3.1. Multiple Linear Regression (MLR)

This research demonstrates the sustained findings by assessing and correlating details of appropriate models implementing central rotatable composite design. Multiple linear regression designs have been formulated and exploited to influence numerous predictor responses over measured variables. The design capability was measured through using correlation coefficient (r square), Fisher exact test (F), and Lack of fit (Table 1). This may indicate that significant variables affect every model for predicting its comparable variables (Table 2).

Table 1. F-value, *p*-value and significance of each variable cellulase.

Source	Y_1		Y_2		Y_3		Y_4		Y_5		Y_6		Y_7	
	F-Value	*p*-Value	F-Value	*p*-Value	F-Value	*p*-Value	F-Value	*p*-Value	F-Value	*p*-Value	F-Value	*p*-Value	F-Value	*p*-Value
Model	9.01	<0.0001	8.50	<0.0001	8.63	<0.0001	9.20	<0.0001	3.64260	<0.0001	8.90	<0.0001	7.84	0.0001
X_1	0.07041	0.0075	0.01260	0.2267	0.00770	0.2088	3.64260	0.1694	4.10853	0.0505	0.09375	0.0089	0.00135	0.7220
X_2	0.00426	0.4588	0.00453	0.4611	0.05133	0.0040	4.10853	0.1613	1.52510	0.0393	0.05041	0.0436	0.35526	<0.0001
X_3	0.06406	0.0100	0.06100	0.0142	0.02870	0.0229	1.52510	0.0002	21.6030	0.1891	0.03375	0.0916	0.16666	0.0011
X_4	0.01401	0.1884	0.00050	0.8043	0.00633	0.2521	21.6030	0.5067	17.1226	0.0001	0.35041	<0.0001	0.02801	0.1195
X_1X_2	0.13043	0.0740	0.24591	0.3154	0.01845	0.0053	17.1226	0.1079	19.5895	0.1063	0.38003	0.8096	0.05200	0.2365
X_1X_3	0.15003	0.4610	0.22682	0.3689	0.10607	0.0005	19.5895	0.0580	38.8484	0.2748	0.02503	0.1066	0.02234	0.0736
X_1X_4	0.07985	0.8637	0.04052	0.3978	0.02027	0.6338	38.8484	0.1733	9.23028	0.7490	0.12574	0.0433	0.38543	0.0880
X_2X_3	0.19914	<0.0001	0.28641	0.3978	0.14792	0.0281	9.23028	0.9077	2.37930	0.2093	0.08360	0.0061	0.00600	0.2950
X_2X_4	0.02722	0.0172	0.00855	0.0014	0.04730	0.5830	2.37930	0.7404	1.03530	0.5439	0.00062	0.0433	0.01562	0.0961
X_3X_4	0.00422	0.0822	0.00680	0.0013	0.08555	0.0029	1.03530	0.1137	0.08555	0.7326	0.03062	0.8096	0.03802	0.0805
$X_1{}^2$	0.00022	0.0008	0.00600	<0.0001	0.00105	0.0602	0.08555	0.0006	1.38650	0.0003	0.05062	<0.0001	0.03422	0.0399
$X_2{}^2$	0.2704	0.0004	0.00600	<0.0001	0.02640	0.0002	1.38650	<0.0001	0.31080	0.0002	0.10562	0.0977	0.0121	0.1610
$X_3{}^2$	0.0529	0.0050	0.12075	0.0391	0.00140	0.0501	0.31080	0.0336	0.09765	<0.0001	0.05062	0.0022	0.0324	<0.0001
$X_4{}^2$	0.0256	0.0001	0.12425	<0.0001	0.05640	<0.0001	0.09765	<0.0001	3.64260	<0.0001	0.00062	0.0081	0.0361	0.4566
R^2	0.8938		0.8880		0.8896		0.8957		0.8902		0.8925		0.8798	
Adj. R^2	0.7946		0.7835		0.7865		0.7883		0.7877		0.7946		0.7676	
Pred. R^2	0.4490		0.4158		0.4418		0.4392		0.5148		0.4490		0.3695	
SSE	10.31		8.94		9.52		92.35		12.09		0.16		0.0183	
LOF	NS		NS		NS		NS		NS		NS		NS	

Table 2. F-value, *p*-value and significance of each variable.

Source	Y_8		Y_9		Y_{10}		Y_{11}		Y_{12}		Y_{13}		Y_{14}	
	F-Value	*p*-Value	F-Value	*p*-Value	F-Value	*p*-Value	F-Value	*p*-Value	F-Value	*p*-Value	F-Value	*p*-Value	F-Value	*p*-Value
Model	10.49	<0.0001	10.52	<0.0001	8.78	<0.0001	7.21	1.316667	8.39	<0.0001	7.21	0.0002	6.17	0.0005
X_1	0.03450	0.0718	0.00015	0.8351	0.00041	0.8261	349.988	−0.0125	0.01550	0.8452	0.00375	0.3243	0.01353	0.3871
X_2	0.01870	0.1743	0.01306	0.0667	0.05226	0.0243	0.81770	−0.02083	4.51533	0.0040	0.01041	0.1101	0.03010	0.2039
X_3	0.04593	0.0411	0.03081	0.0083	0.02041	0.1385	2.74050	−0.00417	0.09003	0.6390	0.00041	0.7388	3.75E-0	0.9632
X_4	0.00700	0.3966	0.01126	0.0863	0.10401	0.0030	6.25260	0.004167	1.19260	0.1018	0.00041	0.7388	0.00220	0.7243
X_1X_2	0.18905	0.0426	0.02714	0.0765	0.15087	0.0212	283.930	−0.03125	0.07832	0.3036	0.01574	0.0551	0.17508	0.0056
X_1X_3	0.00771	0.8980	0.20503	0.2447	0.42287	0.8296	4.09425	0.04375	3.16491	0.1202	0.07145	0.0107	0.37000	0.0082
X_1X_4	0.31268	0.0028	0.12574	0.9322	0.10430	0.3400	32.8812	−0.01875	4.36802	0.1017	2.98E-0	0.2311	0.37000	0.0082
X_2X_3	0.07710	<0.0001	1.19E-0	0.0085	0.12190	0.0909	93.2305	−0.06875	0.54966	0.0046	0.01860	0.0004	0.36208	0.6128
X_2X_4	0.04515	0.0005	0.0121	0.4007	0.05522	0.2272	267.567	−0.00625	0.44555	0.0244	0.01562	0.6833	0.17850	0.2191
X_3X_4	0.00015	0.5921	0.0049	0.6114	0.0004	0.0007	13.9689	0.04375	1.06605	<0.0001	0.03062	0.0107	0.15800	0.0831
$X_1{}^2$	0.11730	0.0004	2.5E-05	0.0019	0.0081	0.0007	312.317	−0.00312	1.19355	0.6615	0.00562	0.7891	0.15800	0.0059
$X_2{}^2$	0.43230	0.3742	0.03062	<0.0001	0.02722	<0.0001	8.65830	0.071875	4.35765	0.0124	0.07562	<0.0001	0.00455	0.0003
$X_3{}^2$	0.17850	<0.0001	0.0025	<0.0001	0.01322	0.0030	579.726	0.021875	2.45705	0.0045	0.00062	0.0759	0.02805	0.0003
$X_4{}^2$	0.00275	0.0111	0.0009	0.3697	0.1521	0.0017	99.9500	0.046875	21.3213	0.2552	0.03062	0.0010	0.05880	0.0003
R^2	0.9073		0.9076		0.8913		0.8706		0.8867		0.8706		0.8574	
Adj. R^2	0.8209		0.8214		0.7898		0.7499		0.7810		0.7499		0.7243	
Pred. R^2	0.5896		0.5421		0.4708		0.3438		0.4650		0.3438		0.2429	
SSE	0.14		0.050		0.13		203.85		5.89		8.69		6.44	
LOF	NS		NS		NS		NS		NS		NS		NS	

3.1.1. Minerals

The cellulase-treated milled rice retained calcium fluctuated around 2.41 to 3.11 mg, whereas xylanase differed from 2.79 to 3.62 mg (Table 3). The maximum retention of calcium for cellulase rice was observed at 80% X_1, 90 min X_2, 30 s X_3, 40 °C X_4, and xylanase rice at 90% X_1, 120 min X_2, 20 s X_3, and 35 °C X_4. The lowest values of calcium retention of cellulase rice were found to be 70% X_1, 120 min X_2, 40 s X_3, and 45 °C X_4, and xylanase rice 80% X_1, 90 min X_2, 30 s X_3, and 40 °C X_4. The coefficient of determination

(R^2 = 89.38 (cellulase) and R^2 = 90.73 (xylanase)) in terms of regression analysis was significant (<0.0001) as well as the lack of fit was nonsignificant. Figures 3a and 4a) show the consequence of predictor parameters over mineral (Ca, P, Fe) retention about enzymatic milled rice. Figure 4a–f initially increased treatment temperature, and enzyme concentration up to 42.5 °C and 85% caused maximum retention of minerals (Ca, P, and Fe) in enzyme-treated bio-milled rice. After, losses of minerals occurred by continuously increasing temperature and enzyme concentration, whereas increased polishing time and treatment time continually caused better retention of minerals (concave shape) in cellulase rice and a reverse trend (convex shape) was observed for xylanase rice. The macronutrients existed inside the bran layer integrated to proteins; such distributions of minerals have been shown to be maximized in the exit stratum against endosperm [27]. Generally, during enzymatic treatment, the cellulase hydrolyzing enzyme acts on a coat around the cellulose [28]. It contains xylan and lignin; xylan invades the mean situation among the spathe of lignin remnants and intertwines through a covalent bond to the sheath in different positions. Covalent linkages around xylan through the lignin scabbard interweave with inter H-bonding imparts off a sheet nearby cellulosic manner. This cellulosic sheath is incorporated with macro elements and proteins. Initially, more enzyme concentration at low temperature causes breakages of covalent and inters H-bonding linkages and leads to moving macronutrients to the inner endosperm; this leads to more retained calcium in bio-polished rice at an initial level during polishing after losses have occurred due to enzyme inactivation by increasing temperature. This reason is due to phosphorus presence in the form of a phosphorus composite matrix with a cellulose stringy chain via hydrogen bonds and Van der Waals forces in the parallel direction. This cellulase reaction upon composite matrix leads to the creation of microfibrils; these were broad as well as forming a crystalline assemblage, leading to a significant release as water-soluble fragments from aleuronic stratum towards endosperm enriches the final products when enzyme concentrations are high its moves opposite to endosperm decreases the final product phosphorus concentration [19,20,29].

Table 3. ANN design (experimental and predicted values) for performance of xylanase.

Y_8		Y_9		Y_{10}		Y_{11}		Y_{12}		Y_{13}		Y_{14}	
Exp.	Pred.	Exp.	Pred.	Exp.	Pred.	Exp.	Pred.	Exp.	Pred.	Exp.	Pred.	Exp.	Pred.
2.84	3.09	2.06	2.00	0.94	0.98	160.67	160.03	61.12	60.66	1.50	1.41	7.65	7.51
2.94	3.18	1.93	2.05	0.77	0.92	162.78	167.10	61.78	60.45	1.40	1.45	7.91	7.75
3.47	3.20	1.97	2.06	0.86	0.79	172.89	170.53	58.11	58.68	1.60	1.41	7.61	7.66
3.62	3.21	2.05	2.13	1.09	0.83	168.95	166.97	59.98	58.73	1.50	1.43	7.92	7.87
3.22	3.07	2.08	2.08	0.59	0.95	147.92	153.31	58.09	59.44	1.30	1.38	7.91	7.63
3.14	3.04	1.98	2.12	0.62	1.11	157.97	162.59	58.14	60.40	1.60	1.44	7.83	7.60
3.00	3.19	1.99	2.08	0.58	0.80	176.89	167.76	57.43	58.53	1.40	1.40	7.64	7.68
3.18	3.10	2.16	2.22	0.63	0.95	162.89	163.10	56.55	57.94	1.30	1.41	8.16	7.80
3.43	3.31	2.09	2.08	0.78	0.73	173.24	169.78	58.57	58.00	1.40	1.55	7.77	8.00
2.98	3.21	2.06	2.12	0.68	0.86	159.47	165.93	57.99	58.37	1.30	1.53	7.9	7.92
3.43	3.16	2.02	2.16	1.03	0.86	172.78	167.95	56.82	57.93	1.60	1.43	7.81	7.78
3.27	3.00	1.97	2.20	0.99	1.13	139.78	160.10	56.45	59.27	1.40	1.43	8.12	7.66
3.52	3.32	1.97	2.08	1.05	0.69	175.78	170.64	59.73	57.73	1.60	1.55	8.13	8.02
3.19	3.20	2.02	2.12	0.78	0.88	168.98	165.59	57.78	58.50	1.60	1.52	7.19	7.90
2.97	3.18	2.22	2.14	0.89	0.84	171.89	169.60	59.89	58.14	1.40	1.43	7.92	7.75
2.89	3.03	2.25	2.27	1.08	1.04	148.99	159.52	60.12	57.78	1.40	1.41	7.84	7.79
3.29	3.18	2.09	2.09	0.79	0.80	172.59	165.58	58.99	58.37	1.30	1.39	7.72	7.70
3.17	3.12	2.11	2.08	0.88	1.01	167.89	164.75	59.78	60.39	1.30	1.45	7.79	7.69
2.94	3.06	1.85	2.13	0.59	1.08	156.78	162.69	61.17	59.76	1.70	1.47	7.87	7.67
2.99	3.12	1.91	2.17	0.68	0.92	154.87	167.09	59.89	58.29	1.50	1.41	7.93	7.71
3.37	3.00	1.87	2.13	0.83	1.16	159.91	162.07	57.17	60.43	1.40	1.44	7.89	7.54
3.28	3.07	2.04	2.22	0.94	0.99	163.59	162.91	57.98	58.05	1.40	1.41	7.91	7.75
3.28	3.21	2.19	2.03	0.77	0.77	161.67	165.10	58.98	59.15	1.50	1.40	7.85	7.71
2.94	3.29	2.26	2.13	0.96	0.74	167.82	167.39	58.23	57.41	1.50	1.55	7.94	8.06
3.05	3.03	2.3	2.16	1.12	1.09	159.23	162.47	58.44	59.54	1.40	1.43	7.52	7.62
2.99	3.03	2.32	2.16	1.06	1.09	156.9	162.47	59.11	59.54	1.30	1.43	7.42	7.62

Table 3. *Cont.*

Y$_8$		Y$_9$		Y$_{10}$		Y$_{11}$		Y$_{12}$		Y$_{13}$		Y$_{14}$	
Exp.	Pred.	Exp.	Pred.	Exp.	Pred.	Exp.	Pred.	Exp.	Pred.	Exp.	Pred.	Exp.	Pred.
2.87	3.03	2.22	2.16	1.21	1.09	158.92	162.47	59.54	59.54	1.60	1.43	7.42	7.62
2.94	3.03	2.28	2.16	1.19	1.09	157.01	162.47	59.40	59.54	1.40	1.43	7.42	7.62
2.89	3.03	2.3	2.16	1.24	1.09	158.15	162.47	58.19	59.54	1.50	1.43	7.42	7.62
2.79	3.03	2.22	2.16	1.09	1.09	156.92	162.47	58.89	59.54	1.20	1.43	7.29	7.62

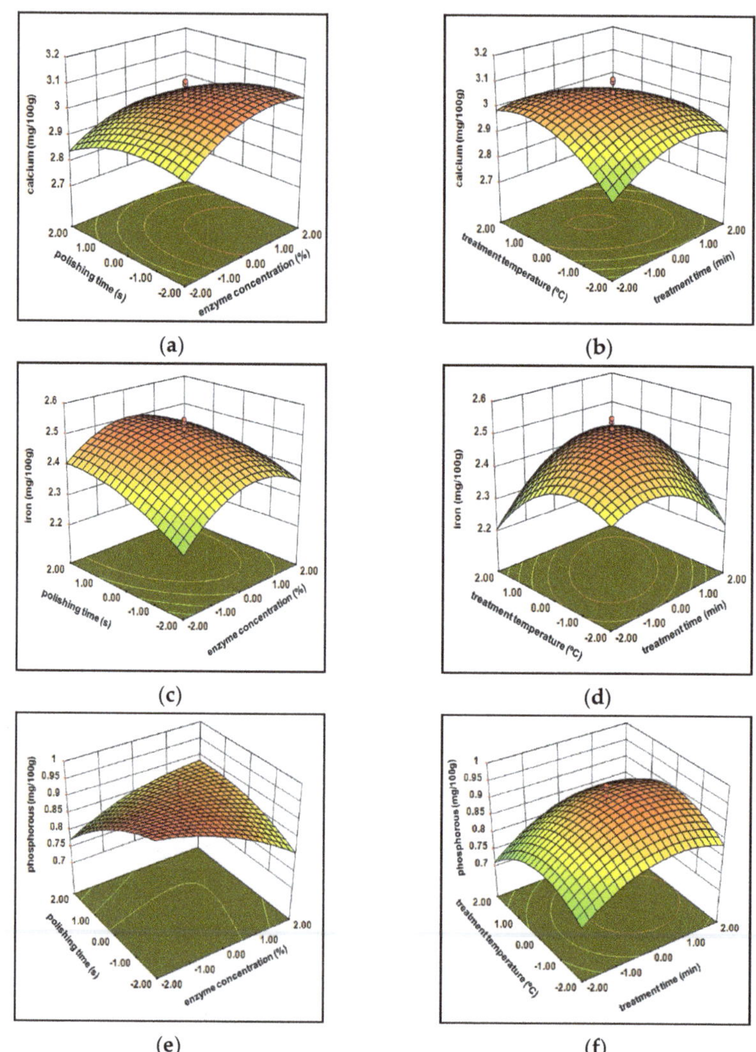

Figure 3. *Cont.*

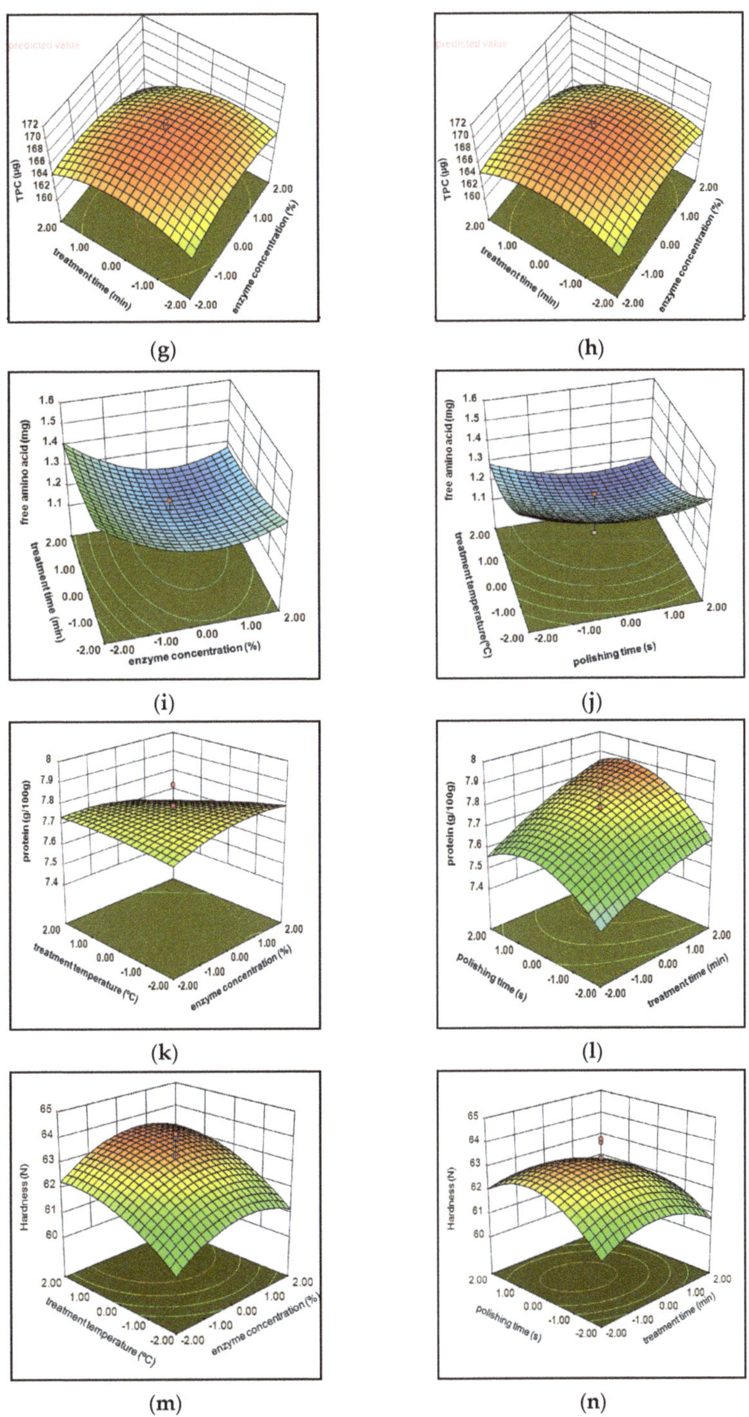

Figure 3. Effect of cellulase enzymatic treatment variables on calcium (**a**,**b**), iron (**c**,**d**), phosphorus (**e**,**f**), TPC (**g**,**h**), free amino acids (**i**,**j**), protein content (**k**,**l**), and hardness (**m**,**n**) of polished rice.

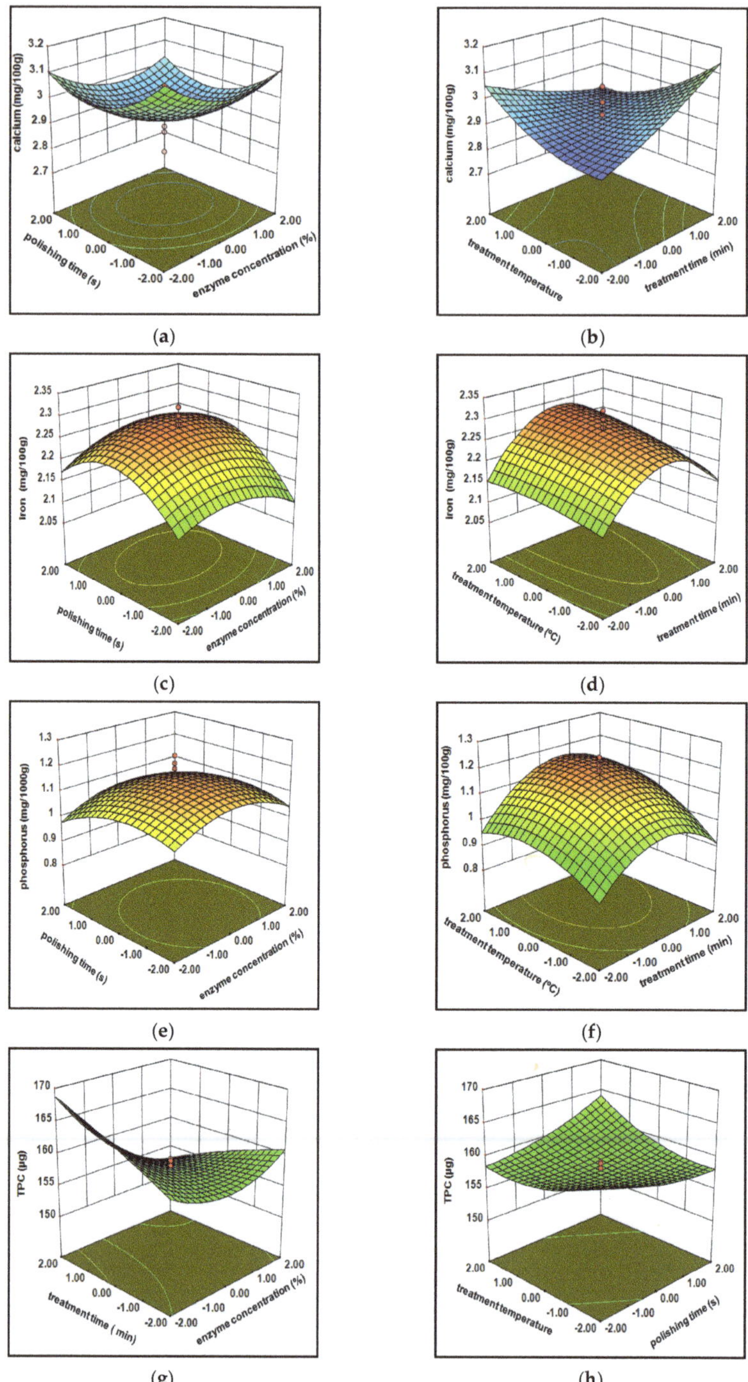

Figure 4. *Cont.*

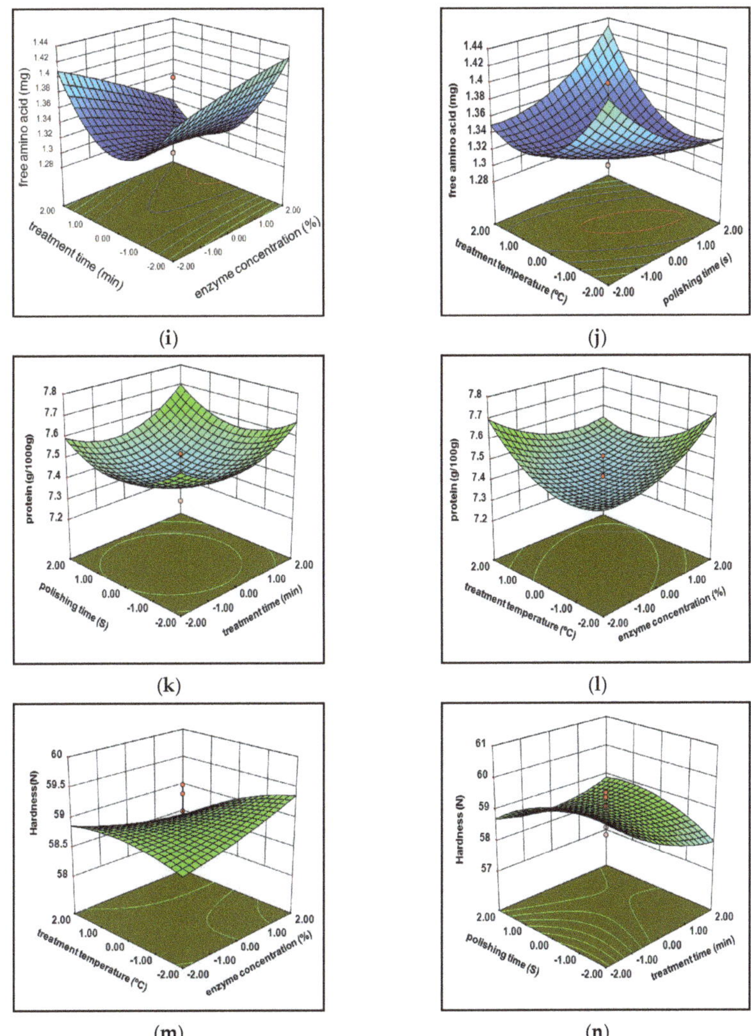

Figure 4. Effect of xylanase enzymatic treatment variables on calcium (**a,b**), iron (**c,d**), phosphorus (**e,f**), TPC (**g,h**), free amino acids (**i,j**), protein content (**k,l**) and hardness (**m,n**) of polished rice.

3.1.2. Total Phenolic Content (TPC)

The observed value for the total phenolic content of cellulase milled rice varied between 151.92 to 169.66 µg, while xylanase milled varied within 139.78 to 176.89 µg. Correlation regression indicates that the phenolic content owned significantly ($p < 0.001$) was influenced via independent variables of cellulose as well as xylanase treated milled rice at a linear form and quadratic terms (Tables 1 and 2). The mathematical formula generated with a variation in the phenolic content under individual parameters (X_1, X_2, X_3, X_4) was considerably suited within second-order polynomial equation (Equations (Y4) and (Y11) in Table 4). The high $R^2 = 0.87$ and 0.87 (Tables 1 and 2) data represent the acquaintance within viewing values and anticipated values and, as a result, improve the paradigm. (Figures 3g,h and 4g,h) 3D surface plots show the effect of the whole phenolic substance against the independent variables. The phenolic importance of cellulase processed milled rice at the beginning increased arbitrarily (160 to 166.21 mg) after that moderately diminish

(166.22 to 164.54 mg) demonstrates as a bulging (convex) shaped, continuous rising enzyme engrossment, time, and temperature, whereas xylanase treated milled rice shows a reverse trend in a concave way (initially decreases after that increases) with all the independent variables. Usually, phenolic acid concentrations are enhanced upon the endosperm towards aleuronic stratum. All phenols existed in an aleuronic layer derivate shape additionally connected to lignin and arabinoxylans compounds. These reactions validate that the enzymes induce the rupture of the attachment among lignin and cell walls and create lesser leaks around phenols derivatives: benzoic, gallic, protocatechuic, and vanillic acids in enzyme-processed milled rice throughout the treatment since phenolic compounds migrate towards the outer layer to endosperm.

Table 4. Quadratic models developed using independent variables from enzymatic treatment experimental design.

Response	Equation
Cellulase Y_1	$Y_1 = 3.05 + 0.0547X_1 + 0.0246X_4 + 0.041X_1X_2$
Y_2	$Y_2 = 0.92 + 0.054X_1X_2 + 0.073X_1X_3 + 0.041X_2X_3 + 9.375 \times 10^{-3}\,X_2X_4$
Y_3	$Y_3 = 2.49 + 0.0237X_1 + 0.014X_2 + 0.050X_3 + 4.583 \times 10^{36}\,X_4 + 0.019X_1X_4 + 0.019X_2X_3 + 0.087X_2X_4$
Y_4	$Y_4 = 168.24 + 1.27X_1X_3 + 0.89X_1X_4 + 1.04X_3X_4$
Y_5	$Y_5 = 63.14 + 0.39X_1 + 0.25X_3 + 0.95X_4 + 0.25X_1X_3 + 0.073X_1X_4$
Y_6	$Y_6 = 1.23 - 0.063X_1 + 0.044X_1X_3 + 0.056X_1X_4 + 0.081X_2X_3 + 0.056X_2X_4 + 0.12X_1{}^2 + 0.034X_2{}^2 + 0.072X_3{}^2 + 0.059X_4{}^2$
Y_7	$Y_7 = +7.77 + 0.12X_2 + 0.083X_3 + 0.031X_1X_2 + 0.027\,X_2X_3 + 0.047\,X_3X_4$
Xylanase Y_8	$Y_8 = 2.92 + 0.028X_2 + 0.053X_1X_2 + 3.125 \times 10^{-0.086}\,X_1X_3 + 0.083X_1{}^2 + 0.017X_2{}^2 + 0.11X_3{}^2 + 0.053X_4{}^2$
Y_9	$Y_9 = 2.27 + 2.500 \times 10^{-3}\,X_1 + 0.023X_2 + 0.036X_3 + 0.022X_4 + 0.028X_1X_2 + 0.018X_1X_3 + 0.044X_2X_3 + 0.013X_2X_4 + 7.500 \times 10^{-3}\,X_3X_4$
Y_{10}	$Y_{10} = 1.15 + 4.167 \times 10^{-3}\,X_1 + 0.047X_2 + 0.066X_4 + 0.059X_1X_2 + 5.000 \times 10^{-3}\,X_1X_3 + 0.029X_2X_4 + 0.098X_3X_4$
Y_{11}	$Y_{11} = 168.24 + 1.27X_1X_3 + 0.89X_1X_4 + 1.04X_3X_4$
Y_{12}	$Y_{12} = 63.14 + 0.39X_1 + 0.25X_3 + 0.95X_4 + 0.073X_1X_4$
Y_{13}	$Y_{13} = 1.23 + 0.044X_1X_3 + 0.056X_1X_4 + 0.081X_2X_3 + 0.12X_1{}^2 + 0.034X_2{}^2 + 0.072X_3{}^2 + 0.059X_4{}^2$
Y_{14}	$Y_{14} = 7.77 - 7.500 \times 10^{-3}\,X_1 + 0.12X_2 + 0.083X_3 - 0.034X_4 - 0.044\,X_1{}^2 + 0.031\,X_1X_2 + 0.027\,X_2X_3 + 0.047\,X_3X_4$

3.1.3. Free Amino Acids

The free amino acid content of cellulase and xylanase milled rice ranging from 1.1 mg to 1.9 mg and 1.2 mg to 1.7 mg across the entire experimental considerations. The experimental and predicted values of xylanase are reported in Table 3. The best free amino acid was obtained by 70% X_1, 60 min X_2, 20 s X_3, 35 X_4 and 80% X_1, 30 min X_2, 30 s X_3, 40 °C X_4. An equilateral polynomial equation was adapted to both enzymes' free amino acid practical information, which evolved a statistical retrogradation portrayed in Table 5 (Equations (Y6) and (Y13) in Table 4). The R^2 (0.89 and 0.87), adjusted R^2 (0.79 and 0.78), and the lack of fit was nonsignificant stated such that the fitness of model acts intimately among observational and anticipated values (Tables 1 and 2). The free amino acid compound of cellulase (1.46 to 1.35 mg) and xylanase (1.41 to 1.33 mg) milled rice at first slightly diminished after they increased (1.21 to 1.42 mg) by an enhancement of all parameters (X_1, X_2, X_3, X_4) continuously from −2 to +2, observed in Figures 3i,j and 4i,j. Free amino acid content is delicate for peak temperature over a lengthy period of cooking; it may cause loss of nutrients easily as a result of processing of the decreased enzyme losses through cooking; the ability of two enzymes in rupturing the glycosidic bonds among proteins with amino acids drives all the compounds (protein plus free amino acids) from the external stratum towards the internal stringy network endosperm. The glycosidic linkage associated compounds (proteins and amino acids) located at the bran layer were smoothly damaged in the middle of mechanical milling owing to the erratic mode through aleurone structure [2,30]. The previous statement reveals that cellulase processed brown rice, at first being slightly diminished, might be because cellulase performing over glycosidic bonds may have taken time. In contrast, xylanase immediately pretends as a result of being verified in the connected linkages (noncovalent bonds).

Table 5. ANN design (experimental and predicted values) for performance of cellulase.

Y_1		Y_2		Y_3		Y_4		Y_5		Y_6		Y_7	
Exp.	Pred.	Exp.	Pred.	Exp.	Pred.	Exp.	Pred.	Exp.	Pred.	Exp.	Pred.	Exp.	Pred.
2.51	2.73	2.04	2.05	0.95	0.79	165.33	163.84	58.14	58.55	1.90	1.72	7.45	7.61
2.65	2.72	2.12	2.06	0.58	0.75	162.88	159.21	59.73	58.53	1.70	1.71	7.31	7.62
2.82	2.73	1.93	2.05	0.76	0.78	165.67	162.60	57.72	58.52	1.80	1.72	7.48	7.61
2.96	2.75	1.82	2.11	0.71	0.74	158.9	157.41	59.19	58.99	1.40	1.67	7.82	7.63
2.88	2.82	2.29	2.19	0.64	0.74	154.91	156.28	58.94	59.32	1.60	1.62	7.46	7.67
2.66	2.87	2.33	2.27	0.67	0.72	156.74	153.89	59.91	59.81	1.50	1.56	7.39	7.70
2.45	2.88	2.25	2.26	0.71	0.75	154.23	157.42	59.19	59.39	1.80	1.59	7.73	7.70
2.76	2.86	2.15	2.25	0.97	0.68	152.21	153.14	58.97	58.67	1.60	1.56	7.91	7.74
2.79	2.91	1.99	2.35	0.91	0.86	160.11	167.67	59.22	62.11	1.80	1.41	7.24	7.65
2.89	2.85	2.22	2.31	0.71	0.80	164.21	163.78	61.15	61.68	1.60	1.38	7.42	7.67
2.92	2.89	2.25	2.33	0.95	0.83	162.7	166.00	59.74	61.92	1.30	1.40	7.46	7.66
3.01	2.73	2.28	2.21	0.79	0.66	155.32	154.87	61.18	60.69	1.30	1.30	7.41	7.71
2.92	2.78	2.05	2.24	0.56	0.71	156.21	158.58	60.18	60.82	1.40	1.33	7.72	7.71
2.91	2.74	1.95	2.21	0.47	0.66	158.9	156.05	63.17	60.37	1.50	1.31	7.44	7.72
2.41	2.75	2.19	2.21	0.59	0.67	151.92	156.63	60.15	60.27	1.30	1.32	7.92	7.73
2.54	2.76	2.25	2.22	0.78	0.67	157.29	157.67	58.77	60.06	1.40	1.33	7.64	7.73
2.65	2.91	2.08	2.35	0.81	0.86	163.67	168.05	59.89	62.04	1.80	1.42	7.65	7.65
2.96	2.75	2.29	2.23	0.79	0.68	157.21	156.95	60.17	60.67	1.50	1.32	7.62	7.71
2.78	2.91	2.15	2.35	0.57	0.86	157.21	167.93	60.91	62.06	1.30	1.42	7.45	7.65
2.79	2.79	2.25	2.25	0.74	0.72	158.78	159.42	58.71	60.84	1.30	1.34	7.94	7.71
2.92	2.90	2.31	2.34	0.76	0.86	165.87	167.79	57.72	62.00	1.50	1.42	7.24	7.65
2.81	2.87	2.51	2.25	0.83	0.74	162.44	167.24	59.14	58.27	1.40	1.44	7.43	7.79
2.77	2.68	2.19	2.00	0.65	0.68	154.92	151.73	58.12	57.22	1.70	1.80	7.88	7.63
2.71	2.72	2.12	2.21	0.57	0.65	152.89	154.64	63.62	60.64	1.10	1.30	7.62	7.72
3.1	2.91	2.54	2.35	0.96	0.86	165.96	167.90	62.14	61.96	1.30	1.42	7.89	7.66
2.96	2.91	2.55	2.35	0.85	0.86	169.01	167.90	63.34	61.96	1.20	1.42	7.79	7.66
3.11	2.91	2.41	2.35	0.97	0.86	167.89	167.90	63.1	61.96	1.30	1.42	7.74	7.66
3.05	2.91	2.49	2.35	0.94	0.86	168.11	167.90	63.98	61.96	1.40	1.42	7.74	7.66
3.03	2.91	2.52	2.35	0.91	0.86	169.66	167.90	64.14	61.96	1.20	1.42	7.74	7.66
3.06	2.91	2.45	2.35	0.87	0.86	168.79	167.90	62.14	61.96	1.10	1.42	7.74	7.66

3.1.4. Protein Composition

The protein compound values of cellulase and xylanase treated milled rice among 7.24 to 7.94 g and 7.29 to 8.16 g. The maximum protein content was shown at 80% X_1, 150 min X_2, 30 s X_3, 40 °C X_4 and 90% X_1, 120 min X_2, 30 s X_3, 45 °C X_4. A simplified equation (Equations (Y7) and (Y14) in Table 4) has been created regarding multinomial polynomial. It shows the result of significant parameters of protein recovery from milled rice. The correlation coefficient (R^2) protein composition is nearly 95%, the F value was immensely significant ($p < 0.0001$), and the lack of fit was nonsignificant ($p < 0.05$) and may imply that the model has been confirmed within observational values and predicted values. The protein complex of cellulase processed rice was enhanced through the convex shape (progressively raised from 7.4 to 7.62 g, then moderately diminished until 7.56 g) by constantly increasing whole parameters (-2 to 2). In contrast, xylanase processed milled rice demonstrates slightly enhanced protein content over an increase of all parameters (Figures 3k,l and 4k,l); this is the reason behind the large number of embryonic proteins tightly attached to a stringy network bran stratum via glycosidic bonds. While traditional milling processes a significant loss of protein due to the above reason, enzyme processing improves bond breakage, stringing of bonds and collisions among enzymes and its bonds lead to relocation of free amino acid outer matrix layer to inner endosperm during specific conditions [30].

3.1.5. Hardness

The hardness of enzymes treated milled rice varied between 57.72 to 64.14 N and 56.45 to 61.78 N. The highest hardness of enzymatic milled rice was sustained at 80% X_1,

90 min X_2, 30s X_3, 40 °C X_4 (cellulase), and 90% X_1, 60 min X_2, 20s X_3, 35 °C X_4 (xylanase). The correlation coefficient (R^2 = 89.9 and 88.6) for the regression model of the hardness of both enzymes treated milled rice was highly significant (<0.0001), and lack of fit has now been recognized as nonsignificant as evidenced in Tables 1 and 2. Figures 3m,n, and 4m,n represents the hardness of cellulase processed milled rice, in the beginning, enhanced randomly after the moderate decline (60.51–60 N) by way of bulging shape (convex form) through enhancing the entire treatment of variables (−2 to 2), considering that xylanase milled rice shows linearly decreased hardness with all treatment conditions. This may be due to the aleuronic stratum removal caused by vulnerability around starch granules by enhanced gelatinization of the starch. Two of the maximum oil and fiber concentrations within the bran stratum inside the brown rice rendered a rigid structure with the hardness, subsequently reducing hardness continually by enzymes reacting over the Huron structure, thereby decreasing both compounds (fibers and oils) around the bran layer and making it smoother [21,31].

3.1.6. Scanning Electron Microscopy Structural Analysis

Microstructure analysis performed by SEM revealed loosening of rice bran from the endosperm due to the splitting activity of cellulase and xylanase on the bran layer of brown rice is observed in Figure 5A,B, which is deviated from regular brown rice followed in Figure 5C.

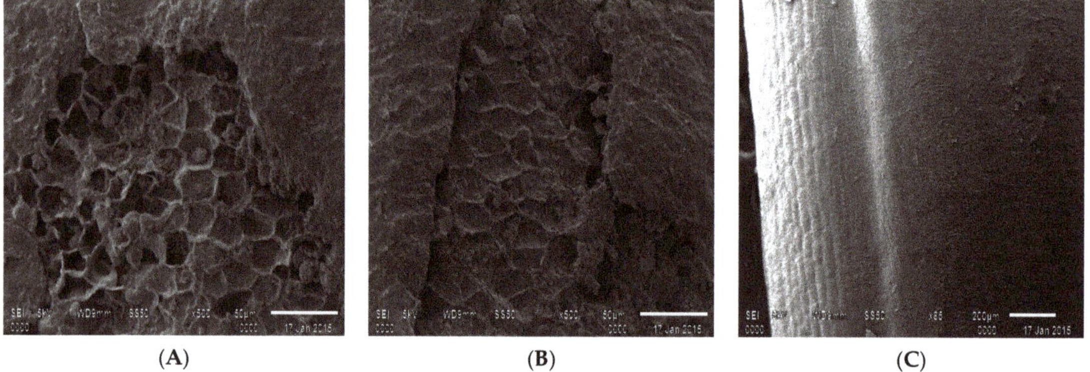

Figure 5. SEM images of cellulase treated rice (**A**), xylanase treated rice (**B**) and normal brown rice (**C**).

3.1.7. Model Optimization

Equation (1) shows the outcome of using multiple linear regression in the design of quadratic multinomial retaliation. Usually, many responses were used to produce the quadratic multinomial regression retaliation for predicting related responses. The models developed through significant variables have been reconstructed with MATLAB's fitlm operation, as displayed in Table 4. It might have identified the novel linear equations that have shortened the significantly influenced parameters, and the achievement has been strengthened. Therefore, the present multivariate models (Table 4) have been endorsed as their potential for creating nutritionally enhanced enzymatic milled rice process parameters.

3.2. Artificial Neural Network

The ANN model was used further for differentiating the execution over the multiple linear regression model. The observed and anticipated assessment ranges were almost to an identical field (0.95 = 1) (Table 5). This interrelation distinguishes data and values expected as projected in Figures 6 and 7. Counterpointing the previous exploration [26] comprises seven neurons at the exit structure while six neurons extracted at hidden stratum were abundant for establishing good achievement for ongoing investigations. Can it be ascertained that R^2 enhanced over a maximum quantity of neurons at the invisible stratum. The complete artificial neural network design performance was better across multiple linear regression designs presented in Table 6. This might be noted that great fit would have been noticed wherein nutrients were retained in enzyme treated milled rice because determined R^2 and MSE values fluctuated upon 0.91 to 0.97 and 0.005 to 6.13. The maximum R^2 and minimum mean square error (MSE) values for ANN showed improved precision and authenticity [13,26]. ANN and MLR designs were advisable to anticipate the xylanase performance, which was superior over cellulase and stated R^2 values of 0.97 and 0.90. Still, the ANN design could be strengthened through implementing maximum concealed neurons over every model; however, on this point, design engorgement is a possiblity. Therefore, a conclusion over quantity about neurons through established models has been applied appropriately. The MSE and correlation coefficient (R^2) of a practiced network subsist approximately at 0.006 and 9.99. Attainment programs based on the aimed network, further to the scatter plot overtraining, validation, and test are depicted at Figure 8. Equating ANN and MLR, the ANN design achievement has been excellent; consequently, it is possible to exploit the speculation of the modeling of enzyme application for enhancing nutrients in cereals.

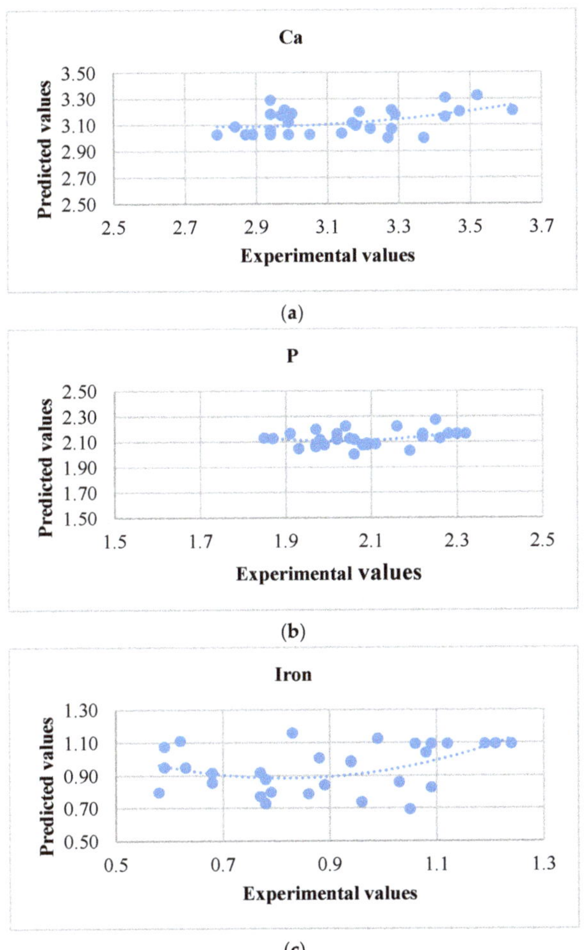

(a)

(b)

(c)

Figure 6. *Cont.*

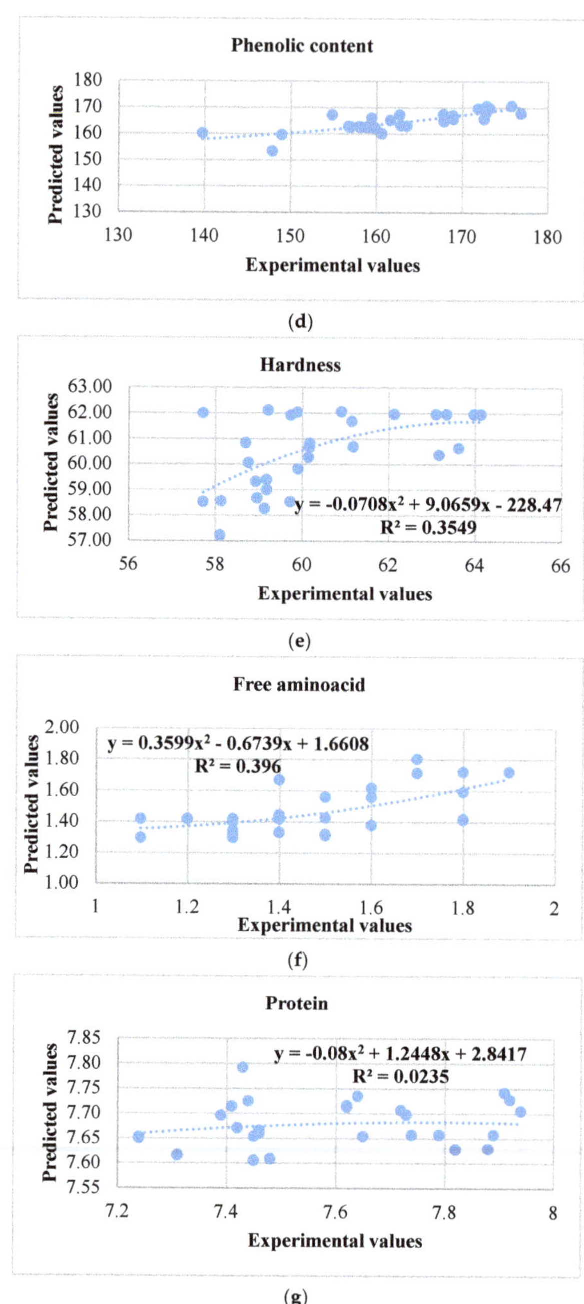

Figure 6. The relationship between experimental values and predicted values for the models developed using artificial neural network. (**a**) Calcium, (**b**) phosphorus, (**c**) iron, (**d**) phenolic content, (**e**) hardness, (**f**) free amino acid, and (**g**) protein of cellulase treated rice.

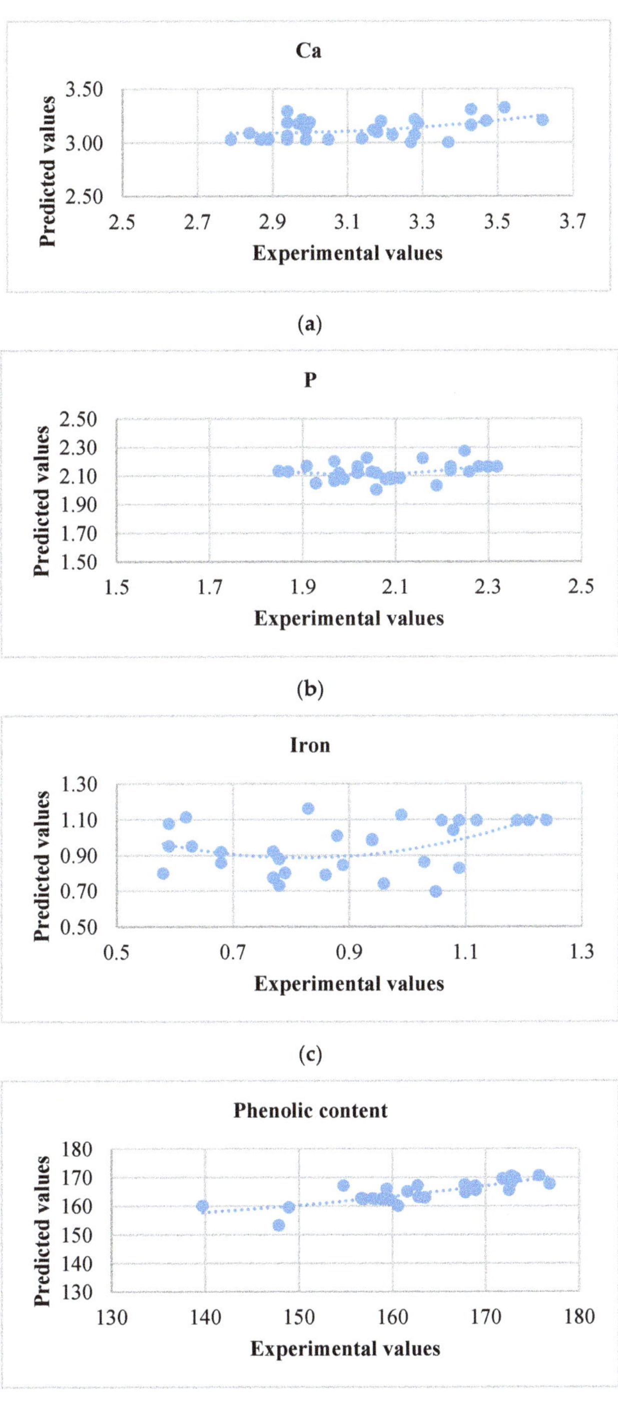

(a)

(b)

(c)

(d)

Figure 7. *Cont.*

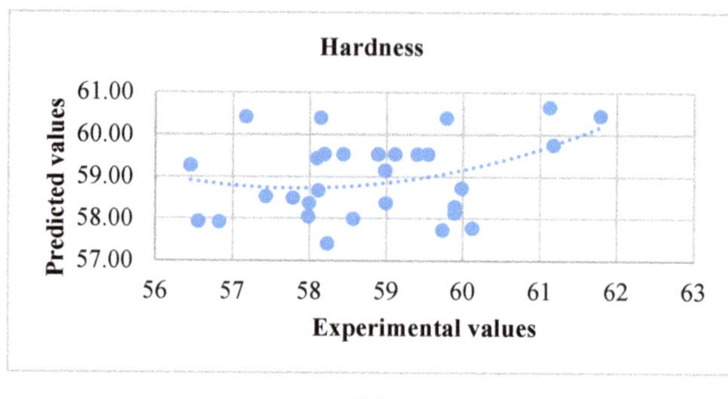

(e)

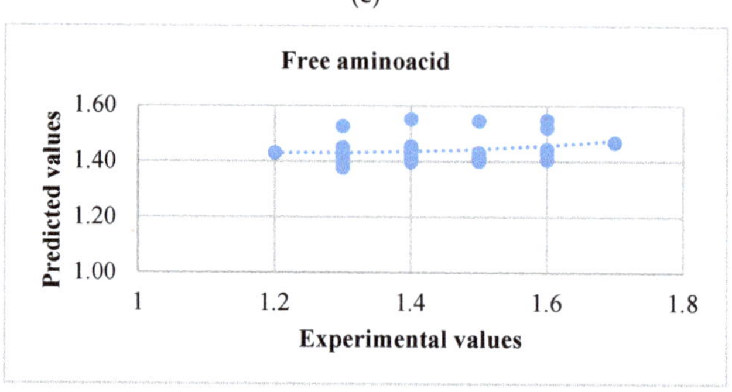

(f)

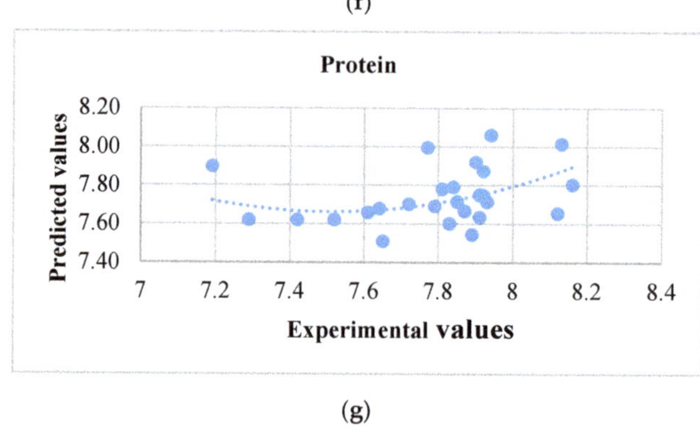

(g)

Figure 7. The relationship between experimental values and predicted values for the models developed using artificial neural network. (**a**) Calcium, (**b**) phosphorus, (**c**) iron, (**d**) phenolic content, (**e**) hardness, (**f**) free amino acid, and (**g**) protein of xylanase treated rice.

Table 6. Performance measure of artificial neural network model.

Variables	Experimental Range	Predicated Range	R^2	MSE
Cellulase Y_1	2.41–3.11	2.68–2.91	0.91	1.1641
Y_2	1.82–2.51	2.00–2.35	0.94	1.1622
Y_3	0.47–0.97	0.65–0.87	0.96	6.1344
Y_4	152.21–169.66	151.92–168.05	0.92	0.971
Y_5	57.72–64.14	57.72–64.14	0.94	2.734
Y_6	1.10–1.90	1.30–1.80	0.93	4.924
Y_7	7.24–7.88	7.62–7.72	0.96	0.229

Table 6. *Cont.*

Variables	Experimental Range	Predicated Range	R^2	MSE
Xylanase Y_8	2.84–3.62	3.00–3.32	0.91	0.677
Y_9	1.93–2.32	2.00–2.27	0.94	0.354
Y_{10}	0.58–1.24	0.69–1.16	0.93	0.005
Y_{11}	147.92–176.89	159.52–170.64	0.90	1.667
Y_{12}	57.45–61.78	57.41–60.66	0.95	2.871
Y_{13}	1.20–1.70	1.39–1.55	0.92	0.195
Y_{14}	7.29–8.16	7.51–8.02	0.97	0.2044

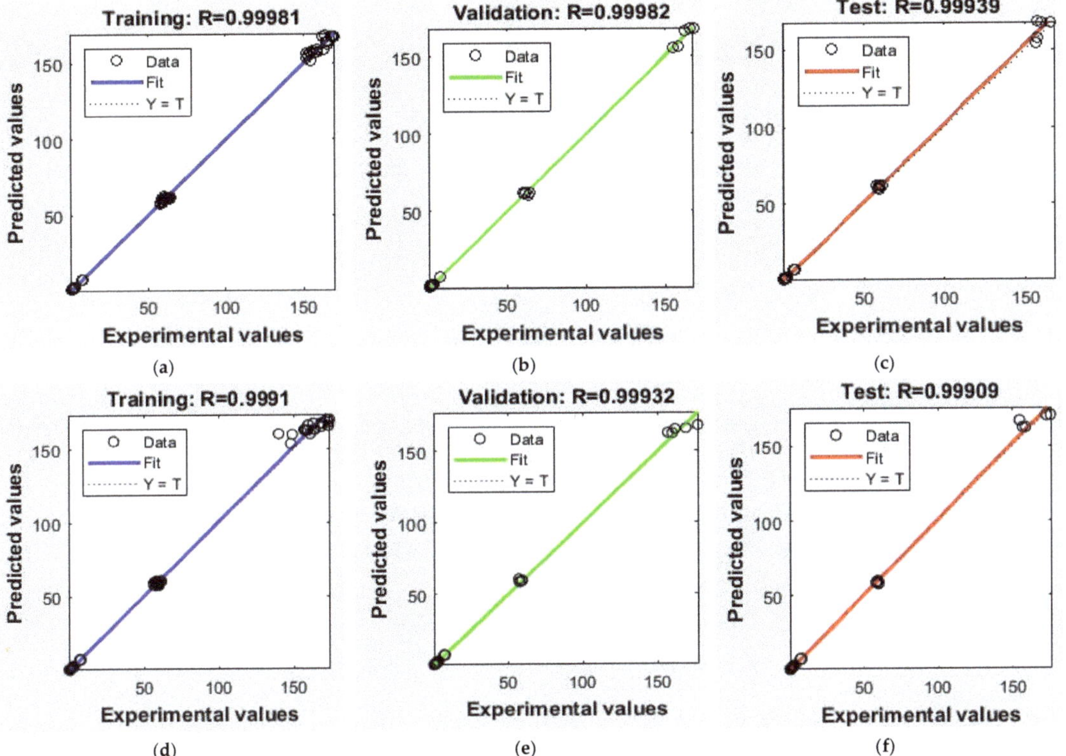

Figure 8. Regression plots for the training (**a,d**), validation (**b,e**) and (**c,f**) test for cellulose enzyme and xylanase treated milled rice ANN mode.

Process Optimization

Economic software (Design-Expert version 11.0) was used for the optimization procedure for mathematical optimization for nutrients in the cellulase and xylanase processed milled rice. Simultaneously, enzyme concentration, treatment time, temperature, and milling time were the independent variables. The optimized values of processing conditions were acquired to create the following characteristics: maximum Ca, P, Fe, phenolic content, hardness, and free amino acid in enzyme processed polished rice. The desirable optimized values for cellulose (87.2% X_1, 80.1 min X_2, 21.8 s X_3 and 33.95 °C) and xylanase (85.7% X_1, 77.1 min X_2, 20 s X_3 and 35 °C X_4) were found for treatment conditions in milled rice. The nutritional enhancement of enzyme treated milled rice under the above optimum conditions of experimental and predicted values can be observed in Table 6. Consequently, the chosen mathematical design exists precisely as well as immediately to create deviations in entire parameters.

4. Conclusions

The current research focused on modeling and optimizing process parameters for enhancing nutrients in enzyme treated milled rice by multiple optimization techniques. During the optimization of enzyme treatment, the consequences from the process assessment revealed that the enzyme concentration, treatment time, temperature, and polishing time had more impact upon the nutritional improvement of milled rice. The optimized cellulase treated milled rice was improved by 66% in calcium, 17% in iron, 64% in phosphorus, 78% in total phenol content, 33% in free amino acid, and 84% in protein content. In contrast, xylanase treated milled rice was improved by 70% in calcium, 15% in iron, 62% in phosphorus., 79% in total phenol content, 34% in free amino acid, and 83% in protein content compared to polished rice. The overall hardness (19–20%) of cooked milled rice was reduced. The xylanase showed better performance than cellulase. The designs were measured based upon the correlation coefficient (R^2), the sum of squared error (SSE), and mean squared error (MSE). The observation outcome of MLR was enhanced by the multiple polynomial retrogression equations despite the predominate multilayer neuromorphic model of ANN obtained six neurons over a "transig" for activating in the hidden stratum. Two designs (MLR and ANN) have been better adapted to optimizing enzyme-treated milled functioning responses. The more significant coefficient and a lesser sum of squared error data of ANN indicate greater anticipation on observational values across MLR.

Author Contributions: A.K. (Anjineyulu Kothakota): Conceptualization, supervisor-lead, writing-original draft preparation, writing-review and editing. R.P.: Visualization, investigation, writing-review draft-lead, writing-review and editing. K.S.: Supervisor, writing-original draft preparation, writing-review and editing. J.P.P.: Software, writing-review and editing. N.S.: Writing-original draft preparation, Writing-review and editing. C.H.S.S.: Software, writing-original draft preparation, writing-review and editing. A.K. (Anil Kumar): Supervisor, writing-original draft preparation, writing-review and editing. A.S.: Writing-original draft preparation, writing-review and editing. S.D.P.: Writing-original draft preparation, writing-review and editing. All authors have read and agreed to the published version of the manuscript.

Funding: This research received no external funding.

Institutional Review Board Statement: Not applicable.

Informed Consent Statement: Not applicable.

Data Availability Statement: Not applicable.

Conflicts of Interest: The authors declare no conflict of interest.

References

1. Fernandes Monks, J.L.; Vanier, N.L.; Casaril, J.; Manica Berto, R.; de Oliveira, M.; Gomes, C.B.; de Carvalho, M.P.; Guerra Dias, A.R.; Elias, M.C. Effect of milling on proximate composition, folic acid, fatty acids and technological properties of rice. *J. Food Compos. Anal.* **2013**, *30*, 73–79. [CrossRef]
2. Chen, H.H.; Chen, Y.K.; Chang, H.C. Evaluation of physicochemical properties of plasma treated brown rice. *Food Chem.* **2012**, *135*, 74–79. [CrossRef]
3. Kennedy, G.; Burlingame, B.; Nguyen, V. Nutritional Contribution of Rice and Impact of Biotechnology and Biodiversity in Rice-Consuming Countries. In Proceedings of the 20th Session of the International Rice Commission, Bangkok, Thailand, 23–26 July 2002; pp. 59–70. Available online: http://www.fao.org/3/Y4751E/y4751e05.htm%0Ahttp://www.fao.org/docrep/006/Y4751E/y4751e05.htm (accessed on 15 October 2021).
4. Lamberts, L.; De Bie, E.; Vandeputte, G.E.; Veraverbeke, W.S.; Derycke, V.; De Man, W.; Delcour, J.A. Effect of milling on colour and nutritional properties of rice. *Food Chem.* **2007**, *100*, 1496–1503. [CrossRef]
5. Thirumdas, R.; Deshmukh, R.R.; Annapure, U.S. Effect of low temperature plasma processing on physicochemical properties and cooking quality of basmati rice. *Innov. Food Sci. Emerg. Technol.* **2015**, *31*, 83–90. [CrossRef]
6. Boluda-Aguilar, M.; Taboada-Rodríguez, A.; López-Gómez, A.; Marín-Iniesta, F.; Barbosa-Cánovas, G.V. Quick cooking rice by hydrostatic pressure processing. *LWT Food Sci. Technol.* **2013**, *51*, 196–204. [CrossRef]
7. Ding, J.; Ulanov, A.V.; Dong, M.; Yang, T.; Nemzer, B.V.; Xiong, S.; Zhao, S.; Feng, H. Enhancement of gamma-aminobutyric acid (GABA) and other health-related metabolites in germinated red rice (*Oryza sativa* L.) by ultrasonication. *Ultrason. Sonochem.* **2018**, *40*, 791–797. [CrossRef]

8. Qian, J.-Y.; Gu, Y.-P.; Jiang, W.; Chen, W. Inactivating effect of pulsed electric field on lipase in brown rice. *Innov. Food Sci. Emerg. Technol.* **2014**, *22*, 89–94. [CrossRef]

9. Das, M.; Banerjee, R.; Bal, S. Evaluation of physicochemical properties of enzyme treated brown rice (Part B). *LWT* **2008**, *41*, 2092–2096. [CrossRef]

10. Sarao, L.K.; Arora, M.; Sehgal, V.K.; Bhatia, S. The Use of Fungal Enzymes viz Protease, Cellulase, And Xylanase for Polishing Rice. *Int. J. Food Saf.* **2011**, *13*, 26–37.

11. Spagna, G.; Barbagallo, R.; Palmeri, R.; Restuccia, C.; Giudici, P. Properties of endogenous β-glucosidase of a Pichia anomala strain isolated from Sicilian musts and wines. *Enzym. Microb. Technol.* **2002**, *31*, 1036–1041. [CrossRef]

12. Kamruzzaman, M.; ElMasry, G.; Sun, D.-W.; Allen, P. Non-destructive prediction and visualization of chemical composition in lamb meat using NIR hyperspectral imaging and multivariate regression. *Innov. Food Sci. Emerg. Technol.* **2012**, *16*, 218–226. [CrossRef]

13. Pandiselvam, R.; Manikantan, M.R.; Sunoj, S.; Sreejith, S.; Beegum, S. Modeling of coconut milk residue incorporated rice-corn extrudates properties using multiple linear regression and artificial neural network. *J. Food Process. Eng.* **2019**, *42*, e12981. [CrossRef]

14. Kothakota, A.; Yadav, K.C.; Panday, J.P. Development of Microwave Baked Potato Chips using Tomato Flavour. *Asian J. Agric. Food Sci.* **2013**, *1*, 1571–2321.

15. Kothakota, A. A study on evaluation and characterization of extruded product by using various by-products. *Afr. J. Food Sci.* **2013**, *7*, 485–497. [CrossRef]

16. Zurera-Cosano, G.; García-Gimeno, R.M.; Rodríguez-Pérez, M.R.; Hervás-Martínez, C. Validating an artificial neural network model of Leuconostoc mesenteroides in vacuum packaged sliced cooked meat products for shelf-life estimation. *Eur. Food Res. Technol.* **2005**, *221*, 717–724. [CrossRef]

17. Sagarika, N.; Prince, M.; Kothakota, A.; Pandiselvam, R.; Sreeja, R.; Mathew, S.M. Characterization and Optimization of Microwave Assisted Process for Extraction of Nutmeg (*Myristica fragrans* Houtt.) Mace Essential Oil. *J. Essent. Oil Bear. Plants* **2018**, *21*, 895–904. [CrossRef]

18. Pakalapati, H.; Tariq, M.A.; Arumugasamy, S.K. Optimization and modelling of enzymatic polymerization of ε-caprolactone to polycaprolactone using Candida Antartica Lipase B with response surface methodology and artificial neural network. *Enzym. Microb. Technol.* **2019**, *122*, 7–18. [CrossRef] [PubMed]

19. Kothakota, A.; Pandey, J.P.; Ahmad, A.H.; Kumar, A.; Ahmad, W. Determination and optimization of Vitamin B complex in xylanase enzyme treated polished rice by response surface methodology. *J. Environ. Biol.* **2016**, *37*, 543–550. [PubMed]

20. Kothakota, A.; Pandiselvam, R.; Rajesh, G.; Gowthami, K. Determination and Optimization of Vitamin B Complex (B1, B2, B3 and B6) in Cellulase Treated Polished Rice by HPLC with UV Detector. *Asian J. Chem.* **2017**, *29*, 385–392. [CrossRef]

21. Bora, P.; Ragaee, S.; Marcone, M. Effect of parboiling on decortication yield of millet grains and phenolic acids and in vitro digestibility of selected millet products. *Food Chem.* **2019**, *274*, 718–725. [CrossRef] [PubMed]

22. Smith, A.J. *Post Column Amino Acid Analysis—Protein Sequencing Protocols*; Humana Press: Totowa, NJ, USA, 2003; Volume 211.

23. Park, J.K.; Kim, S.S.; Kim, K.O. Effect of milling ratio on sensory properties of cooked rice and on physicochemical properties of milled and cooked rice. *Cereal Chem.* **2001**, *78*, 151–156. [CrossRef]

24. AOAC. Official Methods of Analysis of the Association of Analytical Chemists. In *AOAC (1990)*, 15th ed.; A.A.C.C.: Washington, DC, USA, 1990.

25. Nourbakhsh, H.; Emam-Djomeh, Z.; Omid, M.; Mirsaeedghai, H.; Moini, S. Prediction of red plum juice permeate flux during membrane processing with ANN optimized using RSM. *Comput. Electron. Agric.* **2014**, *102*, 1–9. [CrossRef]

26. Srikanth, V.; Rajesh, G.K.; Kothakota, A.; Pandiselvam, R.; Sagarika, N.; Manikantan, M.R.; Sudheer, K.P. Modeling and optimization of developed cocoa beans extractor parameters using box behnken design and artificial neural network. *Comput. Electron. Agric.* **2020**, *177*, 105715. [CrossRef]

27. Itani, T.; Tamaki, M.; Arai, E.; Horino, T. Distribution of Amylose, Nitrogen, and Minerals in Rice Kernels with Various Characters. *J. Agric. Food Chem.* **2002**, *50*, 5326–5332. [CrossRef]

28. Mo, H.; Qiu, J.; Yang, C.; Zang, L.; Sakai, E.; Chen, J. Porous biochar/chitosan composite for high performance cellulase immobilization by g;utaraldehyde. *Enzyme Microb. Technol.* **2020**, *138*, 109561. [CrossRef]

29. Somerville, C.; Bauer, S.; Brininstool, G.; Facette, M.; Hamann, T.; Milne, J.; Osborne, E.; Paredez, A.; Persson, S.; Raab, T.; et al. Toward a Systems Approach to Understanding Plant Cell Walls. *Science* **2004**, *306*, 2206–2211. [CrossRef] [PubMed]

30. Mohapatra, D.; Bal, S. Cooking quality and instrumental textural attributes of cooked rice for different milling fractions. *J. Food Eng.* **2006**, *73*, 253–259. [CrossRef]

31. Arora, G.; Sehgal, V.; Arora, M. Optimization of process parameters for milling of enzymatically presented Basmati rice. *J. Food Eng.* **2007**, *82*, 153–159. [CrossRef]

 *foods*

Article

Increases of Lipophilic Antioxidants and Anticancer Activity of Coix Seed Fermented by *Monascus purpureus*

Haiying Zeng [1,2]**, Likang Qin** [1,2,*]**, Xiaoyan Liu** [3] **and Song Miao** [4,*]

[1] Key Laboratory of Plant Resource Conservation and Germplasm Innovation in Mountainous Region (Ministry of Education), Collaborative Innovation Center for Mountain Ecology & Agro-Bioengineering (CICMEAB), College of Life Sciences/Institute of Agro-Bioengineering, Guizhou University, Guiyang 550025, China; hyzeng1@gzu.edu.cn
[2] School of Liquor and Food Engineering, Guizhou University, Guiyang 550025, China
[3] Zhongkai University of Agriculture and Engineering, Guangzhou 510000, China; lxyan0344@163.com
[4] Teagasc Food Research Centre, Moorepark, Co. Cork, D15 DY05 Fermoy, Ireland
[*] Correspondence: lkqin@gzu.edu.cn (L.Q.); song.miao@teagasc.ie (S.M.)

Abstract: Lipophilic tocols, γ-oryzanol, and coixenolide in coix seed before and after fermentation by *Monascus purpureus* were determined. Antioxidant and anticancer activities of raw and fermented coix seed were evaluated using free-radical-scavenging assays and polyunsaturated fatty acid oxidation model, and human laryngeal carcinoma cell HEp2, respectively. Compared to the raw seed, the tocols, γ-oryzanol, and coixenolide contents increased approximately 4, 25, and 2 times, respectively, in the fermented coix seed. Especially, γ-tocotrienol and γ-oryzanol reached 72.5 and 655.0 μg/g in the fermented coix seed. The lipophilic extract from fermented coix seed exhibited higher antioxidant activity in scavenging free radicals and inhibiting lipid oxidation. The inhibitory concentrations for 50% cell survival (IC$_{50}$) of lipophilic extract from fermented coix seed in inhibiting HEp2 cells decreased by 42%. This study showed that coix seed fermented by *M. purpureus* increased free and readily bioavailable lipophilic antioxidants and anticancer activity. Therefore, fermentation could enhance the efficacy of the health promoting function of coix seeds.

Keywords: coix seed; *Monascus purpureus*; antioxidant; fermentation; HEp2

Citation: Zeng, H.; Qin, L.; Liu, X.; Miao, S. Increases of Lipophilic Antioxidants and Anticancer Activity of Coix Seed Fermented by *Monascus purpureus*. *Foods* **2021**, *10*, 566. https://doi.org/10.3390/foods10030566

Academic Editors: Barbara Laddomada and Weiqun Wang

Received: 19 February 2021
Accepted: 5 March 2021
Published: 9 March 2021

1. Introduction

Coix (*Coix lacryma-jobi* L. var. adlay) is a cereal widely cultivated in Asian countries, including China, Japan, Thailand, Myanmar, Laos, and India [1]. It has been recommended as a nourishing whole food, since coix seeds contain different amino acids, fibers, and phytochemical antioxidants, especially the lipophilic antioxidant vitamin E (tocols) and γ-oryzanol [2]. Numerous studies have reported different health benefits of the consumption of coix seed including decreasing in low-density lipoprotein cholesterol and increasing high-density lipoprotein cholesterol triglycerides, preventing fatty liver, reducing cell inflammation induced by lipoprotein oxidation, inhibiting allergic effects, etc. [3–5]. It has also been used for rheumatism, neuralgia, and diuretic medications and anticancer treatment [1,6]. However, those previous studies mainly focused on raw coix seed. As most of the antioxidants in cereals are in bound or blocked form due to cellulose, they are not readily converted to a free form in normal cooking processes and in the human digestive tract [7]. Thus, the bioavailability and bio-absorption of the bound-form antioxidants in the body are usually much lower than that of the free-form antioxidants.

Recently, the fermentation by probiotic microorganisms as a pre-digestion process has been applied to release the bindings of bound hydrophilic antioxidants and increase free-form antioxidants and their bioavailability [8]. However, the effects of the fermentation process on bound lipophilic antioxidants in cereal have not been reported. In this study, the changes in lipophilic antioxidants and the antioxidant and anticancer activity of coix seed fermented by *Monascus purpureus* were evaluated. The increase in the levels of free-form antioxidants by fermentation could enhance their bioavailability or bio-absorption in the body and increase the efficacy of the health-promoting functions of coix seed.

M. purpureus is a fungus, traditionally used in the preparation of fermented grains in China, such as fermented rice [9,10]. The fermentation process increases organoleptic qualities including desirable pigment and flavor. It also inhibits the growth of pathogenic microorganisms by producing organic acids and other compounds, thereby extending the shelf-life of fermented rice [11]. Previous studies found that the antioxidants of sweet potato, oat, and soybean, especially hydrophilic antioxidants, such as total phenolics and flavonoids, were significantly increased due to fermentation of *Lactobacillus acidophilus*, *Aspergillus oryzae, Monascus anka,* and *Bacillus subtilis* [8,9,12]. The results showed that fermentation was one of the effective and practical ways to increase the hydrophilic antioxidants in those plant foods. Fermentation not only assists in releasing those bound antioxidants to increase their bioavailability, but it could also reduce the sugar level in fermented foods. The fermented food can also help people with small-intestinal bowel infection or colonic infection effectively metabolize and absorb those antioxidants in the fiber-rich cereals [8].

In this study, the antioxidant activities of the lipophilic extracts from raw and fermented coix seed were measured using DPPH (2,2-diphenyl-1-picrylhydrazyl), ABTS (2,2'-azino-bis(3-ethylbenzothiazoline-6-sulfonic acid)) cation, and superoxide anion radical-scavenging assays. The activities of the extracts from raw and fermented coix seed in stabilizing susceptible polyunsaturated fatty acids C20:5 and C22:6 were evaluated as well. As vitamin E is an important lipophilic antioxidant in the human body, the anticancer activities of lipophilic extracts from raw and fermented coix seed were determined using HEp2 laryngeal carcinomatous cells. The results of this study could be helpful in understanding the changes of lipophilic antioxidants in fermented cereals. The fermented cereal or its extract could be used as a healthy processed food or ingredient with highly bioavailable bioactivity and enhanced health promoting function.

2. Materials and Methods

2.1. Chemicals and Materials

HPLC-grade hexane and acetic acid were purchased from Fisher Chemicals (Fair Lawn, NJ, USA). α- and γ-Tocopherols, α- and γ-tocotrienols, γ-oryzanol, 2,2-diphenyl-1-picrylhydrazyl (DPPH), 2,2'-azinobis(3-ethylbenzothiazoline-6-sulfonic acid)(ABTS), EDTA, Trolox, Tween 20, and menhaden fish oil were purchased from Sigma Aldrich (St. Louis, MO, USA). Human laryngeal carcinoma cell HEp2 and normal monkey kidney cell CV-1 lines were purchased from American Type Culture Collection (ATCC, Manassas, VA, USA). Other cell culture reagents, Dulbecco's modified Eagle medium (DMEM), fetal bovine serum (FBS), antibiotic (penicillin–streptomycin), phosphate-buffered saline (PBS), dimethyl sulfoxide (DMSO), and CellTiter-Blue were purchased from Invitrogen (Grand Island, NY, USA). *M. purpureus* (CGMCC 3.4629) was purchased from China General Microbiological Culture Collection Center (Beijing, China). Broken coix seed was provided by Xinlong Green Development Company (Guizhou, China).

2.2. Fermentation of Coix Seed Using M. purpureus Strain

The *M. purpureus* strain was cultivated on potato dextrose agar (PDA) medium at 30 °C for 7 days. The spores were harvested in sterile distilled water and then adjusted to prepare a solution with concentration of 10^6 spores/mL. Coix seed were sterilized at 121 °C and 15 psi for 20 min. Then, the strain solution was mixed with the sterilized seed at a ratio of 1:10 (v/w). The mixture was incubated at 30 °C for 10 days. The pigment extraction and estimation were performed according to the method described by Marič [10]. The fermentation was independently carried out in triplicate.

2.3. Extraction and Determination of Lipophilic Antioxidants and Coixenolide in Raw and Fermented Coix Seed

Lipids in the seed were extracted by the method described by Shen with slight modifications [13]. Briefly, raw or fermented coix seed (10 g) was extracted using 50 mL of hexane at 45 °C for 20 min. The extraction was repeated three times. All collected supernatants were combined and evaporated using a vacuum centrifugal evaporator (CentriVap Mobile System; Labconco, Kansas City, MO, USA) to remove the hexane solvent. The dried extract was weighed (dry weight basis) and dissolved in isopropanol to prepare a stock solution (100 mg/mL).

Tocopherols, tocotrienols, and oryzanol in the extract were determined using a normal-phase HPLC system (1100 series; Agilent, Santa Clara, CA, USA) with Supelcosil LC-Si column (id 250 × 4.60 mm 5 μm, Supelco, Bellefonte, PA, USA) and a series of fluorescence and UV detectors. The HPLC analysis conditions were as described by Jang and Xu [14]. The concentration of each component was calculated based on standard curves. Coixenolide was determined by the method of Yang et al. [15].

2.4. Determination of Free-Radical-Scavenging Activity of the Lipophilic Extract

During fermentation, the DPPH-scavenging activity of lipophilic extract from fermented seed was determined every day. The DPPH-scavenging activity of Trolox was used to prepare a calibration curve of the activity. The DPPH-scavenging activity of lipophilic extract was calculated and converted to μmol Trolox equivalent/gram.

The DPPH assay was performed according to a previous study with slight modifications [16]. Briefly, 1.9 mL of methanolic solution containing 0.1 mM DPPH radicals was mixed with 0.1 mL of different concentrations of sample solutions in a range of 0 to 100 mg/mL and/or methanol used as a blank. The mixtures were vortexed thoroughly and then stood in a dark area for 30 min at room temperature. The absorbance of the reaction mixture was measured at 517 nm using a spectrophotometer. The DPPH free-radical-scavenging activities were calculated as follows:

$$\text{DPPH free-radical-scavenging rate (\%)} = (1 - \text{Abs}_{\text{sample}}/\text{Abs}_{\text{blank}}) \times 100 \qquad (1)$$

where $\text{Abs}_{\text{sample}}$ and $\text{Abs}_{\text{blank}}$ were the absorbance of the mixture of blank and sample with DPPH reagent after reaction, respectively.

For ABTS cation radical-scavenging activity, the measurement method was based on our previous research [4]. Briefly, 9 mL of ABTS solutions was mixed with 3 mL sample solutions with different concentrations (0–100 mg/mL). Methanol was used as a blank. The mixtures were incubated in dark at room temperature for 20 min. The absorbance of each mixture was measured at 734 nm. The ABTS-scavenging activity was calculated according to the equation:

$$\text{ABTS-scavenging rate (\%)} = \{[\text{Abs}_0 - (\text{Abs}_1 - \text{Abs}_2)]/\text{Abs}_0\} \times 100 \qquad (2)$$

where Abs_0 was the absorbance of ABTS solution; Abs_1 was the absorbance of the reacted mixture of ABTS with tested sample; Abs_2 was the absorbance of the methanol with tested sample.

The superoxide anion radical-scavenging activity of the extract was measured based on the method in previous study [17]. Different concentrations of the sample solutions (0.1 mL) were mixed with 4.5 mL of 0.05 mol/L Tris-HCl buffer (pH 8.2) containing 2 mmol/L EDTA. The mixtures were reacted at 25 °C for 20 min. Then 0.4 mL of 25 mmol/L 1,2,3-phentriol was added and allowed to react at 25 °C for 5 min more. Finally, to the mixture was added 1 mL of HCl (8 mol/L) to stop the reaction. The absorbance was recorded at 325 nm. Methanol was used as a blank. The superoxide anion radical-scavenging activity was calculated as follows:

$$\text{Superoxide anion radical-scavenging rate (\%)} = (1 - \text{Abs}_{\text{sample}}/\text{Abs}_{\text{blank}}) \times 100 \qquad (3)$$

where Abs_{blank} was the absorbance of the control group and Abs_{sample} was the absorbance of mixtures with sample solutions.

2.5. Determination of Anti-Lipid-Oxidation Activity Using Fatty Acid Model

The anti-lipid-oxidation activity of lipophilic extract from raw or fermented seed was determined according to a previous study by Zhang et al. [18]. Briefly, 2 mL sample solution (100 mg/mL) was mixed with 20 mL of fish oil emulsion, which consisted of 10 g menhaden fish oil, 10 mL Tween 20, and phosphate buffer. Then, the mixed emulsion was incubated at 37 °C for 5 days with continuously stirring. The concentrations of fatty acid EPA (C20:5) and DHA (C22:6) in the mixed emulsion were measured at day 1, 3, and 5. The fatty acid analysis was performed on a GC system equipped with an FID detector with analysis condition described in the study of Zhang et al. [18]. The emulsion without sample solution was used as a blank control. Anti-lipid-oxidation activities were expressed by the retained rate of DHA and EPA and were calculated as follows:

$$\text{Retained DHA or EPA rate (\%)} = (C_t/C_0) \times 100 \tag{4}$$

where C_0 was the original concentration of EPA or DHA; C_t was the concentration of EPA or DHA at different incubation time.

2.6. Determination of Anticancer Activity of Lipophilic Extract on HEp2 Cells

Anticancer activities were determined based on the survival rate of HEp2 cells treated with different concentrations of lipophilic extract from raw or fermented coix seed. The procedure was described in our previous study [13]. The cells were maintained in 95% DMEM (containing 10% FBS and 1% penicillin−streptomycin) and incubated with 5% CO_2 at 37 °C. The cells were placed in a 96-well plate and incubated for 24 h. Then, the cells were exposed to the medium with different concentrations (0 as blank, 0.625, 1.25, 2.5, 5, and 10 mg/mL) of lipophilic extract from coix seed and incubated for another 24 h at 37 °C. After incubation, the cells were washed three times with PBS and mixed with new medium containing 20% CellTiter-Blue. They were incubated at 37 °C for 4 h. By using a FluoStar Optima microplate reader, the cell viability or survival rate in each well was determined at excitation and emission wavelengths of 570 and 615 nm, respectively. Meanwhile, the normal monkey kidney cells, CV-1, were used as the control, and were treated with the same concentration extract of fermented coix seed. The survival rate of each concentration's treatment group relative to that of the blank group was used to express anticancer activity.

2.7. Data Analysis

The determinations of lipophilic antioxidants and the evaluations of antioxidation activities were carried out in triplicate and expressed as mean with standard deviation values. The SPSS 22.0 software (IBM Company, New York, NY, USA) was used for statistical analysis. The significant differences between the two groups were determined by ANOVA (SAS, 9.1.3, Cary, NY, USA). Difference between two groups was determined at a significant difference $p < 0.05$ or at an extremely significant difference $p < 0.01$. The determinations of anticancer activities of extracts were repeated five times and analyzed by GraphPad Prism (version 6.0; GraphPad Software Inc., La Jolla, CA, USA).

3. Results and Discussion

3.1. Lipophilic Antioxidants and Coixenolide in Raw and Fermented Coix Seed

Figure 1 shows the color changes of coix seeds at different fermentation times. The original color of raw coix seeds was yellowish white and changed to a brownish color in the middle of fermentation (Figure 1a,b). Eventually, the fermented coix seeds had reddish brown color and were readily crushed to a fine powder (Figure 1c). At this time, color value units (CVUs) of pigments in fermented coix seeds reached 687.2 CVU/g. Usually, the growth of *Monascus* sp. and the change of substrate fermentation could be preliminarily judged by the accumulation of pigments. The *Monascus* pigments were the secondary metabolites of polyketides and were biosynthesized by malonyl-CoA catalysis from tetraketide and pentaketide to hexaketide [10,19]. They were accumulated in the solid-state aerobic fermentation of *Monascus* sp.

Figure 1. Images of raw coix seeds (**a**), coix seeds in the middle of fermentation (**b**), and coix seed powder after fermentation (**c**).

The yield of lipophilic extract from raw coix seeds was 4.2%, while it increased to 8.6% from fermented coix seeds after 10 days of fermentation. The fermentation significantly released bound-form lipophilic compounds in coix seeds, which were likely contributed by enzymatic hydrolysis induced by *M. purpureus* during fermentation [20]. As fermentation could produce a significant amount of acid and lower pH in the fermented medium, acid hydrolysis would also contribute to the release of bound or blocked forms of lipophilic compounds [8].

Four tocols—α-tocopherol, α-tocotrienol, γ-tocopherol, and γ-tocotrienol—were found in both raw and fermented coix seeds (Table 1 and Figure 2). Although vitamin E has eight different tocols—α, β, γ, and δ-tocopherols and α, β, γ, and δ-tocotrienols, α-tocopherol, α-tocotrienol, and γ-tocopherol are the common tocols in most cereals, beans, and grains [21]. In raw coix seeds, γ-tocopherol was the leading tocol at a level of 21.2 µg/g DW, followed by γ-tocotrienol, while α-tocopherol and tocotrienol were at a very low level in raw coix seeds (Table 1). Typically, α-tocopherol and α-tocotrienol were the dominant tocols in beans and grain oils, followed by γ-tocopherol. Therefore, the profile of tocols in coix seeds was different from that of most cereals, but similar to rice bran, which contains a high level of γ-tocotrienol [14]. After fermentation, α-tocopherol and γ-tocotrienol were significantly increased ($p < 0.01$), approximately 160 (from 0.1 to 17.9 µg/g) and 16 (from 4.4 to 72.5 µg/g) times, respectively (Table 1). It indicated that raw coix seeds had a high level of the bound or blocked form α-tocopherol and γ-tocotrienol. γ-Tocotrienol recently has been reported to have greater health promoting function than other tocols [22,23]. Therefore, the high level of γ-tocotrienol in coix seeds fermented by *M. purpureus* could significantly enhance the health benefits of coix seeds.

Table 1. Lipophilic antioxidants and coixenolide in raw and fermented coix seeds.

Compound	Raw Coix Seed	Fermented Coix Seed
α-Tocopherol (µg/g DW)	0.1 ± 0.1 [A]	17.9 ± 7.1 [B]
α-Tocotrienol (µg/g DW)	2.4 ± 0.6 [A]	4.2 ± 3.5 [A]
γ-Tocopherol (µg/g DW)	21.2 ± 3.9 [A]	25.4 ± 3.2 [A]
γ-Tocotrienol (µg/g DW)	4.4 ± 1.5 [A]	72.5 ± 10.8 [B]
Total tocols (µg/g DW)	28.1	120.0
γ-Oryzanol (µg/g DW)	26.2 ± 4.1 [A]	655.0 ± 30.1 [B]
Coixenolide (mg/g DW)	4.0 ± 0.2 [A]	8.7 ± 0.8 [B]

Difference between data with different letters in a row is extremely statistically different ($p < 0.01$).

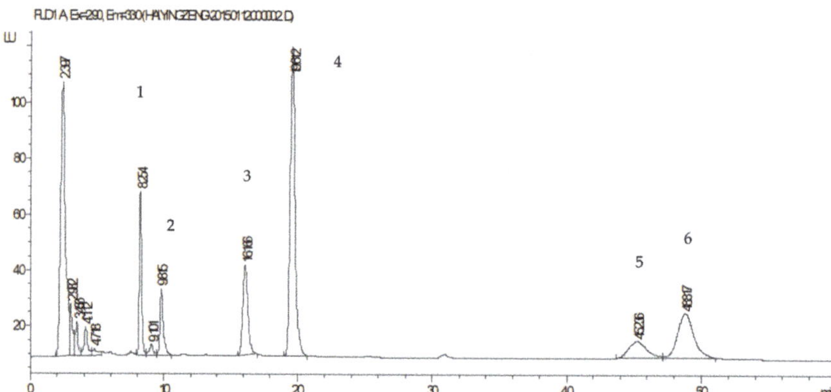

Figure 2. HPLC chromatogram of lipophilic extract from fermented coix seeds: (1) α-tocopherol, (2) α-tocotrienol, (3) γ-tocopherol, (4) γ-tocotrienol, (5 and 6) γ-oryzanols.

The level of γ-oryzanol in fermented coix seeds increased about 25 times to 655.0 μg/g from the level of 26.2 μg/g in raw coix seeds (Table 1 and Figure 2). γ-Oryzanol is usually rich in rice bran and not commonly present in most cereals. It is a type of plant sterol and contains ferulic acid in its structure [24]. Thus, γ-oryzanol is an antioxidant phytosterol because ferulic acid is a strong antioxidant phenolic. It has a potent antioxidant activity in preventing cholesterol oxidation, which could generate toxic cholesterol oxidation products in foods and human body, resulting in development of cardiovascular diseases [25]. Although the level of γ-oryzanol in raw coix seeds was much lower compared to that of most rice varieties (26.2 vs. 200–300 μg/g), the level in fermented coix seeds (655.0 μg/g) was much higher than the level in rice [26]. It was reported that rice bran fermentation by *Rhizopus oryzae* could also increase the free γ-oryzanol level but only by 1.5 times that of raw rice bran. [27]. During fermentation, the microorganisms could degrade the plant cell wall through a variety of self-generated enzymes that stimulate the release of intracellular compounds, especially fiber-bound compounds [28].

Coixenolide is a diol lipid uniquely present in coix seeds. It is a long carbon chain with two long-chain fatty acid esters (Figure 3). Although it may not have antioxidant function based on its chemical structure, previous studies reported that it has outstanding anticancer activity [1,6]. After the fermentation by *M. purpureus*, coixenolide in fermented coix seed increased to 8.7 mg/g, which was about 2.2 times higher than that in raw coix seed (Table 1). The increase of coixenolide could assist in the enhancement of health benefits of coix seeds.

Figure 3. Chemical structure formula of coixenolide in coix seeds ($C_{38}H_{70}O_4$).

3.2. Scavenging Activities on Different Free Radicals of Lipophilic Extracts from Raw and Fermented Coix Seed

DPPH-scavenging activity is widely used as an index of antioxidant activity of a test sample in vitro. The activity of lipophilic extract in fermented coix seeds at different fermentation times was monitored. The scavenging activity rose slightly within the first 7 days of fermentation (Figure 4a), then it dramatically increased from 0.4 to 3.3 μmol/g of Trolox equivalent after 9 days of fermentation. The increase of the activity began to slow down from day 9 to day 10 of fermentation. After 10 days of fermentation, the scavenging activity of the lipophilic extract reached 3.8 µmol/g of Trolox equivalent. Tocols and γ-oryzanol were monitored simultaneously during fermentation. The results displayed that the level of total tocols and γ-oryzanol was extremely significantly enhanced during fermentation ($p < 0.01$), especially after the 6th day of fermentation (Figure 4b,c). Correlation analysis showed that total tocols had a significantly positive correlation with DPPH-scavenging activity at a correlation coefficient of 0.9. The correlation coefficient of γ-oryzanol was only 0.8. Thus, tocols could be the major contributor in scavenging DPPH free radicals.

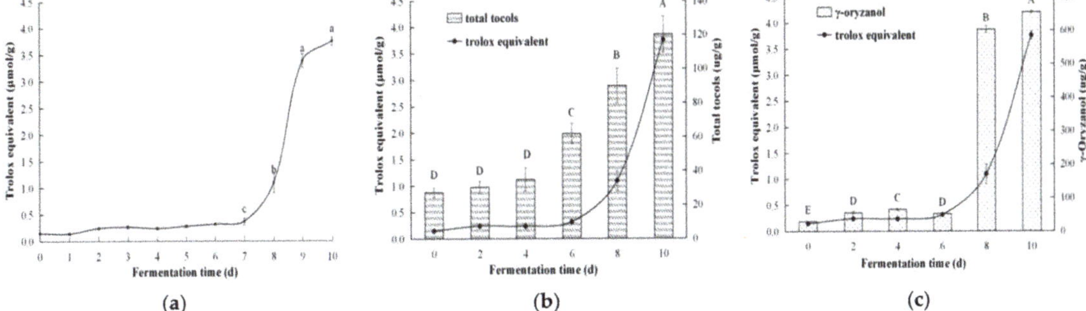

Figure 4. Changes of DPPH-scavenging activity (**a**), total tocols (**b**) and γ-oryzanol (**c**) of lipophilic extract from coix seeds during fermentation. Bars with different letters indicate significant differences (lowercase letters, $p < 0.05$; capital letters, $p < 0.01$).

Free radicals are a group of very reactive species that not only cause rancid deterioration in food products but also result in oxidative inflammation in mammalian cells, leading to eventual development of different chronic diseases, such as cardiovascular diseases and cancers [29]. The harmful free-radical initiators can be effectively annihilated by antioxidants to prevent their negative impacts. Antioxidants from natural sources, such as cereals, vegetables, and fruits are considered as safe antioxidants compared to synthetic antioxidant, such as BHA (butylated hydroxyanisole), BHT (butylated hydroxytoluene), and PG (propyl gallate) [30]. Raw and fermented coix seeds with high levels of tocols and γ-oryzanol could be a rich source of natural antioxidants destined for use as ingredients in food products.

Figure 5a shows the scavenging activities of lipophilic extracts from raw and fermented coix seeds at 0–100 mg/mL. The scavenging activities increased with the concentration increase of lipophilic extracts. It also revealed dose–response relationships between the additive concentration of extracts and scavenging activities. The extract from fermented coix seeds on scavenging DPPH, ABTS cation, or superoxide anion free radicals was significantly higher than that of the extract from raw coix seeds at 100 mg/mL (Figure 5b). Moreover, in three different scavenging tests, the lipophilic extract of fermented coix seeds had the most prominent scavenging effect on ABTS cation. The scavenging activity (60.79%) increased 1.7 times after fermentation, followed by superoxide anion and DPPH (about 1.3 times). The effective concentration for 50% of scavenging DPPH rate of pure α-tocopherol is 120 µmol/L or 51.6 µg/mL [30]. α-Tocopherol was normally considered to be the most active tocol in scavenging different free radicals and had 1.5 times higher activity than γ-tocopherol [31]. The increased α-tocopherol in fermented coix seeds could be an important factor for the higher free-radical-scavenging activity, in addition to γ-tocotrienol and γ-oryzanol.

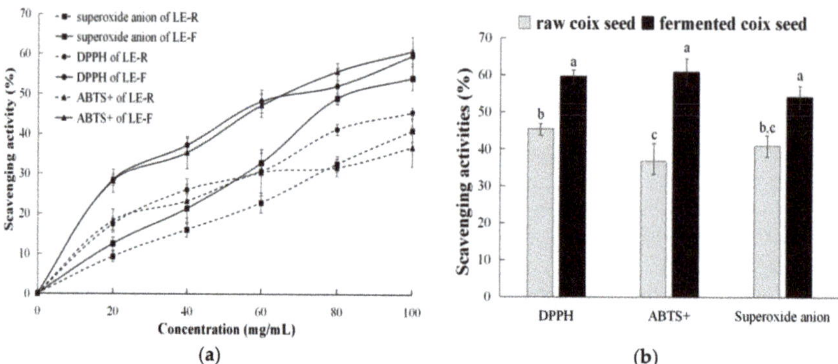

Figure 5. DPPH, ABST cation, and superoxide anion free-radical-scavenging activities of lipophilic extracts from raw and fermented coix seeds at 0–100 mg/mL (**a**) and at 100 mg/mL (**b**). LE-R, lipophilic extract from raw coix seeds. LE-F, lipophilic extract from fermented coix seeds.

Traditionally, the antioxidant activity of a compound is assessed by measuring its scavenging activity against DPPH or other free radicals and through reducing power assays [32]. However, a good result in these chemical-based spectrophotometric assays cannot predict whether the sample has a good antioxidant capability in inhibiting lipid oxidation in an emulsion. In other words, the higher free-radical-scavenging activity of an antioxidant may not be closely correlated with the actual antioxidant activity in inhibiting the lipid oxidation in a lipid-rich food or oxidative inflammation in mammalian cell in which fluid is a complicated emulsion [30]. Therefore, the antioxidant activity of the lipophilic extract of fermented coix seeds was further evaluated using an oil emulsion model.

3.3. Anti-Lipid-Oxidation Activities of Lipophilic Extracts from Raw and Fermented Coix Seeds

In this study, fatty acids, the key component of lipids, were used as the substrates to evaluate antioxidant activity of lipophilic extracts from coix seeds. Compared with free-radical-scavenging chemical assays, the activity obtained in this method is closely correlated to the capability of the antioxidant in stabilizing lipids in a food matrix to extend its shelf life or in a biological system to prevent oxidative stress [18,30]. EPA (C20:5) and DHA (C22:6), polyunsaturated long-chain fatty acids, are the most vulnerable fatty acids to oxidation. Fatty acids could be oxidized to produce a group of short-chain aldehydes, alcohols, and other components. Therefore, the retention of EPA and DHA in oil emulsion with or without the lipophilic extract from coix seeds was monitored. The retained rates of these susceptible polyunsaturated fatty acids could be more accurate to indicate the oxidation status in the emulsion [30]. For example, although α-tocopherol was considered to be the most active tocopherol in reacting with free radicals, in a liposomal membrane model, tocotrienols had higher antioxidant activity than tocopherols [23]. In a recent study, the activity of γ-tocotrienol was also higher than that of α-tocopherol in oil systems [31].

The retention rates of EPA and DHA in the control group decreased to below 20% in 24 h and almost totally oxidized after 120 h (Figure 6a,b). Both of the lipophilic extracts from raw and fermented coix seeds still had much higher retention rate for EPA or DHA after 120 h of incubation. Although the extract from fermented raw coix seeds had slightly lower retention rate of EPA or DHA than the extract from raw coix seeds at 24 h, its retention rate at 72 and 120 h of incubation was more than two times higher than that of the extract from raw coix seeds (Figure 6a,b). It still retained approximately 40% and 32% for EPA or DHA after 120 h of incubation, respectively (raw coix seed, only 15% and 11%). The results were similar to that in a previous study that reported anti-lipid-oxidation activity of tocols extract from rice bran against EPA and DHA oxidation [18]. Overall, the results indicated that fermented coix seeds had higher anti-lipid-oxidation activity than raw coix seeds in inhibiting lipid oxidation. Recently, γ-oryzanol as a new lipophilic antioxidant has been studied to stabilize corn oil and fish oil in yogurt [33,34]. Fermented coix seeds could be a good food preservative to reduce lipid oxidation in food. It could also have a higher antioxidant function to lower oxidative stress in a biological system.

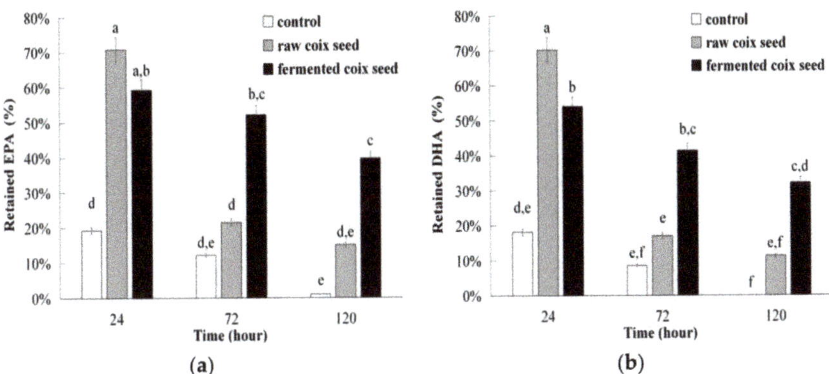

Figure 6. Retention rates of EPA (20:5) (**a**) and DHA (22:6) (**b**) of lipophilic extracts from raw and fermented coix seeds. Bars with different letters indicate significant differences at $p < 0.05$.

3.4. Anticancer Activities of Lipophilic Extracts from Raw and Fermented Coix Seeds

The anticancer activities of raw and fermented coix seed were evaluated by using the HEp2 cell line, which is a human laryngeal carcinomatous cell. Laryngeal carcinoma is one of the most common respiratory cancers and accounts for 25% of head and neck carcinomas [35]. Figure 7a shows the images of HEp2 cell morphology in control and treatment groups after 24 h of incubation. The inhibitory concentrations for 50% of cell survival rate (IC_{50}) of lipophilic extracts from raw and fermented seeds were 3.67 and 2.13 mg/mL, respectively (Figure 7b). The IC_{50} of lipophilic extract from fermented coix seeds in inhibiting HEp2 cell decreased by 42%. The anticancer activity of lipophilic extract was significantly improved after the fermentation, but its cytotoxicity toward normal CV-1 cells was relatively low. Even at the concentration of 10 mg/mL, only 30.5% of the normal cells were inhibited. The IC_{50} values of lipophilic extracts from raw and fermented seeds for HEp2 cells were higher than that of the lipophilic extract from butterfly pea seeds, for which the IC_{50} for HEp2 cells is 8 mg/mL [13]. In a previous study that compared anticancer activity for T24 cell, a human urinary bladder cancer cell, of coix seed oil extracts from 11 different varieties, all of them had IC_{50} values higher than 0.33% or 3.3 mg/mL in medium [6]. Thus, the anticancer activity of coix seeds could be improved through the fermentation by *M. purpureus*.

Tocols and γ-oryzanol could be the primary compounds responsible for the anticancer activity of the lipophilic extract from coix seeds, besides coixenolide. For example, tocols are involved in the DR5 (death receptor 5) protein upregulation, which stimulates tumor necrosis and restricts its proliferation [36]. For antioxidant phytosterols, such as γ-oryzanol, they could inhibit cancer cell proliferation by meddling with protein phosphatase 2A (PP2A) in the sphingomyelin cycle and blocking the cell cycle at G0/G1 phase in different cancer cells, such as prostate cancer, hepatocyte, and breast cancer cells [37,38]. Therefore, the increase of tocols and γ-oryzanol could assist in the improvement of anticancer activity of the lipophilic extract from fermented coix seeds.

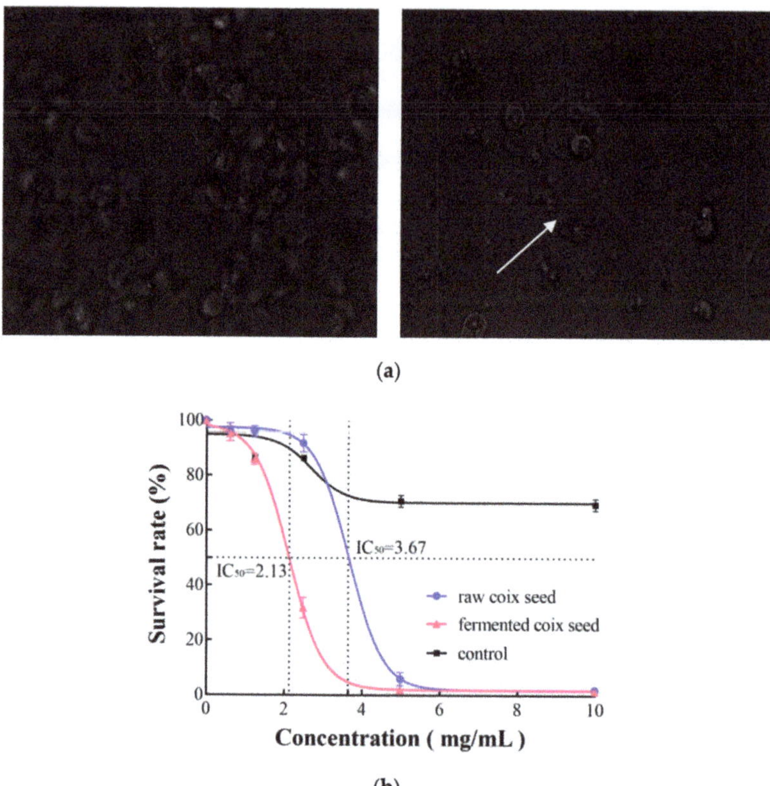

(a)

(b)

Figure 7. Images of HEp2 cell morphology in medium in the control (**a-left**) and treatment (**a-right**) groups, and survival rates of HEp2 cells incubated with lipophilic extracts from raw and fermented coix seeds (**b**).

4. Conclusions

In summary, this study revealed positive changes in lipophilic antioxidants and anticancer activity of coix seeds after fermentation by *Monascus purpureus*, a typical mold used in the preparation of fermented grains. The fermented coix seeds had increased tocols (vitamin E), γ-oryzanol, and coixenolide contents. The levels γ-tocotrienol or γ-oryzanol in fermented coix seeds were much higher than the level found in most cereals. The increased antioxidants enhanced the antioxidant activity in scavenging different free radicals and stabilizing susceptible polyunsaturated fatty acids. The anticancer activity of coix seeds was significantly improved after the fermentation by *M. purpureus*. Therefore, the fermentation by *M. purpureus* is a good approach in increasing the lipophilic bioactivity and bioavailability of health-promoting compounds in cereals. The fermented coix seed or its extract could be used as a food, food ingredient, or nutritional supplement to provide added health benefits to humans.

5. Patents

A new product, coix seed tea fermented by *Monascus purpureus*, has been described in China patent no. CN 201810372538.3.

Author Contributions: Conceptualization, H.Z.; methodology, H.Z. and L.Q.; software, H.Z.; validation, L.Q.; formal analysis, H.Z., L.Q., X.L. and S.M.; investigation, H.Z. and X.L.; resources, H.Z. and L.Q.; data curation, H.Z., L.Q. and S.M.; writing—original draft preparation, H.Z.; writing—review and editing, H.Z., L.Q. and S.M.; Visualization, H.Z. and S.M.; supervision, H.Z. and L.Q.; project administration, H.Z. and L.Q.; Funding acquisition, H.Z. All authors have read and agreed to the published version of the manuscript.

Funding: This research was funded by the Natural Science Foundation of Guizhou Province, grant number (2019) 1111, and the Agriculture Committee of Guizhou Province, grant number (2017) 106 & (2018) 81.

Institutional Review Board Statement: Not applicable.

Informed Consent Statement: Not applicable.

Data Availability Statement: Data is contained within the article.

Conflicts of Interest: The authors declare no conflict of interest.

References

1. Tseng, Y.; Yang, J.; Chang, H.; Mau, J. Taste quality of monascal adlay. *J. Agric. Food Chem.* **2004**, *52*, 2297–2300. [CrossRef]
2. Zhu, F. Coix: Chemical composition and health effects. *Trends Food Sci. Technol.* **2017**, *61*, 160–175. [CrossRef]
3. Huang, D.; Kuo, Y.; Lin, F.; Lin, Y.; Chiang, W. Effect of Adlay (*Coix lachryma-jobi* L. var. *Ma-yuen Stapf*) Testa and its phenolic components on Cu2+-treated Low-Density Lipoprotein (LDL) oxidation and Lipopolysaccharide (LPS)-induced inflammation in RAW 264.7 macrophages. *J. Agric. Food Chem.* **2009**, *57*, 2259–2266. [CrossRef] [PubMed]
4. Wen, A.; Qin, L.; Zeng, H.; Zhu, Y. Comprehensive evaluation of physicochemical properties and antioxidant activity of *B. Subtili*-fermented polished adlay subjected to different drying methods. *Food Sci. Nutr.* **2020**, *8*, 2124–2133. [CrossRef] [PubMed]
5. Chen, H.; Chung, C.; Chiang, W.; Lin, Y. Anti-inflammatory effects and chemical study of a flavonoid-enriched fraction from adlay bran. *Food Chem.* **2011**, *126*, 1741–1748. [CrossRef]
6. Xi, X.; Zhu, Y.; Tong, Y.; Yang, X.; Tang, N.; Ma, S.; Li, S.; Cheng, Z. Assessment of the genetic diversity of different Job's tears (*Coix lacryma-jobi* L.) accessions and the active composition and anticancer effect of its seed oil. *PLoS ONE* **2016**, *11*, e153269. [CrossRef] [PubMed]
7. Xu, Z.; Howard, L.R. *Analysis Methods of Phenolic Acids*; John Wiley Sons, Ltd.: Hoboken, NJ, USA, 2012. [CrossRef]
8. Shen, Y.; Sun, H.; Zeng, H.; Prinyawiwatukul, W.; Xu, W.; Xu, Z. Increases in phenolic, fatty acid, and phytosterol contents and anticancer activities of sweet potato after fermentation by *Lactobacillus acidophilus*. *J. Agric. Food Chem.* **2018**, *66*, 2735–2741. [CrossRef]
9. Chen, G.; Liu, Y.; Zeng, J.; Tian, X.; Bei, Q.; Wu, Z. Enhancing three phenolic fractions of oats (*Avena sativa* L.) and their antioxidant activities by solid-state fermentation with *Monascus anka* and *Bacillus subtilis*. *J. Cereal Sci.* **2020**, *93*, 102940. [CrossRef]
10. Marič, A.; Skočaj, M.; Likar, M.; Sepčić, K.; Cigić, I.K.; Grundner, M.; Gregori, A. Comparison of lovastatin, citrinin and pigment production of different *Monascus purpureus* strains grown on rice and millet. *J. Food Sci. Technol.* **2019**, *56*, 3364–3373. [CrossRef] [PubMed]
11. Hur, S.J.; Lee, S.Y.; Kim, Y.; Choi, I.; Kim, G. Effect of fermentation on the antioxidant activity in plant-based foods. *Food Chem.* **2014**, *160*, 346–356. [CrossRef]
12. Handa, C.L.; de Lima, F.S.; Guelfi, M.F.; da Silva Fernandes, M.; Georgetti, S.R.; Ida, E.I. Parameters of the fermentation of soybean flour by *Monascus purpureus* or *Aspergillus oryzae* on the production of bioactive compounds and antioxidant activity. *Food Chem.* **2019**, *271*, 274–283. [CrossRef]
13. Shen, Y.; Du, L.; Zeng, H.; Zhang, X.; Prinyawiwatkul, W.; Alonso Marenco, J.R.; Xu, Z. Butterfly pea (*Clitoria ternatea*) seed and petal extracts decreased HEp-2 carcinoma cell viability. *Int. J. Food Sci. Technol.* **2016**, *51*, 1860–1868. [CrossRef]
14. Jang, S.; Xu, Z. Lipophilic and hydrophilic antioxidants and their antioxidant activities in purple rice bran. *J. Agric. Food Chem.* **2009**, *57*, 858–862. [CrossRef]
15. Yang, J.; Tseng, Y.; Chang, H.; Lee, Y.; Mau, J. Storage stability of monascal adlay. *Food Chem.* **2005**, *90*, 303–309. [CrossRef]
16. Du, L.; Shen, Y.; Zhang, X.; Prinyawiwatkul, W.; Xu, Z. Antioxidant-rich phytochemicals in miracle berry (*Synsepalum dulcificum*) and antioxidant activity of its extracts. *Food Chem.* **2014**, *153*, 279–284. [CrossRef] [PubMed]
17. Zhang, Z.H.; Fan, S.T.; Huang, D.F.; Yu, Q.; Liu, X.Z.; Li, C.; Wang, S.; Xiong, T.; Nie, S.P.; Xie, M.Y. Effect of *Lactobacillus plantarum* NCU116 fermentation on Asparagus officinalis polysaccharide: Characterization, antioxidative, and immunoregulatory activities. *J. Agric. Food Chem.* **2018**, *66*, 10703–10711. [CrossRef] [PubMed]
18. Zhang, X.; Shen, Y.; Prinyawiwatkul, W.; King, J.M.; Xu, Z. Comparison of the activities of hydrophilic anthocyanins and lipophilic tocols in black rice bran against lipid oxidation. *Food Chem.* **2013**, *141*, 111–116. [CrossRef]
19. Chen, G.; Yang, S.; Wang, C.; Shi, K.; Zhao, X.; Wu, Z. Investigation of the mycelial morphology of *Monascus* and the expression of pigment biosynthetic genes in high-salt-stress fermentation. *Appl. Microbiol. Biotechnol.* **2020**, *104*, 2469–2479. [CrossRef]

20. Masiero, S.S.; Peretti, A.; Trierweiler, L.F.; Trierweiler, J.O. Simultaneous cold hydrolysis and fermentation of fresh sweet potato. *Biomass Bioenergy* **2014**, *70*, 174–183. [CrossRef]
21. Moreau, R.A.; Lampi, A.M. *Analysis Methods for Tocopherols and Tocotrienols*; Xu, Z., Howard, L.R., Eds.; Wiley-Blackwell: Oxford, UK, 2012; pp. 353–386. [CrossRef]
22. Nesaretnam, K.; Yew, W.W.; Wahid, M.B. Tocotrienols and cancer: Beyond antioxidant activity. *Eur. J. Lipid Sci. Technol.* **2007**, *109*, 445–452. [CrossRef]
23. Yoshida, Y.; Niki, E.; Noguchi, N. Comparative study on the action of tocopherols and tocotrienols as antioxidant: Chemical and physical effects. *Chem. Phys. Lipids* **2003**, *123*, 63–75. [CrossRef]
24. Nyström, L. *Analysis Methods of Phytosterols*; Xu, Z., Howard, L.R., Eds.; Wiley-Blackwell: Oxford, UK, 2012; pp. 313–351. [CrossRef]
25. Lesma, G.; Luraghi, A.; Bavaro, T.; Bortolozzi, R.; Rainoldi, G.; Roda, G.; Viola, G.; Ubiali, D.; Silvani, A. Phytosterol and γ-Oryzanol conjugates: Synthesis and evaluation of their antioxidant, antiproliferative, and anticholesterol activities. *J. Nat. Prod.* **2018**, *81*, 2212–2221. [CrossRef] [PubMed]
26. Heinemann, R.J.; Xu, Z.; Godber, J.S.; Lanfer-Marquez, U.M. Tocopherols, tocotrienols, and γ-oryzanol contents in Japonica and Indica subspecies of rice (*Oryza sativa* L.) cultivated in Brazil. *Cereal Chem.* **2008**, *85*, 243–247. [CrossRef]
27. Massarolo, K.C.; de Souza, T.D.; Collazzo, C.C.; Furlong, E.B.; Souza Soares, L.A. The impact of *Rhizopus oryzae* cultivation on rice bran: Gamma-oryzanol recovery and its antioxidant properties. *Food Chem.* **2017**, *228*, 43–49. [CrossRef]
28. Cho, J.; Lee, H.J.; Kim, G.A.; Kim, G.D.; Lee, Y.S.; Shin, S.C.; Park, K.; Moon, J. Quantitative analyses of individual γ-oryzanol (steryl ferulates) in conventional and organic brown rice (*Oryza sativa* L.). *J. Cereal Sci.* **2012**, *55*, 337–343. [CrossRef]
29. Kruk, J.; Aboul-Enein, H.Y.; Kładna, A.; Bowser, J.E. Oxidative stress in biological systems and its relation with pathophysiological functions: The effect of physical activity on cellular redox homeostasis. *Free Radic. Res.* **2019**, *53*, 497–521. [CrossRef]
30. Zhang, Y.; Shen, Y.; Zhu, Y.; Xu, Z. Assessment of the correlations between reducing power, DPPH scavenging activity and anti-lipid-oxidation capability of phenolic antioxidants. *LWT Food Sci. Technol.* **2015**, *63*, 569–574. [CrossRef]
31. Seppanen, C.M.; Song, Q.; Csallany, A.S. The antioxidant functions of tocopherol and tocotrienol homologues in oils, fats, and food systems. *J. Am. Oil Chem. Soc.* **2010**, *87*, 469–481. [CrossRef]
32. Dudonné, S.; Vitrac, X.; Coutière, P.; Woillez, M.; Mérillon, J.M. Comparative study of antioxidant properties and total phenolic content of 30 plant extracts of industrial interest using DPPH, ABTS, FRAP, SOD, and ORAC assays. *J. Agric. Food Chem.* **2009**, *57*, 1768–1774. [CrossRef]
33. Yi, B.; Lee, J.; Kim, M. Increasing oxidative stability in corn oils through extraction of γ-oryzanol from heat treated rice bran. *J. Cereal Sci.* **2020**, *91*, 102880. [CrossRef]
34. Zhong, J.; Yang, R.; Cao, X.; Liu, X.; Qin, X. Improved physicochemical properties of yogurt fortified with fish oil/γ-oryzanol by nanoemulsion technology. *Molecules* **2018**, *23*, 56. [CrossRef] [PubMed]
35. Abbasi, R.; Ramroth, H.; Becher, H.; Dietz, A.; Schmezer, P.; Popanda, O. Laryngeal cancer risk associated with smoking and alcohol consumption is modified by genetic polymorphisms in ERCC5, ERCC6 and RAD23B but not by polymorphisms in five other nucleotide excision repair genes. *Int. J. Cancer* **2009**, *125*, 1431–1439. [CrossRef] [PubMed]
36. Kannappan, R.; Gupta, S.C.; Ji, H.K.; Aggarwal, B.B. Tocotrienols fight cancer by targeting multiple cell signaling pathways. *Genes Nutr.* **2011**, *7*, 43–52. [CrossRef] [PubMed]
37. Woyengo, T.A.; Ramprasath, V.R.; Jones, P.J. Anticancer effects of phytosterols. *Eur. J. Clin. Nutr.* **2009**, *63*, 813–820. [CrossRef]
38. Awad, A.B.; Fink, C.S. Phytosterols as anticancer dietary components: Evidence and mechanism of action. *J. Nutr.* **2000**, *130*, 2127–2130. [CrossRef]

 *foods*

Article

Germinated Buckwheat: Effects of Dehulling on Phenolics Profile and Antioxidant Activity of Buckwheat Seeds

Andrej Živković, Tomaž Polak, Blaž Cigić and Tomaž Požrl *

Department of Food Science and Technology, Biotechnical Faculty, University of Ljubljana, SI-1111 Ljubljana, Slovenia; andrej.zivkovic@bf.uni-lj.si (A.Ž.); tomaz.polak@bf.uni-lj.si (T.P.); blaz.cigic@bf.uni-lj.si (B.C.)
* Correspondence: tomaz.pozrl@bf.uni-lj.si; Tel.: +386-1-3203716

Abstract: The aim was to investigate the effects of the cold dehulling of buckwheat seeds on their germination, total phenolic content (TPC), antioxidant activity (AA) and phenolics composition. Cold dehulling had no negative effects on germination rate and resulted in faster rootlet growth compared to hulled seeds. Although the dehulling of the seeds significantly decreased TPC and AA, the germination of dehulled seeds resulted in 1.8-fold and 1.9-fold higher TPC and AA compared to hulled seeds. Liquid chromatography coupled to mass spectrometry identified several phenolic compounds in free and bound forms. Rutin was the major compound in hulled seeds (98 µg/g dry weight), orientin and vitexin in 96-h germinated dehulled seeds (2205, 1869 µg/g dry weight, respectively). During germination, the increases in the major phenolic compounds were around two orders of magnitude, which were greater than the increases for TPC and AA. As well as orientin and vitexin, high levels of other phenolic compounds were detected for dehulled germinated seeds (e.g., isoorientin, rutin; 1402, 967 µg/g dry weight, respectively). These data show that dehulled germinated seeds of buckwheat have great potential for use in functional foods as a dietary source of phenolic compounds with health benefits.

Keywords: buckwheat; dehulling; germination; LC-MS; free phenolic; bound phenolic; antioxidant activity

Citation: Živković, A.; Polak, T.; Cigić, B.; Požrl, T. Germinated Buckwheat: Effects of Dehulling on Phenolics Profile and Antioxidant Activity of Buckwheat Seeds. *Foods* **2021**, *10*, 740. https://doi.org/10.3390/foods10040740

Academic Editors: Barbara Laddomada and Weiqun Wang

Received: 26 February 2021
Accepted: 27 March 2021
Published: 1 April 2021

Publisher's Note: MDPI stays neutral with regard to jurisdictional claims in published maps and institutional affiliations.

1. Introduction

The germination of edible seeds is a recognized method for improving the nutritional value of seeds [1–4]. Malted grains have traditionally been used in beer production and in recent years, as a functional ingredient in the baking industry, as groats and flour. The use of the germinated seeds of many plant species is increasing due to the often high contents and availability of nutrients compared to dry seeds, and to the associated beneficial effects on human health. In recent years, there have been many studies on the effects of the germination of cereals and pseudocereals, and their various health benefits for the prevention of chronic diseases, such as heart disease, cancers and diabetes [5–9]. Dynamic changes during germination can lead to the breakdown of macronutrients, such as carbohydrates, proteins and lipids. In addition, the levels of polyphenols, vitamins and other bioactive compounds rapidly increase during germination due to de novo synthesis and transformation, thus enhancing the health-promoting effects of seeds [10–13].

Buckwheat (*Fagopyrum esculentum*) is a pseudocereal that is highly adaptable to harsh environmental conditions, has a short growth period, and shows greater resistance to pests compared to other cereals. Nowadays, buckwheat is recognized as an important gluten-free functional food ingredient due to its balanced composition of macronutrients and high content of bioactive compounds. The protein content in buckwheat seeds is similar to that of wheat grain, but with a higher biological value due to the balanced, lysine-rich amino-acid composition and the low content of storage prolamins [14]. Buckwheat seeds have a high levels of phenolic compounds, and especially flavonoids, such as rutin [15], orientin, vitexin and isovitexin, which have been reported to have several health benefits [16–19]. Different studies in humans have shown associations between the consumption of buckwheat seed products with antihypertensive effects [20] and cholesterol-lowering effects [21]. However, the total phenolic content (TPC) in cereals has often been underestimated, as significant proportions of these phenolic compounds are covalently bound to cell wall materials, and thus need to be extracted by alkaline hydrolysis. For the correct estimation of TPC, this portion of bound phenolics must also be considered [22].

Buckwheat seeds are usually consumed in the form of products made from their flour, or as the dehulled seeds, known as groats. Almost all of the data available on the TPC of germinated buckwheat seeds refer to hulled seeds, and although the hull is not edible, it is an important contributor to TPC [23,24]. The usual processing method used for groats production is thermal dehulling, where the seeds are first boiled or steamed at high temperatures of 130 °C to 160 °C, under high pressure. They are then cooled, conditioned and dried prior to dehulling. During this process, the seed endosperm swells and the husk breaks so that it can be easily removed [25,26]. The thermal processing used in the production of groats significantly reduces the nutritional value of the seeds, and from the point of view of sprout production, this reduces the seed germination. Several studies have shown the effects of thermal dehulling, in terms of decreases in TPC, antioxidant activity (AA) and the major flavonoids, rutin and quercetin [27,28].

Buckwheat has been used previously for the preparation of sprouts and microgreens [29], which are generally considered to be a rich source of bioactive compounds. For the production of buckwheat sprouts ready for consumption, it is nevertheless essential to obtain the groats, with the need to thus remove the seed hull. To retain high germination rates, the typical thermal processing method for the seeds is not an option. In comparison to other hulled cereals, buckwheat seeds are easily damaged during dehulling, and hence the obtainment of intact groats with high germination rates is a challenging task.

However, the many reported advantages of hull-less buckwheat seeds justify the efforts needed to obtain this product for use as a functional food ingredient. A previous study on oat [30] showed that germination of the dehulled grain resulted in higher TPC and AA in comparison to the grain with the hulls. It can be assumed that similar effects will apply to buckwheat seeds, which would justify the efforts to prepare hull-less seeds that retain high germination rates and can be used as a functional food ingredient.

Germination is initiated by increasing the moisture content of grain or seeds to 43% to 45% by soaking in water [31]. Upon the initiation of germination, enzyme synthesis and kernel modification take place, and the seeds rapidly undergo metabolic activities that include respiration, nutrient degradation and secondary metabolite synthesis [32,33]. Several studies have investigated the effects of germination on the TPC, AA and phenolics composition of buckwheat seeds [34,35].

In our study, a new processing approach of mechanical—the cold dehulling method— was used for the buckwheat, which maintained a high germination rate. Previous studies and other scientific articles have focused on the effects of germination on phenolic content and antioxidant activity of hulled germinated buckwheat. However, to the best of our knowledge, there have not been any studies into the effects of buckwheat seed dehulling on the germination bioactivities. Furthermore, for the Slovenian domestical cultivar of common buckwheat (*Fagopyrum esculentum*) "Čebelica" [36] used here, there appears to be no information on their phenolic profile.

The objectives of this study were: (1) to determine the effects of the dehulling of buckwheat seeds on their growth (dry weight, rootlet length); (2) to study the dynamic changes in the TPC and AA of extracts of hulled and dehulled seeds during germination; and (3) to characterise the dynamic changes in the content of individual phenolic compounds in their free and bound forms at different stages of germination.

2. Materials and Methods

2.1. Materials

The buckwheat seeds were purchased from a local producer (Krasinec, Slovenia) and grown under organic growing conditions. Methanol, sodium hydroxide, hydrochloric acid, sodium bicarbonate, 2,2′-azino-bis (3-ethylbenzothiazolin-6-sulfonic acid diammonium salt) (ABTS reagent), 1,1-diphenyl-2-picrylhydrazyl (DPPH), and gallic acid were from Sigma-Aldrich (Steinheim, Germany). Folin–Ciocalteu reagent was from Merck (Darmstadt, Germany), and manganese dioxide was from Kemika (Zagreb, Croatia). The analytical standards of rutin (PN:78095-25MG-F), orientin (PN:09765-1MG), isoorientin (PN:78109-5MG), vitexin (PN:49513-10MG-F), catechin (PN:43412-10MG), epicatechin (PN:68097-10MG) and p-coumaric acid (PN:C9008-10G) were from Sigma–Aldrich (Steinheim, Germany), and hyperoside (PN:0018-05-85) was from HWI Analytik (Rülzheim, Germany). Along with the compounds identified, eight other phenolic standards were used: caffeic acid (PN:C0625-5G), gallic acid (PN:91215-100MG), luteolin-7-glucoside (PN:49968-10MG), naringenin (PN:N5893-5G), morin (PN:M4008-5G), quercetin (PN:Q4951-10G) and kaempferol (PN:K0133-10MG), which were all from Sigma-Aldrich (Steinheim, Germany), and chlorogenic acid (PN:0050-05-90), which was from HWI Analytik (Rülzheim, Germany). All of the standards used were analytical or HPLC grade. All aqueous solutions were prepared using Milli-Q purified water (Merck Millipore, Bedford, MA, USA).

2.2. Cold Dehulling of Buckwheat Seeds

A "cold dehulling" method was used to dehull the seeds in this study, whereby the groats obtained appeared to show good germination rates. Prior to dehulling, the hulled seeds were fractionated using a series of square-holed sieves with mesh sizes defined as 5.5, 6.0 and 6.5 (3.70, 3.35, 3.00 mm). Approximately 15% of the seeds (diameter >3.7 mm, <2.7 mm) were discarded, and so not used further in this study. The remaining seed fractions of different sizes were dehulled separately using a stone mill (TYP A130; Osttiroler Grain Mills, Austria) [37] with a gap between the stones of 3.0 mm, 3.4 mm and 3.8 mm, respectively. The hulls were separated from the groats using a flow of air. The dehulled seeds were separated using the respective sieves, and for the remaining hulled seeds, the dehulling was repeated.

2.3. Germination

Before germination, the hulled and dehulled seeds were soaked in water for 8 h, with a 15-min period out of the water every hour. After soaking, all of the seeds were placed in a thin layer on moistened filter paper in glass Petri dishes. Germination was carried out in a 20 °C thermostated growth chamber, with moistened filter papers to ensure high humidity (relative humidity, >95%). The filter papers were kept moist by spraying with distilled water as needed. The sprouts were harvested for analysis at 24, 48, 72 and 96 h after the start of soaking.

2.4. Determination of Total Dry Mass Loss during Germination

Previous studies have shown that during germination, the energy reserves in the form of starch, lipids, and to a certain extent, proteins are mobilised, and become depleted [38,39]. As all of the data here are presented on a dry weight basis, the characterisation of these losses was of crucial importance. For this purpose, a method for estimating the loss of dry matter during germination was developed. Briefly, 100 seeds from each batch with the determined moisture content were placed in glass Petri dishes, and the weight of each batch was determined using laboratory scales. Germination was initiated as described above, and after 24, 48, 72 and 96 h the germinated seeds in the Petri dishes were dried to constant mass in a laboratory dryer. The loss of dry matter was expressed as the difference between the calculated dry matter mass of the initial seeds and the mass determined for the dry germinated seeds.

2.5. Extraction of Free Phenolic Compounds

Samples of the seeds and the germinated seeds (i.e., sprouts) were frozen in liquid nitrogen and milled in a laboratory mill (A11 Basic; IKA-Werke, Staufen, Germany). The free phenolic compounds were extracted using 70% aqueous methanol. Briefly, 0.5 g of ground seeds were mixed with 3.0 mL of 70% aqueous methanol, and the mixture was shaken in the dark at room temperature for 40 min, at 200 rpm (EV-403; Tehtnica Železniki, Slovenia). After centrifugation at $8709 \times g$ for 8 min at 10 °C (Avanti JXN-26; Beckman Coulter, Krefeld, Germany), the supernatant was removed and saved, and the extraction was repeated twice more. The three supernatants were pooled, diluted to 10 mL with 70% aqueous methanol, filtered using 0.45-µm pore size syringe filters (Chromafil A-45/25; cellulose acetate, hydrophilic membrane; Macherey-Nagel, Düren, Germany), and stored at 2 °C until the determination of the TPC, and DPPH and ABTS analysis, within 24 h.

2.6. Extraction of Bound Phenolic Compounds

After the methanol extraction, the solid residues were hydrolysed with sodium hydroxide, as described previously [10]. Then, 20 mL of 2 M NaOH was added to the reaction tube, and the mixture was shaken in the dark at room temperature for 4 h, at 200 rpm (Tehtnica Železniki EV-403, Slovenia). The hydrolysed mixture was acidified to pH 3.2 to 3.4 by the addition of 3.5 mL of concentrated formic acid. After centrifugation at $8709 \times g$ for 8 min at 10 °C (Avanti JXN-26; Beckman Coulter, Krefeld, Germany), the supernatant was removed and filtered through 0.45-µm pore size syringe filters (Chromafil A-45/25; cellulose acetate, hydrophilic membrane; Macherey-Nagel, Düren, Germany) and stored at 2 °C until the determination of the TPC, and DPPH and ABTS analysis, within 24 h.

2.7. Total Phenolic Content

The TPC of the seed extracts was determined using the Folin–Ciocalteu spectrophotometric method described previously [41], with some modifications. Briefly, 0.1 mL of each extract was dispensed into 2.0 mL microcentrifuge tubes and mixed with 1.3 mL of Milli-Q water and 0.3 mL 1:3 diluted of Folin–Ciocalteu phenolic reagent, and allowed to react for 5 min. Then, 0.3 mL of 20% (*w/v*) aqueous Na_2CO_3 was added, and after 1 h at room temperature, the absorbances were measured at 765 nm (UV–Vis spectrophotometer; model 8453; Agilent Technologies, Santa Clara, CA, USA). The measurements were compared with a standard curve of a gallic acid (GA) solution, and the TPC is expressed as mg gallic acid equivalents (GAE) per g dry weight of the seed sample (mg GAE/g DW).

2.8. DPPH Radical Scavenging Activity

The DPPH radical scavenging activity of the extracts was determined according to a method described previously [41], with some modifications. Briefly, 50.0 µL of each extract was dispensed into 2.0 mL microcentrifuge tubes, and 250 µL of acetic buffer was added, and the volume was adjusted to 1.0 mL with methanol. Finally, 1 mL of 0.2 mM methanol solution of DPPH was added. The mixture was shaken and then left in the dark for 1 h. The absorbance was then measured at 517 nm (UV–Vis spectrophotometer; model 8453; Agilent Technologies, Santa Clara, CA, USA). A lower absorption of the reaction mixture indicates higher free radical scavenging activity. The absorbance of the control was achieved by replacing the seed sample with methanol. The measurement was compared to a standard curve of a Trolox solution, and the radical scavenging activity is expressed as mg Trolox equivalents per g dry matter (mg TE/g DW).

2.9. ABTS Radical Cation Scavenging Activity

The radical scavenging activities of the seed extracts against the ABTS radical cation were determined according to a method described previously [34], with some modifications. The ABTS stock solution was prepared by reacting ABTS reagent with manganese dioxide as the oxidising agent. Before analysis, 10 mL of ABTS stock solution was diluted with 25 mL of 0.325 M phosphate buffer and 65 mL of Milli-Q water. An aliquot of each extract

(0.05 mL) was mixed with 0.5 mL 0.325 M of phosphate buffer and 1.0 mL of diluted ABTS radical cation solution, and 0.45 mL Milli-Q water was added, to give the final volume of 2 mL. The mixture was shaken and left in the dark for 1 h. The absorbance was measured at 734 nm (UV–Vis spectrophotometer; model 8453; Agilent Technologies, Santa Clara, CA, USA). A lower absorption of the reaction mixture indicates higher free radical scavenging activity. The absorbance of the control was achieved by replacing the seed sample with methanol. The measurement was compared to a standard curve of a Trolox solution, and the radical scavenging activity is expressed as mg TE/g DW.

2.10. Purification of the Seed Extracts

For the purification of the crude seed extracts, 100 mg Strata-X RP cartridges were used (Phenomenex, Torrance, CA, USA). The seed extracts were initially filtered (0.45 μm pore size syringe filters; Chromafil A-45/25; cellulose acetate, hydrophilic membrane; Macherey-Nagel, Düren, Germany), to remove any solid residue. The cartridges were conditioned with 3.0 mL of methanol followed by 3.0 mL of Milli-Q water. Then, 30 mL of the diluted methanol seed extracts (extract:water, 1:9) or 3.0 mL of the hydrolysed seed extracts were applied to the cartridges, and allowed to pass through them. The phenolic compounds remained bound to the cartridges, and the co-extracted compounds were washed from the columns with 4.0 mL of Milli-Q water. The cartridges were then dried using a vacuum pump. The compounds bound to the cartridges were eluted with 2.0 mL of 70% (*v/v*) aqueous methanol. The resulting extracts were filtered through 0.20 μm pore size syringe filters (Chromafil Xtra-20/13; cellulose acetate; Macherey-Nagel, Düren, Germany), and then stored at $-80\,^{\circ}$C until the liquid chromatography–mass spectrometry (LC-MS) analysis.

2.11. Liquid Chromatography–Mass Spectrometry Analysis

Reversed-phase LC-MS analysis was used to separate and quantify the individual phenolic acids in the seed extracts. The LC system used (1100 chromatography system; Agilent Technologies, Santa Clara, CA, USA) included of a thermostated auto-sampler (G1330B), a thermostated column compartment (G1316A), a diode array detector (G1315B), and a binary pump (1312A). The LC system was coupled with a mass spectrometer (Quattro micro API; Waters, Milford, MA, USA). Chromatographic separation was carried out using a C18 column (2.7 μm, 150 mm × 2.1 mm; Ascentis Express) with an C18 guard column (2.7 μm, 5 mm × 2.1 mm; Ascentis Express; Supelco, Bellefonte, PA, USA). The conditions used were: column temperature, 30 °C; injection volume, 20 μL; and mobile phase flow rate, 250 μL/min. The components of the mobile phase were 0.1% aqueous formic acid (solution A) and acetonitrile (solution B). The mobile phase gradient was programmed as follows (%B): 0–2 min, 10%; 2–18 min, 10–60%; 18–18.2 min, 60–80%; 18.2–20 min, 80%; 20–20.2 min, 80–10%; 20.2–26 min, 10%; Detection was performed with the scanning diode array spectra from 240 nm to 650 nm.

The mass spectrometer was operated in negative ionisation mode, and the operating conditions were as follows: electrospray capillary voltage, 3.5 kV; cone voltage, 20 V; extractor voltage, 2 V; source block temperature, 100 °C; desolvation temperature, 350 °C; cone gas flow rate, 30 L/h, and desolvation gas flow rate, 350 L/h. The data signals were acquired and processed on a PC using the MassLynx software (V4.1 2005; Waters Corporation).

The identification of the individual compounds was achieved by comparing their retention times and both the spectroscopic and mass spectrometric data, with quantification according to peak areas, as compared to previously determined calibration curves. The recoveries of the different compounds were determined using the standard addition method (Supplementary Materials Table S1). The samples were spiked with all of the analysed compounds at four spiking levels (1, 10, 200, 2000 μg/g DW of sample) by adding different volumes of a methanolic solution of the analytes.

2.12. Statistical Analysis

All of the experiments were performed as six independent replicates, using a complete randomisation method. The data are reported as the means ± standard deviation (SD) for three analyses for each extract. The results were subjected to two-way ANOVA, and the significances of the differences between the mean values were determined using Tukey's multiple comparison tests. Pearson's correlation analysis was used to define correlations between the means. All of the tests were performed using the SPSS Statistics software (version 24; IBM, New York, NY, USA). Statistical significance was defined at the level of $p < 0.05$.

3. Results and Discussion

3.1. Effects of Dehulling and Germination on Growth Rates and Total Dry Weight

All of the samples of the hulled and cold dehulled seeds reached high germination rates of over 95%. The dry weights of the sprouts were monitored for 96 h, at 24-h intervals, as shown in Figure 1.

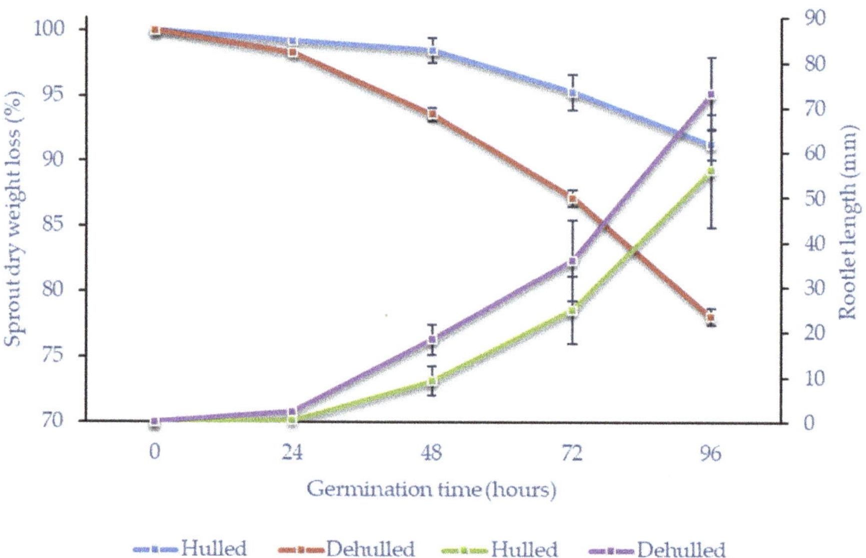

Figure 1. Buckwheat sprout dry weights and rootlet lengths over the 96-h germination period.

As also shown in Figure 1, the mean dry weight losses of the sprouts after 48 h were 1.5% for hulled seeds and 6.4% for dehulled seeds. After 96 h, these losses had reached 8.7% and 21.9%, respectively. Compared to the sprouts from the hulled seeds, the dry weight losses of those from the dehulled seeds were significantly higher at all of the sampled stages of germination ($p < 0.05$).

To determine the growth of the seedlings, the lengths of the rootlets were measured. After 24 h of germination, the roots were visible for most of the seeds. At this time, the rootlets of hulled seeds were visible as white spots at the tips of the seeds, and were less than 1 mm long, while those of the dehulled seeds were significantly longer, with a mean length of 2.3 mm ($p < 0.05$) (Figure 2) For the rest of the germination period analysed, the rootlets of dehulled seeds were always significantly longer than those of hulled seeds.

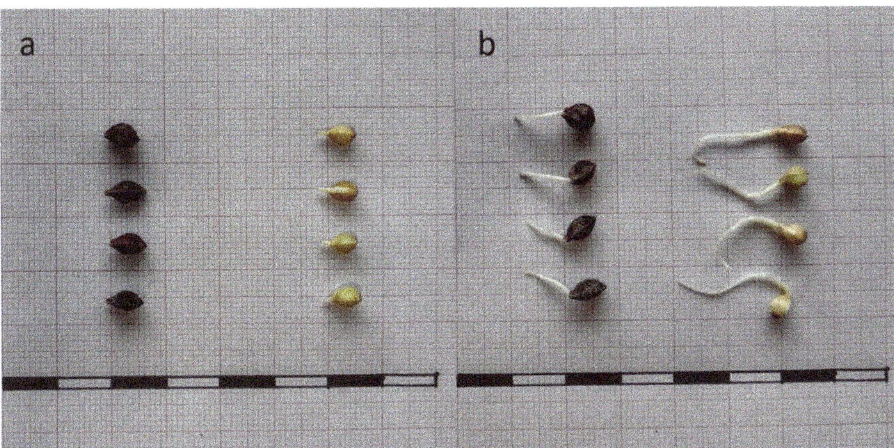

Figure 2. Germinated of representative hulled (left) and dehulled (right) buckwheat seeds after 24 h (**a**) and 48 h (**b**) of germination. Scale measures, 1.0 cm.

This difference in rootlet growth rates between hulled and dehulled seeds is probably due to the higher respiration rate and easier O_2/CO_2 transfer for dehulled seeds, compared to hulled seeds. It is also possible that dehulled seeds had better water absorption over the first few hours of germination [42]. A study performed on barley [43] showed that hull-less barley malted more quickly than hulled barley, with significant benefits in terms of reductions in the consumption of water and energy. For these buckwheat seeds, it can be concluded here that the removal of the hulls results in quicker germination, but not significantly greater, germination rates.

It can be assumed that the dry weights of these sprouts will have continued to decrease beyond the 96 h. However, the aim of this study was to determine the dynamic changes in the early germination stages for the production of buckwheat malt that can be used in baking and brewing processes.

3.2. Effects of Germination on Total Phenolic Content

The changes in the content of free and bound phenolic compounds at the different germination stages are shown in Table 1. The total and the free and bound fractions of the phenolic compounds showed dynamic changes. In the first 24 h of germination, no significant increases in the TPC were detected. It is possible that, although the growth and bioactivation of the seeds was observed, some of the water-soluble bioactive compounds might have been lost during the soaking [44]. For the rest of the germination process, gradual and significant increases in TPC were observed, with the maximum values obtained at the end of the 96 h period of monitoring, which is in agreement with a previous study on germinating buckwheat seeds [34]. Compared to hulled seeds, for dehulled seeds the increase in TPC was faster, probably as a consequence of the earlier onset of germination. Germination had the greatest effects on the phenolic content of the free fractions (hulled vs. dehulled, 2.93-fold vs. 5.40-fold increases, respectively). On the other hand, smaller increases were seen for the bound fractions of the phenolic compounds (24% vs. 41% increases, respectively).

Table 1. Total phenolic content (Folin-Ciocalteu) and antioxidant activities (1,1-diphenyl-2-picrylhydrazyl (DPPH), 3-ethylbenzothiazolin-6-sulfonic acid diammonium salt (ABTS)) for the free and bound extraction fractions of the buckwheat seeds during germination.

Analysis	Seed Preparation	Measure	Germination Time (h)				
			0	**24**	**48**	**72**	**96**
			Total phenolic content (mg GAE/g dry weight)				
Folin-Ciocalteu	Hulled	Free	4.60 ± 0.42 [ab]	4.21 ± 0.09 [a]	5.61 ± 0.38 [b]	8.01 ± 1.09 [c]	13.46 ± 0.89 [d]
		Bound	5.47 ± 0.29 [a]	5.87 ± 0.09 [ab]	6.13 ± 0.32 [bc]	6.57 ± 0.50 [c]	7.28 ± 0.37 [d]
		Total	10.06 ± 0.38 [a]	10.08 ± 0.19 [a]	11.74 ± 0.46 [b]	14.58 ± 1.50 [c]	20.74 ± 1.20 [d]
	Dehulled	Free	4.32 ± 0.12 [a]	4.46 ± 0.15 [a]	6.36 ± 0.24 [b]	14.28 ± 1.05 [c]	23.31 ± 2.09 [d]
		Bound	3.65 ± 0.12 [a]	3.91 ± 0.27 [a]	4.40 ± 0.12 [b]	4.50 ± 0.25 [b]	6.19 ± 0.22 [c]
		Total	7.98 ± 0.21 [a]	8.37 ± 0.35 [a]	10.76 ± 0.19 [b]	18.78 ± 1.03 [c]	29.50 ± 2.17 [d]
			Antioxidant activity (TE/g dry weight)				
DPPH	Hulled	Free	4.61 ± 0.19 [a]	3.99 ± 0.08 [a]	5.45 ± 0.56 [a]	7.60 ± 0.75 [b]	14.00 ± 2.14 [c]
		Bound	4.15 ± 0.38 [a]	4.04 ± 0.42 [a]	5.64 ± 0.66 [b]	6.10 ± 1.19 [b]	8.53 ± 0.53 [c]
		Total	8.76 ± 0.49 [a]	8.03 ± 0.38 [a]	11.08 ± 1.03 [b]	13.70 ± 1.83 [c]	22.53 ± 2.49 [d]
	Dehulled	Free	4.45 ± 0.15 [ab]	3.84 ± 0.18 [a]	6.42 ± 0.71 [b]	15.00 ± 1.06 [c]	24.07 ± 2.39 [d]
		Bound	2.01 ± 0.22 [a]	2.54 ± 0.19 [a]	3.37 ± 0.27 [b]	3.99 ± 0.66 [c]	5.80 ± 0.16 [d]
		Total	6.46 ± 0.33 [a]	6.38 ± 0.31 [a]	9.79 ± 0.64 [b]	18.99 ± 1.41 [c]	29.87 ± 2.29 [d]
ABTS	Hulled	Free	7.02 ± 0.18 [a]	7.53 ± 0.36 [a]	9.09 ± 0.35 [b]	12.34 ± 0.82 [c]	19.20 ± 1.11 [d]
		Bound	5.94 ± 0.65 [b]	4.72 ± 0.47 [a]	4.64 ± 0.54 [a]	5.06 ± 0.46 [a]	6.65 ± 0.33 [b]
		Total	12.96 ± 0.61 [a]	12.26 ± 0.48 [a]	13.74 ± 0.71 [a]	17.4 ± 1.17 [b]	25.85 ± 1.31 [c]
	Dehulled	Free	7.19 ± 0.22 [a]	7.73 ± 0.38 [ab]	10.44 ± 1.22 [b]	23.22 ± 2.28 [c]	35.02 ± 2.74 [d]
		Bound	3.14 ± 0.26 [a]	2.59 ± 0.25 [a]	2.45 ± 0.23 [a]	2.80 ± 0.26 [a]	4.71 ± 1.02 [b]
		Total	10.34 ± 0.46 [a]	10.32 ± 0.6 [a]	12.89 ± 1.28 [a]	26.01 ± 2.36 [b]	39.73 ± 3.51 [c]

Data are means ± SD, from three independent replicates. Means with different letters indicate statistically significant differences between the different stages of germination ($p < 0.05$). GAE: gallic acid equivalents; TE: Trolox equivalents.

For the seeds prior to germination, the contents of free phenolic compounds were similar in hulled and dehulled seeds (Table 1, 0 h). Initially here, as percentages of TPC, the free phenolic compounds were 45.7% in hulled seeds and 54.1% in dehulled seeds. In comparison to cereals (i.e., the *Poaceae* family), the contribution of free phenolic compounds to the TPC in buckwheat seeds is higher, as in cereals most of the phenolic compounds are in a bound form [22].

Throughout the germination period, the bound phenolic content remained significantly higher for hulled seeds compared to dehulled seeds. These data are in agreement with previous studies that reported higher bound phenolic content in buckwheat seed hulls compared to the rest of the seeds [23,45].

From Figure 1, it can be seen that the dry mass of seeds decreases during germination. Here, 10% to 20% of the dry mass was lost during the respiratory metabolic processes required for seed germination and sprout growth to 96 h. The increases in all of the fractions of the phenolic compounds in these sprouts at 96 h will be a combination of de novo synthesis and, to a lesser extent, the concentration of the tissues rich in phenolic compounds, due to the loss of the starchy endosperm [45].

3.3. Effect of Germination on Antioxidant Activity

The total AA measured by both the DPPH and ABTS assays gradually increased for both hulled and dehulled sprouts over the 96-h germination period, as summarised in Table 2.

Table 2. Contents of the individual phenolic compounds (μg/g dry weight (DW)) in the free and bound fractions of the germinating buckwheat seeds.

Phenolic Compound	Phenolic Content (μg/g DW) during Germination (h) for the Hulled and Dehulled Seeds									
	0		24		48		72		96	
	Hulled	Dehulled	Hulled	Dehulled	Hulled	Dehulled	Hulled	Dehulled	Hulled	Dehulled
Free fraction										
Orientin	11.82 ± 0.45 [b]	2.51 ± 0.12 [a]	13.54 ± 1.63 [b]	2.54 ± 0.31 [a]	24.96 ± 3.97 [a]	96.78 ± 7.57 [b]	171.98 ± 17.67 [a]	501.34 ± 33.12 [b]	865.85 ± 91.02 [a]	2205.06 ± 213.18 [b]
Isoorientin	7.79 ± 0.37 [b]	1.28 ± 0.07 [a]	9.43 ± 0.98 [b]	1.25 ± 0.17 [a]	11.88 ± 1.02 [a]	35.33 ± 4.46 [b]	76.26 ± 4.97 [a]	339.68 ± 34.49 [b]	725.37 ± 35.5 [a]	1402.5 ± 106.51 [b]
Rutin	98.05 ± 8.66 [b]	65.39 ± 5.67 [a]	85.17 ± 5.42 [b]	60.96 ± 5.63 [a]	106.04 ± 8.11 [a]	138.46 ± 7.76 [b]	194.76 ± 13.56 [a]	347.6 ± 51.67 [b]	664.32 ± 67.18 [a]	967.66 ± 81.34 [b]
Vitexin	33.33 ± 3.3 [b]	3.35 ± 0.33 [a]	52.13 ± 2.73 [b]	7.99 ± 0.68 [a]	59.91 ± 4.55 [a]	51.55 ± 3.72 [b]	385.83 ± 19.31 [a]	723.34 ± 60.09 [b]	1180.89 ± 157.33 [a]	1869.17 ± 185.14 [b]
Catechin	23.15 ± 1.32 [b]	10.67 ± 0.95 [a]	18.55 ± 1.48 [a]	27.39 ± 2.5 [b]	64.5 ± 8.26 [a]	148.41 ± 10.65 [b]	134.95 ± 6.68 [a]	285.1 ± 17.96 [b]	243.85 ± 21.19 [a]	472.5 ± 33.23 [b]
Epicatechin	17.88 ± 1.15 [b]	7.48 ± 1.22 [a]	12.05 ± 1.23 [a]	32.84 ± 1.94 [b]	46.89 ± 3.76 [a]	148.59 ± 6.72 [b]	89.44 ± 2.70 [a]	413.11 ± 15.19 [b]	150.16 ± 20.53 [a]	527.6 ± 39.57 [b]
Hyperin	13.48 ± 1.33 [b]	1.72 ± 0.09 [a]	8.15 ± 1.66 [b]	0.87 ± 0.06 [a]	10.00 ± 0.94 [b]	1.78 ± 0.27 [a]	9.90 ± 0.83 [a]	8.23 ± 1.47 [b]	20.14 ± 2.68 [a]	19.66 ± 1.41 [a]
p-Coumaric acid	ND	ND	1.64 ± 0.11 [a]	2.93 ± 0.27 [b]	5.33 ± 0.35 [a]	13.26 ± 1.22 [b]	11.97 ± 1.29 [a]	16.8 ± 2.05 [b]	18.62 ± 1.62 [a]	31.38 ± 2.06 [b]
Bound fraction										
Orientin	2.12 ± 0.18	ND	1.09 ± 0.13	ND	1.17 ± 0.09 [b]	0.05 ± 0.01 [a]	1.48 ± 0.13 [b]	0.17 ± 0.01 [a]	1.84 ± 0.14 [a]	3.89 ± 0.22 [b]
Isoorientin	0.85 ± 0.07 [b]	0.21 ± 0.03 [a]	0.94 ± 0.08 [b]	0.39 ± 0.03 [a]	1.06 ± 0.09 [b]	0.49 ± 0.03 [a]	1.25 ± 0.09 [a]	1.52 ± 0.11 [b]	3.07 ± 0.41 [a]	11.63 ± 1.13 [b]
Rutin	0.34 ± 0.03 [b]	0.19 ± 0.01 [a]	0.12 ± 0.01	ND	0.34 ± 0.04 [a]	0.51 ± 0.06 [b]	0.49 ± 0.04 [a]	0.74 ± 0.09 [b]	0.61 ± 0.05 [a]	0.86 ± 0.11 [b]
Vitexin	17.99 ± 1.24 [b]	0.47 ± 0.03 [a]	13.74 ± 0.97 [b]	0.53 ± 0.04 [a]	22.86 ± 2.55 [b]	0.93 ± 0.07 [a]	25.41 ± 2.31 [b]	5.31 ± 0.49 [a]	35.95 ± 2.33 [b]	18.47 ± 1.14 [a]
Catechin	ND	ND	ND	ND	ND	ND	ND	ND	ND	ND
Epicatechin	ND	ND	ND	ND	ND	ND	ND	ND	ND	ND
Hyperin	1.25 ± 0.03	ND	0.98 ± 0.07	ND	1.66 ± 0.15	ND	3.07 ± 0.23	ND	4.24 ± 0.57	ND
p-Coumaric acid	31.32 ± 0.25 [b]	2.64 ± 0.14 [a]	32.79 ± 2.15 [b]	3.39 ± 0.32 [a]	36.82 ± 2.87 [b]	4.19 ± 0.37 [a]	37.91 ± 2.86 [b]	5.06 ± 0.41 [a]	60.36 ± 7.24 [b]	7.36 ± 0.63 [a]

Data are means ± SD, from three independent replicates. Means with different letters in rows indicate statistically significant differences between contents in the same stage of germination (hulled vs. dehulled) ($p < 0.05$); not detected (ND).

Before germination, the total AA was higher in hulled seeds compared to dehulled seeds. For both hulled and dehulled seeds, the germination significantly increased AA. The highest measured AA in these methanol extracts was at 96 h of germination for dehulled seeds, and the lowest was for the bound fraction from before germination and after 24 h of germination of dehulled seeds (Table 1). Compared with the Folin–Ciocalteu method for phenolic content (see also Table 1), the AA measures showed similar trends during the course of germination.

In the DPPH test for AA of the free fractions of these seed extracts, up to 48 h after the start of germination there were only minor increases compared to the seeds before germination, with no significant difference between the hulled and dehulled seed extracts. However, after 48 h of germination, AA here increased rapidly, with significantly higher values measured as the maxima that were reached after 96 h.

These data for the DPPH tests showed significant correlations ($p < 0.01$) between the TPC and AA in the free fractions of both hulled ($r = 0.974$) and dehulled ($r = 0.988$) seeds. In the bound fractions, significant correlations were still observed ($p < 0.05$) for both these hulled ($r = 0.781$) and dehulled ($r = 0.926$) seed extracts.

The AA obtained using the ABTS assay showed similar trends, with some small differences. Overall, the AA in the free fractions of these seed extracts measured by the ABTS method was significantly higher than that by the DPPH method, which might be explained by the higher activities of the phenolic compounds contributing to the antioxidant potential of the seed extracts towards the ABTS radical [46,47]. It was also noted that the ABTS/DPPH ratio changed, reaching a maximum at 24 h of germination, at around 2.0, and then decreasing to around 1.4 for the 96-h germinated seed extracts. This indicates that during germination, the chemical compositions of the antioxidants were changing, as well as their total concentrations. The different reactivities of particular antioxidants toward the ABTS/DPPH radicals and the changes in their relative compositions will be reflected in these different AA values obtained by the two methods. Indeed, Floegel et al. reported that antioxidant capacities determined for various foods in in vitro assays differed significantly across these two assays (i.e., ABTS and DPPH) [48].

With the ABTS assay, the maximum values of the AA in the free fraction were reached after 96 h of germination, as 19.20 mg TE/g DW and 35.02 mg TE/g DW in hulled and dehulled seeds, respectively. As with the DPPH and Folin-Ciocalteu methods, the lowest values were detected before germination and after 24 h of germination. These data for the ABTS assay also showed significant correlations ($p < 0.01$) between the TPC and AA for the free fractions of both hulled ($r = 0.976$) and dehulled ($r = 0.992$) seeds. For the bound fractions, significant correlations were again observed ($p < 0.05$) for the hulled ($r = 0.450$) and dehulled ($r = 0.781$) extracts.

In general, the AA in the bound fractions of these seed extracts were lower compared to the free fractions. In the bound fractions, the differences between the extracts before germination and after 96 h of germination were lower in comparison to the free fractions, where strong increases in the AA were seen during the course of germination.

For the DPPH assay, the lowest values in the bound fractions were seen up to 24 h after the start of germination. Afterwards, the AA began to increase, to reach the maxima at 96 h. In the ABTS assay, the AA of the bound phenolic compounds before germination decreased significantly over the first 24 h of germination (20%, 17% for hulled and dehulled seeds, respectively). Afterwards, the AA slowly increased and reached starting values again after 72 h of germination. These data for the AA in the bound fractions of these seed extracts confirmed the significant differences between these two assays. The lower levels of AA measured using ABTS can be explained by the low content of phenolic compounds that contributed to the antioxidant potential towards the ABTS radical [47]. This is also confirmed by the lower correlations between the AA in the bound fraction in the ABTS assay and the Folin–Ciocalteu and DPPH assays.

In general, these data for AA also concur with previous studies on buckwheat seeds and germinated buckwheat seeds [24,49–51]. However, it needs to be noted that detailed comparisons of the data in the present study with the data in the literature can be difficult, as previous studies were mainly related to buckwheat seeds either before germination [52] or as sprouts [53], without consideration of the effects of dehulling on the levels of the bioactive compounds and the antioxidative potential.

3.4. Phenolic Characterisation by Liquid Chromatography–Mass Spectrometry

Liquid chromatography coupled with mass spectrometry is a powerful tool for the analysis of a wide range of compounds. It is now commonly used for the analysis of phenolics and other bioactive compounds in biological samples. In the present study, LC-MS analysis of the extracts of the germinated seeds (i.e., sprouts) identified several phenolic compounds, most of which were flavonoids in glycosylated and aglycone forms. In contrast to cereals (i.e., the *Poaceae* family) where the majority of phenolic compounds are bound to cell walls [54], the buckwheat seed phenolic compounds are mostly in free forms [52].

The characterisation of the methanol extracts showed that the predominant phenolic compound in the seeds before germination was rutin, with significantly higher levels in hulled seeds compared to dehulled seeds (98.05, 65.39 µg/g, respectively) (Table 2). The other important flavonoids in these seeds were vitexin (33.33, 3.35 µg/g, respectively), catechin (23.15, 10.67 µg/g, respectively) and epicatechin (17.88, 7.48 µg/g, respectively). Three other major phenolic compounds in the germinated seeds, orientin, isoorientin and vitexin, were present at low concentrations before germination, and then greatly increased during germination (Table 2). The higher levels of the phenolic compounds overall in hulled seeds indicate that the majority of these bioactive compounds are stored in the hull and the outer parts of the seed, such that dehulling significantly decreases their levels. Similar data for higher levels of phenolic compounds in buckwheat seed hulls have also been reported in previous studies [23,55].

In the first 24 h of germination, no significant differences were seen for the majority of these phenolic compounds analysed. Interestingly, the rutin levels were even decreased after 24 h, by 13.1% in hulled seeds and 6.7% in dehulled seeds. This decrease can be explained by the leaching of water-soluble phenolic compounds into the steeping water during the early seed soaking [45,56]. Despite the loss of some of these seed bioactive compounds during soaking, with longer germination times there were increased contents of all of these compounds analysed in these sprouts, which is in agreement with the data here for TPC and AA (Table 1).

For the dehulled seeds, in the first 24 h of germination they already showed increased vitexin, catechin and epicatechin levels. These increases would indicate the faster onset of de novo synthesis of these bioactive compounds in the dehulled seeds. As already noted, compared to the hulled seeds, this was probably due to their faster water intake, and consequently more vigorous physiological activity [42,43]. However, the greatest increases in the majority of the compounds analysed here occurred after 48 h of germination. Comparing the levels of phenolics going from 48 h to 72 h of germination, over this 24-h period, rutin levels increased by 1.8-fold and 2.5-fold, vitexin by 6.4-fold and 14-fold, and orientin by 6.9-fold and 5.2-fold in hulled and dehulled seeds, respectively. As can be seen, these relative increases in most of the phenolic compounds were lower in hulled seeds compared to dehulled seeds. During the final 24 h of the germination (i.e., to 96 h), the levels of these bioactive compounds continued to increase, with the maximum values for all of the compounds analysed seen after 96 h of germination. With the exception of hyperin, the levels of all of the compounds analysed were significantly higher in dehulled sprouts ($p < 0.05$).

Overall, the 96-h germination increased the contents of the major phenolics in these seeds as follows: epicatechin, 70-fold; catechin, 44-fold; and rutin, 15-fold. The levels of orientin, isoorientin and vitexin were low in the seeds before germination, and so during germination these levels increased by 878-fold, 1095-fold and 558-fold, respectively. These increases might appear to be particularly large, but it must be noted that these compounds were at particularly low concentrations in the seeds before germination, and their levels were greatly increased by germination. Similar rutin, orientin, isoorientin and vitexin levels in seeds and germinated sprouts have been reported in previous studies [57,58].

Based on the specific reactivities of the phenolic compounds identified [59,60] and their levels in these seed extracts, their contributions to the AA measured can be evaluated. It can be estimated that in the seeds before germination, the contribution of the sum of the phenolic compounds to total AA was relatively low, at only about 5% to 8% of AA determined using DPPH. Other redox-active compounds therefore contributed most to the AA before germination. Upon germination, the levels of the major phenolic compounds increase significantly, which coincided with the increased AA (Figure 3). The calculated relative contributions of the major phenolic compounds after 96 h of germination to the total AA was increased from the few percent beforehand, to approximately 60%. These increases in AA during the germination of these seeds can therefore be attributed preferentially to newly synthesised bioactive phenolic compounds. This is also very important from the nutritional point of view, as the compounds identified here are recognized as having health benefits [16–19].

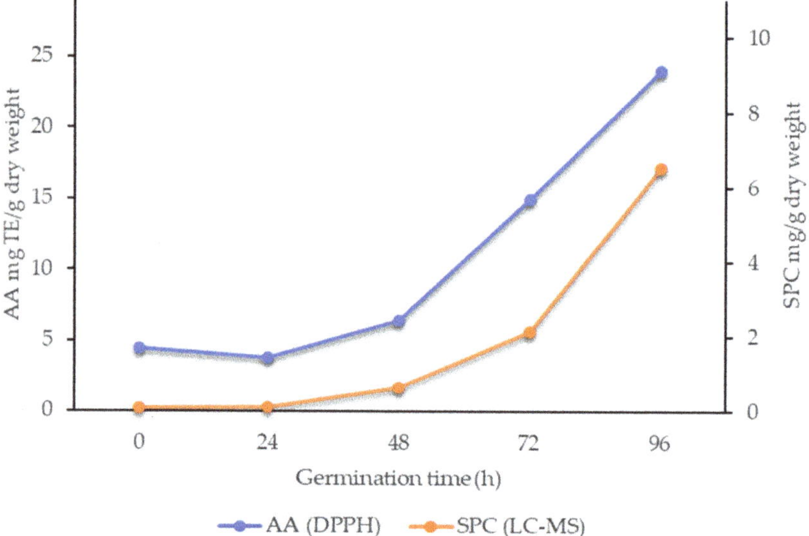

Figure 3. Comparison between antioxidant activity (AA) determined by the DPPH method and the sum of the phenolic compounds (SPC) determined by liquid chromatography-mass spectrometry (LC-MS) for the free fraction of the germinating buckwheat seed extracts.

Analysis of the bound fractions of these seed extracts showed that p-coumaric acid and vitexin were the major phenolics. Orientin, isoorientin, rutin and hyperin were also present in the bound fractions, although at lower levels than in the free fractions (Table 2). It is of note that the dehulling reduced the bound phenolic content in these seeds, which is in agreement with data from Li et al. [61], where it was shown that the highest levels of bound phenolic compounds are in the seed hulls. The accumulation of bound phenolic compounds during the course of germination was nevertheless similar for hulled and dehulled seeds. As the hull is an inedible part of the seeds, and as it is always removed during processing and preparation for human consumption, this initial loss of phenolic compounds due to the dehulling is not relevant in nutritional terms.

According to the data in the present study, the dehulling of the seeds prior to germination had significant effects on the levels of these individual phenolic compounds, as well as the phenolic profile in the edible part of these seeds. Most importantly, the content of the major health-promoting compounds, such as orientin, isoorientin, vitexin and rutin, were enhanced more rapidly in dehulled seeds. These data thus show that the combination of cold dehulling and seed germination is a promising method for implementation for the production of new functional food ingredients.

4. Conclusions

In the present study, the germination of buckwheat seeds is demonstrated to be an excellent way for increasing their content of phenolic compounds as well as their AA. A new processing approach of the mechanical cold dehulling of buckwheat seeds was used that maintained the high germination rates through seed dehulling. The goal was to determine the effects of this dehulling before germination on seed germination, growth, the TPC, AA and individual phenolic contents. The data obtained show that in comparison with hulled seeds, this cold dehulling maintained high germination rates and promoted faster growth of the buckwheat groats, which resulted in greater increases in TPC, AA and all of the phenolics determined. Some of the important health-promoting compounds of the germinated seeds, including orientin, isoorientin, rutin and vitexin, were detected at greater concentrations in the sprouts from the dehulled seeds. As the inedible part of the seeds was removed (i.e., the hull), such dehulled germinated buckwheat seeds can be used directly in various technological food preparation processes. These improved TPC and antioxidant properties thus further demonstrate that germinated dehulled buckwheat groats provide an excellent raw material for the preparation of functional food products.

Supplementary Materials: The following are available online at https://www.mdpi.com/2304-8158/10/4/740/s1, Table S1. Characteristics of the phenolic compounds in buckwheat extracts as analysed by LC-MS, Figure S1: Representative mass chromatogram of (A) nongerminated hulled buckwheat extracts, (B) nongerminated hulled buckwheat extracts spiked with standards and (C) 96 h germinated hulled buckwheat extracts.

Author Contributions: A.Ž.: conceptualisation and methodology, data curation, formal analysis, investigation, methodology, writing—original draft preparation, writing—review and editing. T.P. (Tomaž Polak): formal analysis, writing—review and editing, methodology, supervision, data curation, validation, software. B.C.: writing—review and editing, methodology, validation, supervision. T.P. (Tomaž Požrl): conceptualisation and methodology, data curation, project administration, supervision, visualisation, writing—original draft, writing—review and editing. All authors have read and agreed to the published version of the manuscript.

Funding: The work was financially supported by the research programs funded by the Slovenian Research Agency (P4-0234).

Institutional Review Board Statement: Not applicable.

Informed Consent Statement: Not applicable.

Data Availability Statement: Not applicable.

Acknowledgments: The authors thank Chris Berrie for the critical review of the manuscript and language editing.

Conflicts of Interest: The authors declare no conflict of interest.

References

1. Gan, R.Y.; Lui, W.Y.; Wu, K.; Chan, C.L.; Dai, S.H.; Sui, Z.Q.; Corke, H. Bioactive compounds and bioactivities of germinated edible seeds and sprouts: An updated review. *Trends Food Sci. Technol.* **2017**, *59*, 1–14. [CrossRef]
2. Rico, D.; Peñas, E.; García, M.d.C.; Martínez-Villaluenga, C.; Rai, D.K.; Birsan, R.I.; Frias, J.; Martín-Diana, A.B. Sprouted barley flour as a nutritious and functional ingredient. *Foods* **2020**, *9*, 296. [CrossRef] [PubMed]
3. Kim, H.; Kim, O.-W.; Ahn, J.-H.; Kim, B.-M.; Oh, J.; Kim, H.-J. Metabolomic analysis of germinated brown rice at different germination stages. *Foods* **2020**, *9*, 1130. [CrossRef] [PubMed]
4. Arouna, N.; Gabriele, M.; Pucci, L. The impact of germination on sorghum nutraceutical properties. *Foods* **2020**, *9*, 1218. [CrossRef] [PubMed]

5. Nelson, K.; Stojanovska, L.; Vasiljevic, T.; Mathai, M. Germinated grains: A superior whole grain functional food? *Can. J. Physiol. Pharmacol.* **2013**, *91*, 429–441. [CrossRef]

6. Penãs, E.; Martínez-Villaluenga, C. Advances in production, properties and applications of sprouted seeds. *Foods* **2020**, *9*, 9. [CrossRef]

7. Cid-Gallegos, M.S.; Sánchez-Chino, X.M.; Juárez Chairez, M.F.; Álvarez González, I.; Madrigal-Bujaidar, E.; Jiménez-Martínez, C. Anticarcinogenic activity of phenolic compounds from sprouted legumes. *Food Rev. Int.* **2020**, 1–16. [CrossRef]

8. Baenas, N.; Piegholdt, S.; Schloesser, A.; Moreno, D.; García-Viguera, C.; Rimbach, G.; Wagner, A. Metabolic activity of radish-sprout-derived isothiocyanates in *Drosophila melanogaster*. *Int. J. Mol. Sci.* **2016**, *17*, 251. [CrossRef]

9. Li, R.; Song, D.; Vriesekoop, F.; Cheng, L.; Yuan, Q.; Liang, H. Glucoraphenin, sulforaphene, and antiproliferative capacity of radish sprouts in germinating and thermal processes. *Eur. Food Res. Technol.* **2017**, *243*, 547–554. [CrossRef]

10. Wang, H.; Wang, J.; Guo, X.; Brennan, C.S.; Li, T.; Fu, X.; Chen, G.; Liu, R.H. Effect of germination on lignan biosynthesis, and antioxidant and antiproliferative activities in flaxseed (*Linum usitatissimum* L.). *Food Chem.* **2016**, *205*, 170–177. [CrossRef]

11. Hübner, F.; Arendt, E.K. Germination of cereal grains as a way to improve the nutritional value: A review. *Crit. Rev. Food Sci. Nutr.* **2013**, *53*, 853–861. [CrossRef] [PubMed]

12. Chaturvedi, N.; Sharma, P.; Shukla, K.; Singh, R.; Yadav, S. Cereals nutraceuticals, health ennoblement and diseases obviation: A comprehensive review. *J. Appl. Pharm. Sci.* **2011**, *1*, 6–12.

13. Marton, M.; Mandoki, Z.; Caspo, J.; Caspo-Kiss, Z.; Marton, M.; Mándoki, Z.; Marton, M.; Mandoki, Z.; Caspo, J.; Caspo-Kiss, Z. The role of sprouts in human nutrition. A review. *Aliment. Hung. Univ. Transylvania* **2010**, *3*, 81–117.

14. Gálová, Z.; Rajnincová, D.; Chňapek, M.; Balážová, Ž.; Špaleková, A.; Vivodík, M.; Hricová, A. Characteristics of cereals, pseudocereals and legumes for their coeliac active polypeptides. *J. Microbiol. Biotechnol. Food Sci.* **2019**, *9*, 390–395. [CrossRef]

15. Bai, C.Z.; Feng, M.L.; Hao, X.L.; Zhong, Q.M.; Tong, L.G.; Wang, Z.H. Rutin, quercetin, and free amino acid analysis in buckwheat (*Fagopyrum*) seeds from different locations. *Genet. Mol. Res.* **2015**, *14*, 19040–19048. [CrossRef] [PubMed]

16. Rauf, A.; Imran, M.; Patel, S.; Muzaffar, R.; Bawazeer, S.S. Rutin: Exploitation of the flavonol for health and homeostasis. *Biomed. Pharmacother.* **2017**, *96*, 1559–1561. [CrossRef] [PubMed]

17. Lam, K.Y.; Ling, A.P.K.; Koh, R.Y.; Wong, Y.P.; Say, Y.H. A review on medicinal properties of orientin. *Adv. Pharmacol. Sci.* **2016**, *2016*, 1–9. [CrossRef]

18. Peng, Y.; Gan, R.; Li, H.; Yang, M.; McClements, D.J.; Gao, R.; Sun, Q. Absorption, metabolism, and bioactivity of vitexin: Recent advances in understanding the efficacy of an important nutraceutical. *Crit. Rev. Food Sci. Nutr.* **2021**, *61*, 1049–1064. [CrossRef]

19. He, M.; Min, J.-W.; Kong, W.-L.; He, X.-H.; Li, J.-X.; Peng, B.-W. A review on the pharmacological effects of vitexin and isovitexin. *Fitoterapia* **2016**. [CrossRef]

20. Lu, C.-L.; Zheng, Q.; Shen, Q.; Song, C.; Zhang, Z.-M. Uncovering the relationship and mechanisms of Tartary buckwheat (*Fagopyrum tataricum*) and Type II diabetes, hypertension, and hyperlipidemia using a network pharmacology approach. *PeerJ* **2017**, *5*, e4042. [CrossRef]

21. Wieslander, G.; Fabjan, N.; Vogrincic, M.; Kreft, I.; Janson, C.; Spetz-Nyström, U.; Vombergar, B.; Tagesson, C.; Leanderson, P.; Norbäck, D. Eating buckwheat cookies is associated with the reduction in serum levels of myeloperoxidase and cholesterol: A double-blind crossover study in day-care centre staff. *Tohoku J. Exp. Med.* **2011**, *225*, 123–130. [CrossRef] [PubMed]

22. Terpinc, P. Bound phenolic compounds of whole cereals grain as a functional food component: Part one. *Acta Agric. Slov.* **2019**, *114*, 269–278. [CrossRef]

23. Zhang, W.; Zhu, Y.; Liu, Q.; Bao, J.; Liu, Q. Identification and quantification of polyphenols in hull, bran and endosperm of common buckwheat (*Fagopyrum esculentum*) seeds. *J. Funct. Foods* **2017**, *38*, 363–369. [CrossRef]

24. Dziadek, K.; Kopeć, A.; Pastucha, E.; Piątkowska, E.; Leszczyńska, T.; Pisulewska, E.; Witkowicz, R.; Francik, R. Basic chemical composition and bioactive compounds content in selected cultivars of buckwheat whole seeds, dehulled seeds and hulls. *J. Cereal Sci.* **2016**, *69*, 1–8. [CrossRef]

25. Janeš, D.; Prosen, H.; Kreft, S. Identification and quantification of aroma compounds of Tartary buckwheat (*Fagopyrum tataricum* Gaertn.) and some of its milling fractions. *J. Food Sci.* **2012**, *77*. [CrossRef] [PubMed]

26. Keriene, I.; Mankeviciene, A.; Bliznikas, S.; Cesnuleviciene, R.; Janaviciene, S.; Jablonskyte-Rasce, D.; Maiksteniene, S. The effect of buckwheat groats processing on the content of mycotoxins and phenolic compounds. *CYTA J. Food* **2016**, *14*, 565–571. [CrossRef]

27. Amudha Senthil, S.P. Effect of hydrothermal treatment on the nutritional and functional properties of husked and dehusked buckwheat. *J. Food Process. Technol.* **2015**, *06*. [CrossRef]

28. Klepacka, J.; Najda, A. Effect of commercial processing on polyphenols and antioxidant activity of buckwheat seeds. *Int. J. Food Sci. Technol.* **2021**, *56*, 661–670. [CrossRef]

29. Kou, L.; Luo, Y.; Yang, T.; Xiao, Z.; Turner, E.R.; Lester, G.E.; Wang, Q.; Camp, M.J. Postharvest biology, quality and shelf life of buckwheat microgreens. *LWT Food Sci. Technol.* **2013**, *51*, 73–78. [CrossRef]

30. Aparicio-García, N.; Martínez-Villaluenga, C.; Frias, J.; Peñas, E. Changes in protein profile, bioactive potential and enzymatic activities of gluten-free flours obtained from hulled and dehulled oat varieties as affected by germination conditions. *LWT* **2020**, *134*, 109955. [CrossRef]

31. Muñoz-Insa, A.; Selciano, H.; Zarnkow, M.; Becker, T.; Gastl, M. Malting process optimization of spelt (*Triticum spelta* L.) for the brewing process. *LWT Food Sci. Technol.* **2013**. [CrossRef]

32. Mazzottini-dos-Santos, H.C.; Ribeiro, L.M.; Oliveira, D.M.T.; Paiva, E.A.S. Ultrastructural aspects of metabolic pathways and translocation routes during mobilization of seed reserves in *Acrocomia aculeata* (Arecaceae). *Rev. Bras. Bot.* **2020**, *43*, 589–600. [CrossRef]

33. Erbaş, S.; Tonguç, M.; Karakurt, Y.; Şanli, A. Mobilization of seed reserves during germination and early seedling growth of two sunflower cultivars. *J. Appl. Bot. Food Qual.* **2016**, *89*, 217–222. [CrossRef]

34. Terpinc, P.; Cigić, B.; Polak, T.; Hribar, J.; Požrl, T. LC-MS analysis of phenolic compounds and antioxidant activity of buckwheat at different stages of malting. *Food Chem.* **2016**, *210*, 9–17. [CrossRef] [PubMed]

35. Beitane, I.; Krumina-Zemture, G.; Sabovics, M. Effect of germination and extrusion on the phenolic content and antioxidant activity of raw buckwheat (*Fagopyrum esculentum* Moench). *Agron. Res.* **2018**, *16*, 1331–1340. [CrossRef]

36. Grahić, J.; Dikić, M.; Gadto, D.; Šimon, S.; Kurtović, M.; Pejić, I.; Gaši, F. Assessment of genetic relationships among common buckwheat (*Fagopyrum esculentum* Moench) varieties from western Balkans using morphological and SSR molecular markers. *Genetika* **2018**, *50*, 791–802. [CrossRef]

37. Cajzek, F.; Bertoncelj, J.; Kreft, I.; Poklar Ulrih, N.; Polak, T.; Požrl, T.; Pravst, I.; Polišenská, I.; Vaculová, K.; Cigić, B. Preparation of β-glucan and antioxidant-rich fractions by stone milling of hull-less barley. *Int. J. Food Sci. Technol.* **2020**, *55*, 681–689. [CrossRef]

38. Tian, B.; Xie, B.; Shi, J.; Wu, J.; Cai, Y.; Xu, T.; Xue, S.; Deng, Q. Physicochemical changes of oat seeds during germination. *Food Chem.* **2010**, *119*, 1195–1200. [CrossRef]

39. Yiming, Z.; Hong, W.; Linlin, C.; Xiaoli, Z.; Wen, T.; Xinli, S. Evolution of nutrient ingredients in tartary buckwheat seeds during germination. *Food Chem.* **2015**, *186*, 244–248. [CrossRef]

40. Mencin, M.; Abramovič, H.; Jamnik, P.; Mikulič Petkovšek, M.; Veberič, R.; Terpinc, P. Abiotic stress combinations improve the phenolics profiles and activities of extractable and bound antioxidants from germinated spelt (*Triticum spelta* L.) seeds. *Food Chem.* **2020**, *344*, 128704. [CrossRef] [PubMed]

41. Abramovič, H.; Košmerl, T.; Ulrih, N.P.; Cigi, B. Contribution of SO_2 to antioxidant potential of white wine. *Food Chem.* **2015**, *174*, 147–153. [CrossRef]

42. Crabb, D.; Kirsop, B.H. Water-sensitivity in barley I. Respiration studies and the influence of oxygen availability. *J. Inst. Brew.* **1969**, *75*, 254–259. [CrossRef]

43. Agu, R.C.; Bringhurst, T.A.; Brosnan, J.M.; Pearson, S. Potential of hull-less barley malt for use in malt and grain whisky production. *J. Inst. Brew.* **2009**, *115*, 128–133. [CrossRef]

44. Liu, Y.; Cai, C.; Yao, Y.; Xu, B. Alteration of phenolic profiles and antioxidant capacities of common buckwheat and tartary buckwheat produced in China upon thermal processing. *J. Sci. Food Agric.* **2019**, *99*, 5565–5576. [CrossRef]

45. Kalinová, J.P.; Vrchotová, N.; Tříska, J. Phenolics levels in different parts of common buckwheat (*Fagopyrum esculentum*) achenes. *J. Cereal Sci.* **2019**, *85*, 243–248. [CrossRef]

46. Nenadis, N.; Wang, L.-F.; Tsimidou, M.; Zhang, H.-Y. Estimation of scavenging activity of phenolic compounds using the ABTS•+ assay. *J. Agric. Food Chem.* **2004**, *52*, 4669–4674. [CrossRef] [PubMed]

47. Abramovič, H.; Grobin, B.; Ulrih, N.P.; Cigic, B. The methodology applied in DPPH, ABTS and Folin-Ciocalteau assays has a large influence on the determined antioxidant pPotential. *Acta Chim. Slov.* **2017**, *64*, 491–499. [CrossRef] [PubMed]

48. Floegel, A.; Kim, D.O.; Chung, S.J.; Koo, S.I.; Chun, O.K. Comparison of ABTS/DPPH assays to measure antioxidant capacity in popular antioxidant-rich US foods. *J. Food Compos. Anal.* **2011**, *24*, 1043–1048. [CrossRef]

49. Wiczkowski, W.; Szawara-Nowak, D.; Sawicki, T.; Mitrus, J.; Kasprzykowski, Z.; Horbowicz, M. Profile of phenolic acids and antioxidant capacity in organs of common buckwheat sprout. *Acta Aliment.* **2016**, *45*, 250–257. [CrossRef]

50. Tsurunaga, Y.; Takahashi, T.; Katsube, T.; Kudo, A.; Kuramitsu, O.; Ishiwata, M.; Matsumoto, S. Effects of UV-B irradiation on the levels of anthocyanin, rutin and radical scavenging activity of buckwheat sprouts. *Food Chem.* **2013**, *141*, 552–556. [CrossRef]

51. Swieca, M. Potentially bioaccessible phenolics, antioxidant activity and nutritional quality of young buckwheat sprouts affected by elicitation and elicitation supported by phenylpropanoid pathway precursor feeding. *Food Chem.* **2016**, *192*, 625–632. [CrossRef] [PubMed]

52. Zhu, H.; Liu, S.; Yao, L.; Wang, L.; Li, C. Free and bound phenolics of buckwheat varieties: HPLC characterization, antioxidant activity, and inhibitory potency towards α-glucosidase with molecular docking analysis. *Antioxidants* **2019**, *8*. [CrossRef] [PubMed]

53. Kreft, S.; Janeš, D.; Kreft, I. The content of fagopyrin and polyphenols in common and tartary buckwheat sprouts. *Acta Pharm.* **2013**, *63*, 553–560. [CrossRef]

54. Ti, H.; Li, Q.; Zhang, R.; Zhang, M.; Deng, Y.; Wei, Z.; Chi, J.; Zhang, Y. Free and bound phenolic profiles and antioxidant activity of milled fractions of different indica rice varieties cultivated in southern China. *Food Chem.* **2014**, *159*, 166–174. [CrossRef] [PubMed]

55. Ren, S.-C.; Sun, J.-T. Changes in phenolic content, phenylalanine ammonia-lyase (PAL) activity, and antioxidant capacity of two buckwheat sprouts in relation to germination. *J. Funct. Foods* **2014**, *7*, 298–304. [CrossRef]

56. Li, Q.; Singh, V.; de Mejia, E.G.; Somavat, P. Effect of sulfur dioxide and lactic acid in steeping water on the extraction of anthocyanins and bioactives from purple corn pericarp. *Cereal Chem.* **2019**, *96*, 575–589. [CrossRef]

57. Nam, T.G.; Lim, Y.J.; Eom, S.H. Flavonoid accumulation in common buckwheat (*Fagopyrum esculentum*) sprout tissues in response to light. *Hortic. Environ. Biotechnol.* **2018**, *59*, 19–27. [CrossRef]

58. Kiprovski, B.; Mikulic-Petkovsek, M.; Slatnar, A.; Veberic, R.; Stampar, F.; Malencic, D.; Latkovic, D. Comparison of phenolic profiles and antioxidant properties of European *Fagopyrum esculentum* cultivars. *Food Chem.* **2015**, *185*, 41–47. [CrossRef]
59. Peter, K.; Kolodziej, H. Antioxidant properties of phenolic compounds from *Pelargonium reniforme*. *J. Agric. Food Chem.* **2004**, *52*, 4899–4902. [CrossRef]
60. Villaño, D.; Fernández-Pachón, M.; Moyá, M.; Troncoso, A.; García-Parrilla, M. Radical scavenging ability of polyphenolic compounds towards DPPH free radical. *Talanta* **2007**, *71*, 230–235. [CrossRef]
61. Li, F.H.; Yuan, Y.; Yang, X.L.; Tao, S.Y.; Ming, J. Phenolic profiles and antioxidant activity of buckwheat (*Fagopyrum esculentum* Möench and *Fagopyrum tartaricum* L. Gaerth) hulls, brans and flours. *J. Integr. Agric.* **2013**, *12*, 1684–1693. [CrossRef]

Article

Drought and Heat Stress Impacts on Phenolic Acids Accumulation in Durum Wheat Cultivars

Barbara Laddomada [1,*], Antonio Blanco [2], Giovanni Mita [1], Leone D'Amico [1], Ravi P. Singh [3], Karim Ammar [3], Jose Crossa [3] and Carlos Guzmán [4,*]

[1] Institute of Sciences of Food Production (ISPA), National Research Council (CNR), Via Monteroni, 73100 Lecce, Italy; giovanni.mita@ispa.cnr.it (G.M.); leone.damico@ispa.cnr.it (L.D.)

[2] Department of Soil, Plant and Food Sciences, University of Bari 'Aldo Moro', Via G. Amendola 165/A, 70126 Bari, Italy; antonio.blanco@uniba.it

[3] International Maize and Wheat Improvement Center (CIMMYT), Global Wheat Program, Apdo Postal 6-641, Ciudad de México 56237, Mexico; R.Singh@cgiar.org (R.P.S.); k.ammar@cgiar.org (K.A.); J.CROSSA@cgiar.org (J.C.)

[4] Departamento de Genética, Escuela Técnica Superior de Ingeniería Agronómica y de Montes, Edificio Gregor Mendel, Campus de Rabanales, Universidad de Córdoba, CeiA3, 14071 Córdoba, Spain

[*] Correspondence: barbara.laddomada@ispa.cnr.it (B.L.); carlos.guzman@uco.es (C.G.); Tel.: +39-0832-422-613 (B.L.); +34-957-212-575 (C.G.)

Abstract: Droughts and high temperatures are the main abiotic constraints hampering durum wheat production. This study investigated the accumulation of phenolic acids (PAs) in the wholemeal flour of six durum wheat cultivars under drought and heat stress. Phenolic acids were extracted from wholemeals and analysed through HPLC-DAD analysis. Ferulic acid was the most represented PA, varying from 390.1 to 785.6 μg/g dry matter across all cultivars and growth conditions, followed by sinapic acids, *p*-coumaric, vanillic, syringic, and *p*-hydroxybenzoic acids. Among the cultivars, Cirno had the highest PAs content, especially under severe drought conditions. Heat stress enhanced the accumulation of minor individual PAs, whereas severe drought increased ferulic acid and total PAs. Broad-sense heritability was low (0.23) for *p*-coumaric acid but $\geq$0.69 for all other components. Positive correlations occurred between PA content and grain morphology and between test weight and grain yield. Durum wheat genotypes with good yields and high accumulation of PAs across different growing conditions could be significant for durum wheat resilience and health-promoting value.

Keywords: durum grains; phenolic compounds; genetic variability; heritability; climate constraints; yield performance

Citation: Laddomada, B.; Blanco, A.; Mita, G.; D'Amico, L.; Singh, R.P.; Ammar, K.; Crossa, J.; Guzmán, C. Drought and Heat Stress Impacts on Phenolic Acids Accumulation in Durum Wheat Cultivars. *Foods* **2021**, *10*, 2142. https://doi.org/10.3390/foods10092142

Academic Editor: Rosaria Saletti

Received: 30 July 2021
Accepted: 7 September 2021
Published: 10 September 2021

1. Introduction

Durum wheat (*Triticum turgidum* ssp. *durum* (Desf.) Husnot) is one of the most common cereal crops in Mediterranean climates and the tenth most cultivated species in the world. Despite accounting for only 5% of global wheat production, this wheat species is a key commodity for many areas worldwide, especially for the countries surrounding the Mediterranean basin, North America, the desert area of the southwestern United States, North Mexico, and sub-Saharan Africa [1].

Traditionally, durum-based foods have a large diversification across the producing countries. While pasta is the most popular product worldwide and a symbol of the Italian cousin, couscous is the most common durum-based food in North Africa, and durum wheat breads are traditionally important in Southern Italy, Spain, Turkey, and Mid-East Mediterranean regions. Locally in the Mid-East, durum wheat is also used to make bulgur (made from the cracked parboiled grains) and freekeh (a dish made with green, roasted, and rubbed grains); in Turkey and Cyprus, tarhana is a fermented soup made with durum grains and yogurt or milk. Overall, the above products provide a significant slice of calories and proteins to human diets; additionally, they are an important source of bioactive components contributing to a healthy diet [2,3].

Durum wheat is mainly grown under rainfed conditions, often encountering drought and heat stresses that hamper yield potential and influence the qualitative characteristics of the grains [4,5]. Therefore, it is of great importance to investigate the effects of these environmental constraints that have become more and more frequent due to climate change that is also posing a serious challenge for durum wheat growth [6]. Among the drawbacks of water scarcity in durum cultivation is the reduction of plant height, grain size, transpiration rate, and hormonal imbalance [7,8]. Durum wheat is also very sensitive to elevated temperatures, especially when they occur at grain filling and flowering times [9]. High temperatures can affect its physiological traits and reduce seed germination, grain filling, grain number, photosynthetic capacity, chlorophyll content, and typically induce early leaf senescence [10]. An excess of reactive oxygen species (ROS) production in response to drought and heat stresses can cause damages to proteins, carbohydrates, lipids, and nucleic acids [11]. To counteract the injuries of oxidative stress, the wheat plant has developed a complex defense system based on the production of enzymatic and non-enzymatic antioxidants [12]. Among the latter, phenolic acids are the major subclass of polyphenols participating in the scavenging of ROS in cereals [13]. The antioxidant capacity of these compounds depends on both the number and position of the hydroxyl groups in the benzene ring and on ortho-substitution with the methoxy group [14]. In addition to contributing to biotic and abiotic resistance in plants, in humans, phenolic acids contribute to the prevention of a number of non-communicable diseases due to their anti-inflammatory and anti-carcinogenic properties [3,15].

Typically, phenolic acids are classified as hydroxy derivatives of either cinnamic or benzoic acid, the former including ferulic acid that alone accounts for about 90% of total phenolic acids in the wheat grain [16]. In mature durum kernels, the majority of phenolic acids (up to 75–80%) are insolubly bound to cell wall polymers, while the rest (20–25%) are esterified to sugars and other low molecular mass compounds, and only 0.5–2% are soluble and free [15]. The content and composition of phenolic acids may vary in the wheat germplasm as documented by a large body of literature [17]. The extent of variation for PAs is up to 3.6-fold [18,19], and it is influenced by both the genotype and environmental factors [18,19]. To unravel the genetic base of the trait, investigations were carried out to identify and characterize the enzymatic genes and QTLs involved in phenolic acids pathways [20,21]. To date, only a few reports describe the effect of abiotic and biotic factors on individual phenolic acids accumulation in mature wheat kernels and more studies need to be carried out to support the first evidence [11]. Only a few works have investigated the effects of genotype, environment, and their interaction on the content and composition of phenolic acids in durum wheat grains under drought and heat stress [22–24]. Moreover, elite durum wheat cultivars developed from the large gene pool available at the International Maize and Wheat Improvement Center (CIMMYT) grown under abiotic stresses were assessed for yield performances and nutritional and qualitative traits but not for phenolic acids accumulation [25,26]. Some of the above cultivars have been used largely in several durum producing areas and are of great interest due to their high yield performances, disease resistance, grain quality, and tolerance to drought and heat stresses [25].

The objective of this study was to test the impact of drought and heat stresses on phenolic acids content and composition in the mature grains of a set of durum cultivars representative of CIMMYT durum germplasm. In addition to the variability for individual and total phenolic acids, the study was aimed at assessing the effects of genotypes, growth conditions, and their interactions to assess the possibility of breeding for phenolic compounds.

2. Materials and Methods

2.1. Plant Materials/Agronomic Trials

Six elite durum wheat varieties, which are some of the most important in 50 years of breeding at CIMMYT, were considered in this study: Mexicali (released in 1975), Yavaros (released in 1979), Altar (released in 1984), Atil (released in 2000), Jupare (released in 2001), and Cirno (released in 2008) (Table S1). The cultivars were sown in 2015–16 and in 2016–2017 crop seasons in Ciudad Obregon, Sonora, in northwestern Mexico. The experimental trials were planted with two replicates in a randomized complete block design under six different growth conditions: (1) drip irrigation in beds; (2) full irrigation in flat beds; (3) full irrigation in beds; (4) mild drought stress; (5) severe drought stress; and (6) severe heat stress. All the trials were planted in November, except for severe heat stress (planted in February). All the plots (6.5 m^2) had full irrigation (>500 mm) except medium drought stress (300 mm) and severe drought stress (180 mm). Weed, diseases, and insects were all well controlled. In all the trials, N was applied (pre-planting) at a rate of 50 kg of N/ha, and at tillering, 150 additional units of N were applied in all the plots except in severe drought stress (50 N units). At maturity, whole plots were harvested, grain yield was calculated, and 1 kg of grain from each durum line was used for analyzing the quality traits. The meteorology data of the experimental station in Ciudad Obregon was characterized by almost no precipitation during the wheat growing season. Average temperatures were between 12 and 24 °C in March and April, the grain filling time for all treatments, except for plants under heat stress at temperatures between 19 and 28 °C during grain filling in May. Flowering time and physiological maturity in most of the examined cultivars occur at similar times because these genotypes were bred for the same growing area. According to the general growing stages of durum wheat in Ciudad Obregon, drought stress was continuous from stem elongation to grain ripening in moderate and severe drought stress trials. In severe heat stress trial, higher temperatures than in the normal planting time started from shoot elongation and remained in the grain filling stage until ripening.

2.2. Grain Traits

The digital image system SeedCount SC5000 (Next Instruments, Australia) was used to calculate 1000 kernel weight (g) and test weight (kg/hL). With the same device other grain morphological traits such as grain length, width, and thickness were obtained. Grain protein content (%) and moisture content were determined by near-infrared spectroscopy (NIR Systems 6500, Foss Denmark) calibrated based on official AACC methods 39–10 and 46–11A, respectively (AACC, 2010). Protein content was adjusted to a 12.5% moisture basis.

2.3. Phenolic Acids Analysis

Wholegrain samples were milled at particle size ≤1 mm using a 1093 Cyclotec™ Sample mill (FOSS, Hilleroed, Denmark) to produce wholemeal flour. Milled samples were stored at −20 °C until analysis to protect phenolic acids from degradation. Total phenolic acids (sum of soluble and insoluble fractions) were extracted from wholemeal flour samples according to details previously described [19]. In brief, samples were delipidated twice with hexane, hydrolysed with 2 M NaOH, and acidified with HCl 12 M to pH 2 prior to undergoing ethyl acetate extraction. Extracts were dried under nitrogen flux and dissolved in 200 μL of 80:20 methanol/water and quali-quantitatively analyzed using an Agilent 1100 Series HPLC-DAD system (Agilent Technologies, Santa Clara, CA, USA) equipped with a reversed phase C18 (2) Luna column (Phenomenex, Torrance, CA, USA) (5 μm, 250 × 4.6 mm) at a column temperature of 30 °C. A mobile phase consisting of acetonitrile (A) and 10 mL/L water solution of H_3PO_4 (B) was used for the following elution program: isocratic elution, 100% B, 0–30 min; linear gradient from 100% B to 85% B, 30–55 min; linear gradient from 85% B to 50% B, 55–80 min; linear gradient from 50% B to 30% B, 80–82 min; and post time, 10 min before the next injection. The flow rate of the mobile phase was 1 mL/min, and the injection volume was 20 μL. The column temperature was kept at 30 °C. Peaks were identified by comparing their retention times and UV-Vis spectra to those of authentic phenolic standards: *p*-hydroxybenzoic acid, vanillic acid, syringic acid, *p*-coumaric acid, sinapic acid, and ferulic acid (Sigma-Aldrich, Gillingham, U.K.). All phenolic acids were quantified via a ratio to the internal standard (3,5-dichloro-4-hydroxybenzoic acid, Sigma-Aldrich, Gillingham, U.K.) added to every sample and using calibration curves of phenolic acid standards. The wavelengths used for quantification of phenolic acids were 280, 295, and 320 nm. All samples were extracted and analyzed in duplicate, and concentrations of individual phenolic acids were expressed in micrograms per gram of dry matter.

2.4. Statistical Analysis

We obtained the Best Linear Unbiased Estimated (BLUE) of the durum wheat lines for each of the response traits using a linear mixed model with the fixed effect containing the growth conditions (established by the field management), year (which reflects the different environmental conditions in the two years of the trial that are independent of the field management), durum wheat lines, and their interactions. We also computed the broad-sense heritability (repeatability) and the genetic correlation matrices for the various traits through a combined analysis of the evaluated durum wheat lines across growth conditions and years.

For BLUEs estimation, the following linear mixed model was used:

$$Y_{ijkl} = \mu + E_i + Y_k + EY_{ik} + R_j(EY_{ik}) + G_l + EG_{il} + YG_{kl} + EYG_{ikl} + \varepsilon_{ijkl} \tag{1}$$

where μ is the general mean, E_i is the fixed effects of the growth conditions ($I = 1, \ldots,$ s), Y_k represents the fixed effects of the years ($k = 1.2, \ldots, y$), EY_{ik} is the inetraction between growth condition and year, $R_j(EY_{ik})$ is the effects of the replicates ($j = 1, 2$) within growth condition and year assumed to be identically and independently normally distributed with mean zero and variance $\sigma^2_{j(i\,k)}$, the fixed effects of the durum wheat lines are G_l ($l = 1,2, \ldots, m$), EG_{il} is the line by growth condition interaction, the term YG_{kl} is the line by year interaction, and the triple interaction between the growth conditions, year, and durum wheat lines is denoted by EYG_{ikl}. The term ε_{ijkl} is a random residual associated to the *l*th wheat line in the *j*th replicate within the *i*th growth condition and *k*th year combination and assumed to be identically and independently normally distributed with mean zero and variance σ^2_ε. The code used for fitting the linear mixed model of Equation (1) was generated using SAS software, Version 9.

The broad-sense heritability was calculated as:

$$H^2 = \frac{\sigma^2_g}{\sigma^2_g + \sigma^2_{ge}/nloc + \sigma^2_{gy}/nyear + \sigma^2_{gey}/nlocyear\; \sigma^2_\varepsilon/(nloc \times nyear \times nrep)} \tag{2}$$

where σ^2_g, σ^2_{ge}, σ^2_{gy}, σ^2_{gey}, and σ^2_ε are the genotype, genotype by growth condition interaction, genotype by year interaction, genotype by growth condition and by year interaction, and the error variance components, respectively; nloc, nyear, and nrep are the number of growth conditions; and rg is the number of years and number of replicates, respectively.

The genetic correlation matrices among sites were calculated using equations from Cooper and Delacy [27]: $\rho_{g_{ij}} = \frac{\rho_{p_{ij}}}{h_i h_{i\prime}}$ where $\rho_{p_{ij}}$ is the phenotypic correlation among growth condition–year combination i and $i^\prime$, and h_i and $h_{i\prime}$ are the square roots of the growth condition–year combination i and $i^\prime$, respectively.

3. Results

3.1. Grain Yield, Grain Traits, and Phenolic Acid Profile of Durum Wheat Cultivars

Six durum cultivars out the foremost durum varieties developed at CIMMYT were evaluated for grain yield, general grain traits and the content and composition of phenolic acids in wholemeal flour under six growth conditions across two years. The growth conditions varied from optimal to critical due to the effect of moderate to severe drought and severe heat stresses. A summary of the data of grain yield and other grain traits recorded is showed in Table S2. In full irrigation environments, the cultivars produced more than seven tons per hectare, which was reduced to 4 and 3 tons per hectare under severe drought and heat, respectively. Test weight and 1000 kernel weight showed a similar pattern compared to grain yield, while grain protein content was higher in the stressed growth conditions. In overall, the genotype with best performance was Cirno, showing the highest grain yield and size values and the third highest test weight and grain protein content values. A first picture of results of the phenolic acids composition is provided in Table 1 which summarizes the variation of individual phenolic acids of the cultivars grown across all tested conditions and years. Ferulic acid was the most represented phenolic acid with a grand mean of 563.07 µg/g dry matter and a variation range from 390.1 to 785.6 µg/g dry matter across all cultivars and growth conditions. Sinapic acid was the second phenolic acid for abundancy, followed by four minor components (i.e., *p*-coumaric, vanillic, syringic and *p*-hydroxybenzoic acids). Overall, each variety had a typical phenolic acid profile, differing significantly ($p < 0.05$) for almost all individual components (Table 1). Cirno was the cultivar with the highest content of major phenolic acids (i.e., ferulic and sinapic acids) and, so far it had the highest content of total phenolic acids (TPAs). As a second insight into phenolic variation of the cultivars, we looked at the average values of individual phenolic acids as influenced by the six tested growth conditions tested across two years (Table S3). The outcome data revealed an impact of growth conditions that was different on individual phenolic acids (Figure 1, Table S4). While severe heat stress enhanced the accumulation of some minor individual phenolic acids (i.e., *p*-coumaric, syringic and vanillic acids), severe drought had a higher impact on the most abundant phenolic acid, namely ferulic, and consequently on TPAs. By comparing the three main growing conditions used in Mexico for durum wheat (i.e., drip irrigation in beds, full irrigation in flat beds, and full irrigation in beds) with the stress conditions, an increase of individual phenolic acids was found under severe drought (Table S3).

Table 1. Mean values, minimum/maximum values, coefficient of variation (CV), heritability and least significant difference (LSD) for grain phenolic acids content (µg/g dry matter) in six CIMMYT durum wheat cultivars evaluated across two years and six growth conditions. TPAs: total sum of individual phenolic acids. Numbers with the same letter in each column are not significantly different ($p < 0.05$).

	p-Hydroxy-benzoic Acid	Syringic Acid	Vanillic Acid	*p*-Coumaric Acid	Ferulic Acid	Sinapic Acid	TPAs
Altar	4.87 [b]	5.37 [d]	8.15 [c]	14.24 [bc]	540.96 [c]	51.58 [c]	625.18 [cd]
Atil	4.80 [b]	5.16 [d]	8.14 [c]	12.94 [d]	563.69 [b]	55.04 [b]	649.75 [b]
Cirno	5.12 [a]	5.81 [bc]	8.16 [c]	13.85 [c]	611.00 [a]	59.13 [a]	703.09 [a]
Jupare	3.85 [c]	6.10 [ab]	9.30 [b]	14.84 [b]	530.09 [c]	54.62 [b]	618.80 [d]
Mexicali	3.30 [d]	6.14 [a]	9.83 [a]	14.09 [b]	568.34 [b]	54.74 [b]	657.29 [b]
Yavaros	3.99 [c]	5.73 [c]	9.06 [b]	16.38 [a]	559.15 [b]	45.21 [d]	639.52 [bc]
Grand Mean	4.34	5.72	8.79	14.52	563.07	53.2	649.87
Range	1.9–6.8	2.4–8.5	5.4–12.8	7.4–38.3	390.1–785.6	29.4–92.3	444.9–902.2
CV	6.9	9.87	7.48	8.3	4.92	5.16	4.86
Heritability	0.95	0.75	0.89	0.23	0.69	0.85	0.65
LSD	0.17	0.33	0.38	0.74	17.1	2.21	19.51

Moreover, the results showed that sinapic acid was not enhanced neither by water scarcity nor by elevated temperatures, but it increased under the drip irrigation in beds, full irrigation in flat beds, and full irrigation in beds conditions ($p < 0.05$). The outcome results also remarked a rise of the two major phenolic acids (i.e., ferulic and sinapic acids), and of the least abundant *p*-hydroxybenzoic acid under the full irrigation in flat beds condition.

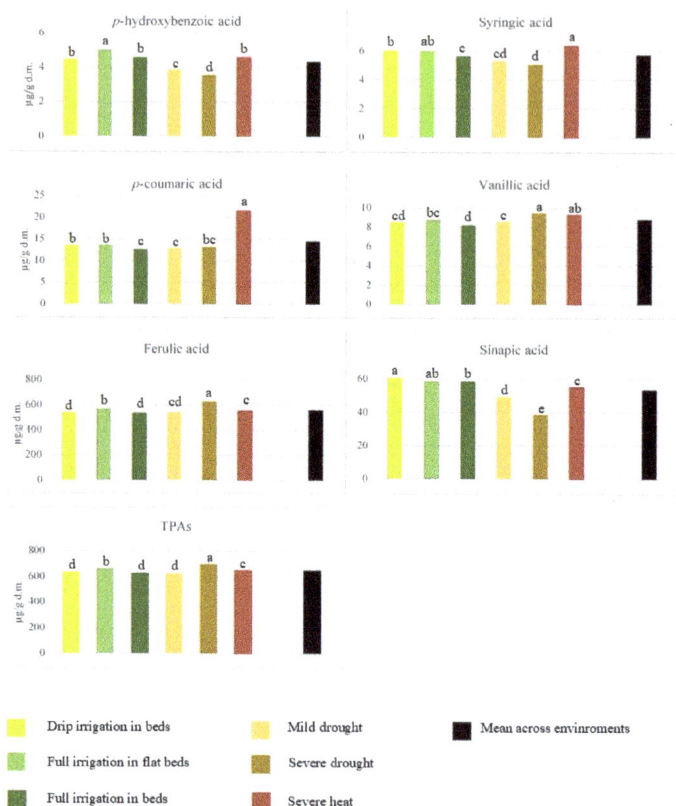

Figure 1. Average values of individual phenolic acids, evaluated in wholemeal flour of six CIMMYT durum cultivars over two years under six different growth conditions. TPAs: total sum of individual phenolic acids. Different letters show significant differences ($p < 0.05$).

3.2. Effects of Genotype, Growth Conditions, Year, and Their Interactions

An ANOVA analysis was carried out to determine the effects of the genotype, growth conditions, year, and their interactions on the grain phenolic acid contents. The analysis showed that almost all the factors involved in the experimental trial had a highly significant effect on phenolic acids ($p < 0.0001$) (Table 2). The year and growth conditions were the most impactful sources of variation, followed by genotype and the E × Y, G × E, E × Y and G × E × Y interactions. The Rep (E × Y) interaction was not significant for all the traits. The variance ascribed to the year, due to the effect of the different climatic conditions occurring across the two years of the experimental trials, was particularly high for almost all individual phenolic acids, with the exception of two minor components (i.e., *p*-hydroxybenzoic acid and *p*-coumaric acid). Conversely, genotype variance was consistent for *p*-hydroxybenzoic acid, while the first source of variation for *p*-coumaric acid was the growth conditions (Table 2). Broad-sense heritability, based on the variance component estimates with combined analysis, varied largely among individual phenolic acids, being low (0.23) for *p*-coumaric acid, above 0.69 for all other components, and 0.65 for TPAs (Table 1).

Table 2. Analysis of variance of individual phenolic acids in six CIMMYT durum wheat cultivars evaluated across two years and six growth conditions. Mean squares values are shown.

Source	DF	*p*-Hydroxy Benzoic Acid	Syringic Acid	Vanillic Acid	*p*-Coumaric Acid	Ferulic Acid	Sinapic Acid	TPAs [1]
Year (Y)	1	0.61 *	59.82 ***	83.31 ***	0.21 ***	220,684.87 ***	6149.96 ***	320,586.19 ***
Growth conditions (E)	5	7.08 ***	5.51 ***	5.26 ***	276.34 ***	27,885.92 ***	1650.95 ***	18,322.21 ***
E x Y	5	4.01 ***	5.11 ***	3.10 ***	32.38 ***	19,815.92 ***	540.06 ***	27,821.01 ***
Rep (E x Y)	12	0.08 ns	0.26 ns	0.44 ns	0.45 ns	214.82 ns	9.10 ns	284.14 ns
Genotype (G)	5	12.07 ***	3.73 ***	12.13 ***	32.43 ***	20507.17 ***	545.30 ***	24,087.22 ***
G × Y	5	0.28 ***	0.214 ***	1.380 *	4.117 *	9046.34 ***	109.01 ***	11,060.94 ***
G × E	25	0.41 *	1.28 ***	1.46 ***	45.12 ***	6831.27 ***	82.69 ***	9043.75 ***
G × E × Y	25	0.95 ***	0.74 **	1.26 ***	9.16 ***	5768.96 ***	95.32 ***	7721.88 ***
Error	60	0.09	0.33 ***	0.43 ***	1.65 ***	876.77 ***	14.71 ***	1141.52 ***

[1]: sum of individual phenolic acids. *: $p < 0.05$, **: $p < 0.01$, ***: $p < 0.001$, ns = not significant, DF = degree of freedom.

3.3. Relationships between Phenolic Acids, Yield Components, and Protein Content

The phenotypic correlation coefficients were calculated among individual phenolic acids, TPAs, and other traits, as shown in Table 3. In particular, we considered the correlations between phenolic acids and grain yield, 1000 kernel weight, test weight, grain length, grain width, and protein content across years and growth conditions. Four of the individual phenolic acids and TPAs were correlated in a positive manner ($p < 0.05$) with grain yield. In addition to this, all individual phenolic acids except vanillic acid showed a positive or neutral correlation with 1000 kernel weight. In the case of test weight, only syringic and sinapic acid showed positive associations with it. TPAs were positively correlated with grain yield and all other kernel morphological traits, except for test weight (Table 3), although, in this case, the correlation was significant but weak ($r = -0.14$). Similarly, the protein content was negatively correlated ($p < 0.05$) with the minor PAs (syringic, vanillic, and *p*-coumaric acids) and positively correlated with the major PAs and TPAs. Moreover, looking at the correlations among individual PAs, *p*-hydroxybenzoic acid had negative correlations with syringic, vanillic, and *p*-coumaric acids and positive correlations with sinapic and ferulic acids and TPAs.

Table 3. Correlation coefficients (*r*) among individual phenolic acids (PAs) contents and other phenotypic traits including yield components protein content and kernel morphology and across year and growth conditions. TPAs: total sum of individual phenolic acids.

	Grain Yield	1000 K. Weight	Test Weight	Protein Content	Grain Length	Grain Width	Grain Thickness	*p*-Hydroxy Benzoic Acid	Syringic Acid	Vanillic Acid	*p*-Coumaric Acid	Ferulic Acid	Sinapic Acid	TPAs
Grain yield	1													
1000 K. weight	0.78	1												
Test weight	−0.16	−0.60	1											
Protein content	NS	−0.13	0.43	1										
Grain length	−0.39	−0.62	0.88	0.69	1									
Grain width	NS	−0.32	0.64	−0.38	0.24	1								
Grain thickness	−0.60	−0.91	0.64	−0.17	0.48	0.66	1							
p-Hydroxybenzoic	0.55	NS	0.24	0.46	NS	NS	NS	1						
Syringic	0.17	0.39	NS	−0.62	−0.25	0.47	NS	−0.64	1					
Vanillic	−0.42	NS	−0.18	−0.51	NS	0.12	NS	−0.98	0.78	1				
p-Coumaric	−0.49	−0.36	NS	−0.84	−0.23	0.53	0.59	−0.64	0.54	0.65	1			
Ferulic	0.32	−0.26	0.87	0.47	0.64	0.59	0.35	0.62	−0.15	−0.52	−0.34	1		
Sinapic	0.77	0.65	0.13	0.55	0.13	−0.13	−0.66	0.43	NS	−0.33	−0.79	0.46	1	
TPAs	0.41	−0.14	0.83	0.49	0.61	0.56	0.24	0.61	NS	−0.50	−0.40	0.99	0.56	1

All correlations are significant ($p < 0.05$) except for NS: not significant.

4. Discussion

Durum wheat-based foods are major components of the human diet in many areas worldwide, and the content of grain functional compounds, such as phenolic acids that are beneficial to human health, has become an important subject of research. While there is a wide knowledge on the genetic variability of phenolic acids in wheat germplasm collections [17], in view of durum breeding programs, the assessment of the genetic stability of phenolic acids under different growth conditions has been reported only in a few investigations [22–24,28]. A recent report released by the Food and Agriculture Organization [29] showed that the increasing frequency and intensity of extreme weather constraints, such as water scarcity and elevated temperatures as a result of climate changes, are having a devastating effect on food security and livelihoods, also posing a challenge to durum wheat cultivation. Our study is the first one affording the effect of drought and heat stress on the phenolic acids profile of a set of durum cultivars developed at CIMMYT over the last 50 years. The examined cultivars have been introduced in many wheat producing areas due to their high yield potential, disease resistance, grain quality, and tolerance to drought and heat stresses, and they have been used by durum breeding programs carried out by different research institutions in several countries [25]. Based on the overall data, we identified six main individual phenolic acids (i.e., ferulic acid, vanillic acid, *p*-coumaric acid, sinapic acid, syringic acid, and *p*-hydroxibenzoic acid), in line with results arisen from durum wheat genetic diversity screen [19] and validating previous findings on the effects of abiotic stress on the accumulation of phenolic acids [23]. The range of variation across the cultivars, years, and growth conditions for phenolic acids was comprised between 444.9 and 902.2 µg/g dry matter. So far, a twofold range of variation was found in this study, which was slightly lower with that observed in the genetic diversity assessment considering a higher number of genotypes [18,19]. This is probably due to the fact that, in our study, we only used modern durum varieties developed by the same breeding program while the cited studies analyzed the variability for phenolic compounds in heterogenous collections with materials of different wheat species and origins.

Compared to previous works which assessed the effects only of elevated temperatures [23,28] or drought [24], this study considered the effects of both drought and heat stress on a set of durum wheat genotypes. Another element of novelty of the present work is that we did not consider the variation for soluble free, soluble bound, and insoluble bound phenolic acids separately, but we considered the variation for all fractions as a whole, as previously shown [19]. This choice was supported by evidence that the amount of total soluble free, soluble bound, and insoluble bound phenolic acids extracted separately is equal to the amount of the three fractions recovered together [15,19]. Moreover, it was shown that the relative proportion of the three fractions in wheat grains (0.5–2% soluble free; 18–22% soluble bound; 77–80% insoluble bound) is common to all wheat genotypes independently of growing conditions and environments [4,19,30]. All phenolic fractions have important biological properties to protect human health. While soluble free phenolic acids are rapidly absorbed by the small intestine and protect against cardiovascular disease and colon cancer due to their antioxidant properties [31], soluble and insoluble bound phenolic acids protect against colon inflammation and cancer [3,32,33]. The occurrence of ester or ether linkages to cell wall polymers or to other low molecular mass components is not a hindrance for phenolic acids to exert their biological activity as they are metabolized locally by the colon microflora [34,35].

Previous works investigating on the effect of water stress on eight durum genotypes found that phenolic acids accumulated differently in the mature grains independently of whether they were resistant, tolerant, or sensitive to stress [24]. Some genotypes did not exhibit any significant change in phenolic acids under water scarcity, while some others had higher concentrations compared to those grown under non-stress conditions [24]. In the current study, the highest TPAs values were found under severe drought conditions, while moderate drought and severe heat stress did not lead in all cases to higher concentrations of PAs compared to the full irrigation growth conditions. This is interesting, particularly in the case of heat stress, because in this growth condition the lowest values of 1000 kernel weight and test weight were registered. One would expect that the smallest and most shrivelled grain would have the highest PAs content because PAs are concentrated in the bran layers.

The absence of this effect under heat stress condition indicates the probable negative effects of heat on the metabolic pathway leading to the production of PAs. A three-way ANOVA was used to evaluate the portion of genetic variation for the individual phenolic acids ascribed to each source of variation. Based on the results, the year and growth conditions had the largest effect on individual and total PAs contents, followed by the genotype and the G × E, G × Y, E × Y and G × E × Y interactions. The results agree with some previous reports and disagree with others. Shamloo et al. [23] found a predominant genotypic effect on the accumulation of total phenolic acids content in mature wheat grains under elevated temperatures. A general increase of phenolic acids was observed, though the different wheat cultivars showed different grades of increase [23]. Other works investigated the impact of terminal heat stress, finding a negative influence on several secondary metabolites of wheat grains, including phenolic acids [28] (Shahid et al., 2017), which agree with our results commented above. Nevertheless, these studies also confirmed a higher effect of the genotype over environmental factors on the observed variation [28]. Conversely, when the effects of several climate parameters were considered, prevalent effects from the environment and the genotype–environment interactions on the total soluble phenolic contents of mature grains were detected [22], similar to what we found. Discrepancies among the different research could be due to the diverse experimental plan of the studies, considering different genetic materials and the highly different climate and agronomical conditions. In addition to this, the heritability values were calculated for all PAs. The heritability estimated for TPAs was 0.65, which confirmed previous reports by studies carried out on durum wheat [19] and higher compared to those observed in bread wheat [30,36]. For ferulic acid, the most abundant PA in the durum grain, heritability was 0.69, indicating that the proportion of the total phenotypic variance that is attributable to the average effects of genes is quite high for these grain compounds, which could make feasible breeding approaches to increase the PA content. Related to this, it was interesting to identify in the correlation analysis that when TPAs increase, the grain length, grain width, and grain thickness also increase and are positively correlated with test weight and grain yield. Indeed, Cirno was the cultivar showing the highest grain yield and TPAs in our trial, indicating that productivity is not at odds with high PA concentration. In previous studies [23,24], it was not clear if PA variation in mature grains caused by heat or drought was an indirect effect of bran to endosperm ratios or grain filling entity or grain size. Our data suggests that, in general, when climate constraints affect grain morphological traits, a concomitant reduction of phenolic accumulation occurs in mature grains.

5. Conclusions

In this study, the effect of water scarcity and elevated temperatures on phenolic acids profile of the wholemeal flour of a set of CIMMYT elite durum wheat cultivars was evaluated. The varieties had a typical phenolic acid profile under different growth conditions, differing significantly for almost all individual PA components. Cirno was the cultivar showing the highest content of phenolic acids across years and growth conditions, especially under severe drought conditions. Overall, severe heat stress enhanced the accumulation of minor phenolic acids (i.e., *p*-coumaric, syringic and vanillic acids) and reduced the main ones (i.e., ferulic acid), whereas severe drought had a higher impact on ferulic acid and the total sum of PAs.

The broad-sense heritability varied largely among the individual phenolic acids, being low for *p*-coumaric acid and ≥ 0.69 for all other components, and significant genotype effects were found for all PAs. In addition, positive correlations were found between total PAs and grain morphology parameters, test weight, and grain yield, suggesting that breeding for high-yielding varieties with high PAs concentration is feasible.

Supplementary Materials: The following are available online at https://www.mdpi.com/2304-815 8/10/9/2142/s1; Table S1. List and pedigree of the CIMMYT durum wheat cultivars; Table S2. Mean and range of values for different traits in six CIMMYT durum wheat cultivars evaluated across two years and six environments; Table S3. Individual phenolic acids (µg/g dry matter) in six CIMMYT durum wheat cultivars evaluated across two years and six growing conditions; Table S4. Means of individual and total phenolic acids in six CIMMYT durum wheat cultivars evaluated across two years and six growing conditions. TPAs: total sum of individual phenolic acids. Different letters between columns indicate significant differences ($p < 0.05$).

Author Contributions: Conceptualization, B.L. and C.G.; methodology, B.L. and C.G.; validation, B.L. and C.G.; formal analysis, B.L., L.D. and C.G.; data curation, B.L., C.G. and J.C.; writing—original draft preparation, B.L., A.B. and C.G.; writing—review and editing, B.L., A.B., C.G., G.M., R.P.S., K.A. and J.C.; resources: R.P.S., B.L. and C.G.; supervision, B.L., C.G. and K.A. All authors have read and agreed to the published version of the manuscript.

Funding: We greatly appreciate financial support from the CRP-Wheat program of CGIAR consortium. Carlos Guzman gratefully acknowledges the European Social Fund and the Spanish State Research Agency (Ministry of Science and Innovation) for financial funding through the Ramon y Cajal Program (RYC-2017-21891).

Institutional Review Board Statement: Not applicable.

Informed Consent Statement: Not applicable.

Data Availability Statement: The data is contained within the article.

Acknowledgments: We would like to acknowledge all CIMMYT scientists, field workers, and lab assistants who helped in the collection of the samples used in this study. Special thanks go to all the members of the CIMMYT Wheat Chemistry and Quality Laboratory. We also acknowledge Mateo Vargas (Universidad Autonoma de Chapingo, Texcoco, Edo. de Mexico, Mexico) for helping with the statistical analyses.

Conflicts of Interest: The authors declare no conflict of interest.

References

1. International Grains Council [IGC]. World Grain Statistics 2016. 2020. Available online: https://www.igc.int/en/subscriptions/subscription.aspx (accessed on 21 May 2020).
2. Ciudad-Mulero, M.; Barros, L.; Fernandes, Â.; Ferreira, I.C.; Callejo, M.J.; Matallana-González, M.C.; Fernández-Ruiz, V.; Morales, P.; Carrillo, J.M. Potential Health Claims of Durum and Bread Wheat Flours as Functional Ingredients. *Nutrients* **2020**, *12*, 504. [CrossRef] [PubMed]
3. Calabriso, N.; Massaro, M.; Scoditti, E.; Pasqualone, A.; Laddomada, B.; Carluccio, M.A. Phenolic extracts from whole wheat biofortified bread dampen overwhelming inflammatory response in human endothelial cells and monocytes: Major role of VCAM-1 and CXCL-10. *Eur. J. Nutr.* **2020**, *59*, 2603–2615. [CrossRef] [PubMed]
4. Li, L.; Shewry, P.R.; Ward, J.L. Phenolic Acids in Wheat Varieties in the HEALTHGRAIN Diversity Screen. *J. Agric. Food Chem.* **2008**, *56*, 9732–9739. [CrossRef] [PubMed]
5. De Leonardis, A.M.; Fragasso, M.; Beleggia, R.; Ficco, D.B.M.; De Vita, P.; Mastrangelo, A.M. Effects of Heat Stress on Metabolite ccumulation and Composition, and Nutritional Properties of Durum Wheat Grain. *Int. J. Mol. Sci.* **2015**, *16*, 30382–30404. [CrossRef]
6. Aberkane, H.; Belkadi, B.; Kehel, Z.; Filali-Maltouf, A.; Tahir, I.S.A.; Meheesi, S.; Amri, A. Assessment of drought and heat tol-erance of durum wheat lines derived from interspecific crosses using physiological parameters and stress indices. *Agronomy* **2021**, *11*, 695. [CrossRef]
7. Abbas, T.; Rizwan, M.; Ali, S.; Adrees, M.; Mahmood, A.; Zia-ur-Rehman, M.; Ibrahim, M.; Arshad, M.; Qayyum, M.F. Biochar application increased the growth and yield and reduced cadmium in drought stressed wheat grown in an aged contam-inated soil. *Ecotoxicol. Environ. Saf.* **2018**, *148*, 825–833. [CrossRef]
8. Raheem, A.; Shaposhnikov, A.; Belimov, A.A.; Dodd, I.C.; Ali, B. Auxin production by rhizobacteria was associated with im-proved yield of wheat (*Triticum aestivum* L.) under drought stress. *Arch Agron. Soil Sci.* **2018**, *64*, 574–587. [CrossRef]
9. Porter, J.R.; Gawith, M. Temperatures and the growth and development of wheat: A review. *Eur. J. Agron.* **1999**, *10*, 23–36. [CrossRef]
10. Poudel, P.B.; Poudel, M.R. Heat Sstress effects and tolerance in wheat: A review. *J. Biol. Today's World* **2020**, *9*, 217.
11. Sharma, A.; Shahzad, B.; Rehman, A.; Bhardwaj, R.; Landi, M.; Zheng, B. Response of Phenylpropanoid Pathway and the Role of Polyphenols in Plants under Abiotic Stress. *Molecules* **2019**, *24*, 2452. [CrossRef]

12. Sharma, P.; Jha, A.B.; Dubey, R.S.; Pessarakli, M. Reactive Oxygen Species, Oxidative Damage, and Antioxidative Defense Mechanism in Plants under Stressful Conditions. *J. Bot.* **2012**, *2012*, 1–26. [CrossRef]
13. Liu, R.H. Whole grain phytochemicals and health. *J. Cereal Sci.* **2007**, *46*, 207–219. [CrossRef]
14. Kikuzaki, H.; Hisamoto, M.; Hirose, K.; Akiyama, K.; Taniguchi, H. Antioxidant Properties of Ferulic Acid and Its Related Compounds. *J. Agric. Food Chem.* **2002**, *50*, 2161–2168. [CrossRef]
15. Laddomada, B.; Durante, M.; Minervini, F.; Garbetta, A.; Cardinali, A.; D'Antuono, I.; Caretto, S.; Blanco, A.; Mita, G. Phytochemical characterization and anti-inflammatory activity of extracts from the whole-meal flour of Italian durum wheat cultivars. *Int. J. Mol. Sci.* **2015**, *16*, 3512–3527. [CrossRef]
16. Lempereur, I.; Rouau, X.; Abecassis, J. Genetic and Agronomic Variation in Arabinoxylan and Ferulic Acid Contents of Durum Wheat (Triticum durumL.) Grain and Its Milling Fractions. *J. Cereal Sci.* **1997**, *25*, 103–110. [CrossRef]
17. Nigro, D.; Grausgruberb, H.; Guzmàn, C.; Laddomada, B. Phenolic compounds in wheat kernels: Genetic and genomic studies of biosynthesis and regulation. In *Wheat Quality for Improving Processing and Health*; Igrejas, G., Ikeda, T., Guzmàn, C., Eds.; Springer Sciences & Business Media: Dordrecht, The Netherlands, 2020; pp. 223–251. ISBN 978-3-030-34162-6.
18. Shewry, P.R. The HEALTHGRAIN programme opens new opportunities for improving wheat for nutrition and health. *BNF* **2009**, *34*, 225–231. [CrossRef]
19. Laddomada, B.; Durante, M.; Mangini, G.; D'Amico, L.; Lenucci, M.S.; Simeone, R.; Piarulli, L.; Mita, G.; Blanco, A. Genetic var-iation for phenolic acids concentration and composition in a tetraploid wheat (*Triticum turgidum* L.) collection. *Genet Resour. Crop Evol.* **2017**, *64*, 587–597. [CrossRef]
20. Rawat, N.; Laddomada, B.; Gill, B.S. Genomics of cereal-based functional foods. In *Cereal Genomics*, 2nd ed.; Varshney, R.K., Gupta, P.K., Eds.; Springer Sciences+Business Media: Dordrecht, The Netherlands, 2013; pp. 247–274. [CrossRef]
21. Nigro, D.; Laddomada, B.; Mita, G.; Blanco, E.; Colasuonno, P.; Simeone, R.; Gadaleta, A.; Pasqualone, A.; Blanco, A. Genome-wide association mapping of phenolic acids in durum wheat. *J. Cereal. Sci.* **2017**, *75*, 25–34. [CrossRef]
22. Branković, G.; Dragičević, V.; Dodig, D.; Zorić, M.; Knežević, D.; Žilić, S.; Denčić, S.; Šurlan, G. Genotype × Environment in-teraction for antioxidants and phytic acid contents in bread and durum wheat as influenced by climate. *Chil. J. Agric. Res.* **2015**, *75*, 139–146.
23. Shamloo, M.; Babawale, E.A.; Furtado, A.; Henry, R.J.; Eck, P.K.; Jones, P.J.H. Effects of genotype and temperature on accumulation of plant secondary metabolites in Canadian and Australian wheat grown under controlled environments. *Sci. Rep.* **2017**, *7*, 9133. [CrossRef]
24. Liu, H.; Bruce, D.R.; Sissons, M.; Able, A.J.; Able, J.A. Genotype-dependent changes in the phenolic content of durum under water-deficit stress. *Cereal Chem. J.* **2018**, *95*, 59–78. [CrossRef]
25. Guzmán, C.; Autrique, J.E.; Mondal, S.; Singh, R.P.; Govindan, V.; Morales-Dorantes, A.; Posadas-Romano, G.; Crossa, J.; Ammar, K.; Peña, R.J. Response to drought and heat stress on wheat quality, with special emphasis on bread-making quality, in durum wheat. *Field Crop. Res.* **2016**, *186*, 157–165. [CrossRef]
26. Hernandez-Espinosa, N.; Laddomada, B.; Payne, T.; Huerta-Espino, J.; Govindan, V.; Ammar, K.; Ibba, M.I.; Pasqualone, A.; Guzman, C. Nutritional quality characterization of a set of durum wheat landraces from Iran and Mexico. *LWT Food Sci. Technol.* **2020**, *124*, 109198. [CrossRef]
27. Cooper, M.; Delacy, I.H. Relationships among analytical methods used to study genotypic variation and geno-type-by-environment interaction in plant breeding multi environment experiments. *Theor. Appl. Genet* **1994**, *88*, 561–572. [CrossRef] [PubMed]
28. Shahid, M.; Saleem, M.F.; Anjum, S.A.; Shahid, M.; Afzal, I. Effect of terminal heat stress on proline, secondary metabolites and yield components of wheat (*Triticum aestivum* L.) genotypes. *Philipp. Agric. Sci.* **2017**, *100*, 278–286.
29. FAO. *The Impact of Disasters and Crises on Agriculture and Food Security*; FAO: Rome, Italy, 2021.
30. Fernandez-Orozco, R.; Li, L.; Harflett, C.; Shewry, P.R.; Ward, J.L. Effects of Environment and Genotype on Phenolic Acids in Wheat in the HEALTHGRAIN Diversity Screen. *J. Agric. Food Chem.* **2010**, *58*, 9341–9352. [CrossRef]
31. Manach, C.; Scalbert, A.; Morand, C.; Rémésy, C.; Jiménez, L. Polyphenols: Food sources and bioavailability. *Am. J. Clin. Nutr.* **2004**, *79*, 727–747. [CrossRef] [PubMed]
32. Andreasen, M.F.; Kroon, P.; Williamson, G.; Conesa, M.T.G. Intestinal release and uptake of phenolic antioxidant diferulic acids. *Free. Radic. Biol. Med.* **2001**, *31*, 304–314. [CrossRef]
33. Vitaglione, P.; Napolitano, A.; Fogliano, V. Cereal dietary fibre: A natural functional ingredient to deliver phenolic com-pounds into the gut. *Trends Food Sci. Tech.* **2008**, *19*, 451–463. [CrossRef]
34. Pandey, K.B.; Rizvi, S.I. Plant Polyphenols as Dietary Antioxidants in Human Health and Disease. *Oxidative Med. Cell. Longev.* **2009**, *2*, 270–278. [CrossRef]
35. Cardona, F.; Andres-Lacueva, C.; Tulipani, S.; Tinahones, F.J.; Queipo-Ortuño, M.I. Benefits of polyphenols on gut microbiota and implications in human health. *J. Nutr. Biochem.* **2013**, *24*, 1415–1422. [CrossRef] [PubMed]
36. Shewry, P.R.; Piironen, V.; Lampi, A.-M.; Edelmann, M.; Kariluoto, S.; Nurmi, T.; Fernandez-Orozco, R.; Ravel, C.; Charmet, G.; Andersson, A.A.M.; et al. The HEALTHGRAIN Wheat Diversity Screen: Effects of Genotype and Environment on Phytochemicals and Dietary Fiber Components. *J. Agric. Food Chem.* **2010**, *58*, 9291–9298. [CrossRef] [PubMed]

Article

Sorghum Phenolic Compounds Are Associated with Cell Growth Inhibition through Cell Cycle Arrest and Apoptosis in Human Hepatocarcinoma and Colorectal Adenocarcinoma Cells

Xi Chen [1,2], Jiamin Shen [1], Jingwen Xu [1,3], Thomas Herald [4], Dmitriy Smolensky [4], Ramasamy Perumal [5] and Weiqun Wang [1,*]

[1] Department of Food Nutrition Dietetics and Health, Kansas State University, Manhattan, KS 66506, USA; xchen@whpu.edu.cn (X.C.); jiamins@ksu.edu (J.S.); jw-xu@shou.edu.cn (J.X.)
[2] School of Food Science and Engineering, Wuhan Polytechnic University, Wuhan 430023, China
[3] College of Food Science and Technology, Shanghai Ocean University, Shanghai 201306, China
[4] USDA-ARS Grain Quality and Structure Research, 1515 College Ave, Manhattan, KS 66502, USA; Tom.Herald@ars.usda.gov (T.H.); dmiriy.smolensky@usda.gov (D.S.)
[5] Agricultural Research Center, Kansas State University, Hays, KS 67601, USA; perumal@ksu.edu
* Correspondence: wwang@ksu.edu; Tel.: +1-(785)-532-5508

Citation: Chen, X.; Shen, J.; Xu, J.; Herald, T.; Smolensky, D.; Perumal, R.; Wang, W. Sorghum Phenolic Compounds Are Associated with Cell Growth Inhibition through Cell Cycle Arrest and Apoptosis in Human Hepatocarcinoma and Colorectal Adenocarcinoma Cells. *Foods* 2021, 10, 993. https://doi.org/10.3390/foods10050993

Academic Editor: Stan Kubow

Received: 10 March 2021
Accepted: 29 April 2021
Published: 1 May 2021

Publisher's Note: MDPI stays neutral with regard to jurisdictional claims in published maps and institutional affiliations.

Abstract: Phenolic compounds in some specialty sorghums have been associated with cancer prevention. However, direct evidence and the underlying mechanisms for this are mostly unknown. In this study, phenolics were extracted from 13 selected sorghum accessions with black pericarp while F10000 hybrid with white pericarp was used as a control, and cell growth inhibition was studied in hepatocarcinoma HepG2 and colorectal adenocarcinoma Caco-2 cells. Total phenolic contents of the 13 high phenolic grains, as determined by Folin–Ciocalteu, were 30–64 mg GAE/g DW in the phenolic extracts of various accessions compared with the control F10000 at 2 mg GAE/g DW. Treatment of HepG2 with the extracted phenolics at 0–200 μM GAE up to 72 h resulted in a dose- and time-dependent reduction in cell numbers. The values of IC_{50} varied from 85 to 221 mg DW/mL while the control of F10000 was 1275 mg DW/mL. The underlying mechanisms were further examined using the highest phenolic content of PI329694 and the lowest IC_{50} of PI570481, resulting in a non-cytotoxic decrease in cell number that was significantly correlated with increased cell cycle arrest at G2/M and apoptotic cells in both HepG2 and Caco-2 cells. Taken together, these results indicated, for the first time, that inhibition of either HepG2 or Caco-2 cell growth by phenolic extracts from 13 selected sorghum accessions was due to cytostatic and apoptotic but not cytotoxic mechanisms, suggesting some specialty sorghums are a valuable, functional food, providing sustainable phenolics for potential cancer prevention.

Keywords: sorghum; phenolic compounds; cell growth inhibition; cell cycle analysis; apoptosis; HepG2; Caco-2

1. Introduction

Comparable with other cereals, the nutritional value of sorghum has been thought to be less than it actually is for many decades. Reference [1] suggests that people still underestimate the nutritional advantages of sorghum. It is well-known that sorghum is a gluten-free food and suitable for people with celiac disease. Some specialty sorghums have recently gained particular interest because of their high level of phenolics and potential health benefits, especially in cancer prevention [2–5]. Previous investigation has shown that an average of 35% of overall human cancer mortality is related to diet [6]. Sorghum can be a part of a plant-based healthy diet as certain accessions of sorghum contain many more bioactive phenolics than other crops, such as wheat, barley, rice, maize, rye, and oats [7]. Numerous studies have reported the cancer-preventive effects of phenolic compounds in fruits and vegetables, although few focused on specialty sorghum whole grain [1,8].

Compelling data from epidemiological and animal studies have suggested that phenolic compounds could potentially contribute to anti-cancer effects through their biological properties including antioxidant activity, induction of cell cycle arrest and apoptosis, and promotion of tumor suppressor proteins, etc. [9–13]. There is plenty of literature on the properties of various phenolic-rich foods such as tea and red wine/grapes in relation to various types of cancer [8], while studies regarding the association between sorghum phenolic compounds and cancers are scarce. Epidemiological studies have reported that sorghum consumption consistently correlates with a low incidence of esophageal cancer in various parts of the world (including several parts of Africa, Russia, India, China, Iran, etc.), whereas wheat and corn consumption correlates with an elevated incidence of esophageal cancer [10–14]. In vivo studies regarding the anti-cancer effects of sorghum phenolics are reported even less. Lewis et al. [15] reported in 2008 that feeding normolipidemic rats a diet containing sorghum bran could significantly reduce the number of aberrant crypts in the rats. More recently, Park et al. [16] found that the metastasis of breast cancer to the lungs was blocked by sorghum extracts in the immune-deficient mouse metastasis model. However, the mechanisms by which sorghum reduced the risk of cancer are unclear. A few in vitro studies published recently, using sorghum extracts to treat several cancer cells, including leukemia (HL-60) [17], breast (MCF-7, MDA-MB 231) [18,19], colon (HT-29) [19], and liver (HepG2) [19] cells, found consistent results in the induction of cell apoptosis, inhibition of cell proliferation, and promotion of the expression of cell cycle regulators.

Given the high potential of the benefits of sorghum phenolics and the lag in research when compared to other plant foods, black sorghum with high phenolic compounds is a candidate that deserves systematic investigation. Therefore, the present study investigated the effects of phenolics extracted from 13 specialty sorghum accessions on cancer cell growth in both hepatocarcinoma HepG2 and colorectal adenocarcinoma Caco2 cell lines. The underlying mechanisms regarding cytotoxicity, cell cycle interruption, and apoptosis induction were further examined.

2. Materials and Methods

2.1. Sorghum Accessions

Thirteen specialty sorghum accessions with black pericarp were selected based upon their high levels of phenolic content, including PI152653, PI152687, PI193073, PI329694, PI559733, PI559855, PI568282, PI570366, PI570481, PI570484, PI570819, PI570889 and PI570993. Another sorghum accession, F10000 with white pericarp, was selected as a control as it contains low levels of phenolics. All the sorghum accessions were provided by the Kansas State University-Agricultural Research Center, Hays, KS, USA.

2.2. Reagents

Reagents including acetone, ethanol, Folin–Ciocalteu reagent, gallic acid, sodium carbonate, HyClone Dulbecco's Modified Eagle Medium (DMEM), fetal bovine serum, phosphate buffered saline (PBS), penicillin/streptomycin, trypsin-EDTA, propidium iodide (PI), and RNase were purchased from Fisher Scientific Co. L.L.C (Pittsburgh, PA, USA). CytoSelect™ LDH Cytotoxicity Assay Kit was purchased from the Cell Biolabs, Inc. (San Diego, CA, USA).

2.3. Phenolic Extraction

Sorghum flour from each accession at 0.2–0.5 g was extracted in 10 mL of 70% aqueous acetone (v/v) for 2 h while shaking at low speed using a 211DS shaking incubator (Labnet International Inc., Edison, NJ 08817, USA) at room temperature, followed by storage at $-20\,^{\circ}\text{C}$ in the dark overnight allowing the phenolics to be completely diffused from the cellular matrix into the solvent. The extract was then equilibrated at room temperature and centrifuged at $2970\times g$ for 10 min. The residue was rinsed with an additional 10 mL of solvents with 5 min of shaking and centrifuged at $2970\times g$ for another 10 min. Both supernatants were combined, and two aliquots were used for either total phenolic content determination or cell culture treatment. The aliquot for cell culture treatment was dried under a stream of nitrogen, and then dissolved in dimethyl sulfoxide (DMSO) to make a stock solution at $-20\,^{\circ}$C. Before its use for cell culture treatment, the stock solution was diluted with a fresh medium to achieve the desired concentration at 0–200 μM GAE. The final DMSO concentration in each treatment was kept at 0.1% *v/v*, which did not alter cell growth or cell cycle measurements significantly when compared with the DMSO-free medium. All extractions and treatments were conducted in triplicate.

2.4. Total Phenolic Content

Total phenolic content was determined using a Folin–Ciocalteu assay [20]. A stock solution of gallic acid at 1 mg/mL in distilled water was prepared and the final concentrations ranging from 12.5 to 200 μg/mL in 70% acetone were diluted for a standard curve. To each of the 96 wells, 75 μL distilled water was added, followed by 25 μL of either an aliquot of extracts or gallic acid as a standard at various concentrations. Folin–Ciocalteu reagent diluted 1:1 with distilled water was added to each well. The reaction was allowed to stand for 10 min at room temperature, and then 100 μL of Na_2CO_3 solution at 7.5% *(w/v)* was added to each well. The plate was covered and left to stand in the dark for 90 min before measuring. Absorbance was read using a microplate reader Synergy HT, BioTek with Gen5™ 2.0 data analysis software (Winnoski, VT, USA). Results were expressed as mg gallic acid equivalent (GAE) per g dry weight (DW).

2.5. Cell Culture

The human hepatocarcinoma HepG2 (HB-8065) and human colorectal adenocarcinoma Caco-2 (HTB-37) were purchased from the American Type Culture Collection, Manassas, VA 20108, USA. Cells were cultured in DMEM supplemented by 10% FBS, 100 μg/mL streptomycin, and 100 units/mL penicillin at $37\,^{\circ}$C in a 5% CO_2 humidified atmosphere. Cells in the exponential growth phase were used for all the experiments.

2.6. Cell Growth Inhibition Assay

Two milliliter cell suspensions (1×10^5 cells/mL) were seeded into 6-well plates and cultured in a humidified incubator to allow adhesion. Cells were then treated with extracted sorghum phenolics at 0–200 μM GAE for up to 72 h. After incubation with each treatment, cells at 24, 48 and 72 h were, respectively, detached by 0.05% trypsin-EDTA solution at $37\,^{\circ}$C and then suspended in PBS. The number of suspended cells was counted with a hemocytometer as described in our previous publication [21,22].

2.7. Cytotoxicity Assay

Cytotoxicity was assessed by lactate dehydrogenase (LDH) leakage into the culture medium. The activity of LDH in the medium was determined using a commercially available kit CytoSelect™ LDH Cytotoxicity Assay Kit from Cell Biolabs, Inc. (San Diego, CA 92126, USA). Cell suspension containing 0.1–1.0×10^6 cells/mL was seeded into a 96-well plate and cultured at $37\,^{\circ}$C and 5% CO_2 with sorghum extracts at 0–200 μM GAE. Negative control was applied using sterile water and positive control was applied by Triton X-100. Aliquots of media and reagents were mixed in a 96-well plate and incubated at $37\,^{\circ}$C for 0.5 h. Absorbance was recorded using a microplate reader SynergyHT, BioTek (Winnoski, VT, USA), and analyzed with Gen5™ 2.0 data analysis software. The % of relative cytotoxicity was calculated using the following equation

$$\frac{OD(experiment\ sample) - OD(negative\ control)}{OD(positive\ control) - OD(negative\ control)} \times 100 = \%\ \text{Relative Cytotoxicity}$$

2.8. Cell Cycle Analysis

Cell cycle analysis was conducted according to our previous publication [21]. After treatment, cells were detached and fixed in 70% ethanol at $4\,^{\circ}$C. Cells were then resuspended in 20 µg/mL of propidium iodide (PI) staining solution with 5 U/mL RNase at 37 °C for 15 min before analysis by a flow cytometry (LSRFortessa X-20 and FACSCalibur, BD, Franklin Lakes, NJ, USA) with excitation at 488 nm and emission at 617 nm.

2.9. Apoptosis Analysis

Apoptosis analysis was conducted according to our previous publication [22]. Briefly, treated cells were collected and fixed by 1% paraformaldehyde and 70% ice cold ethanol at a concentration of 1–2 × 10^6 cells/mL. The cells were stored at −20 °C for several days. Fixed cells were analyzed for apoptosis by FITC annexin V staining protocol according to commercial instructions (BioLegend Inc., San Diego, CA, USA).

2.10. Statistics

Data were analyzed by the SAS statistical system (version 9.2). The significance of the trend for cell growth inhibition at various concentrations for three exposure times was analyzed with linear regression. The effect of extracted phenolic at various concentrations for 72 h on cell cycle and apoptosis was analyzed by one-way ANOVA with Tukey adjustment. Pearson correlation coefficients (r) were used to analyze the relationships between total phenolic contents and IC_{50} values for all sorghum accessions, as well as cell numbers and percentages of cell arrest at G2/M or apoptosis.

3. Results

3.1. Total Phenolic Content

As shown in Table 1, total phenolic contents of the 13 specialty sorghum accessions ranged from 31 to 63.7 mg GAE/g DW, with the highest being PI329694 and the lowest being the control F10000.

Table 1. Phenolic Contents in Various Specialty Sorghum Accessions (mean ± SD, *n* = 3).

Sorghum Accession No.	Phenolic Contents (mg GAE/g DW)
F10000 (control)	2.3 ± 0.2
PI 559855	31.0 ± 0.2
PI 152687	44.6 ± 2.9
PI 570819	48.2 ± 4.4
PI 570993	51.0 ± 1.9
PI 559733	51.5 ± 1.4
PI 570366	54.3 ± 1.7
PI 152653	54.6 ± 1.2
PI 570484	54.8 ± 0.8
PI 193073	55.1 ± 4.5
PI 570889	58.0 ± 2.0
PI 568282	58.3 ± 2.5
PI 570481	58.5 ± 2.5
PI 329694	63.7 ± 2.5

3.2. Cell Growth Inhibition

As shown in Figure 1, treatment of HepG2 with the extracted phenolics at 0–200 µM GAE up to 72 h resulted in a dose- and time-dependent reduction in cell number. The values of IC_{50} varied from 85 to 221 mg DW/mL, while the control of F10000 was 1275 mg DW/mL (Table 2). The lowest IC_{50} value was observed in the phenolic extract from PI570481, suggesting PI570481 to be the most potent in suppressing HepG2 cell growth.

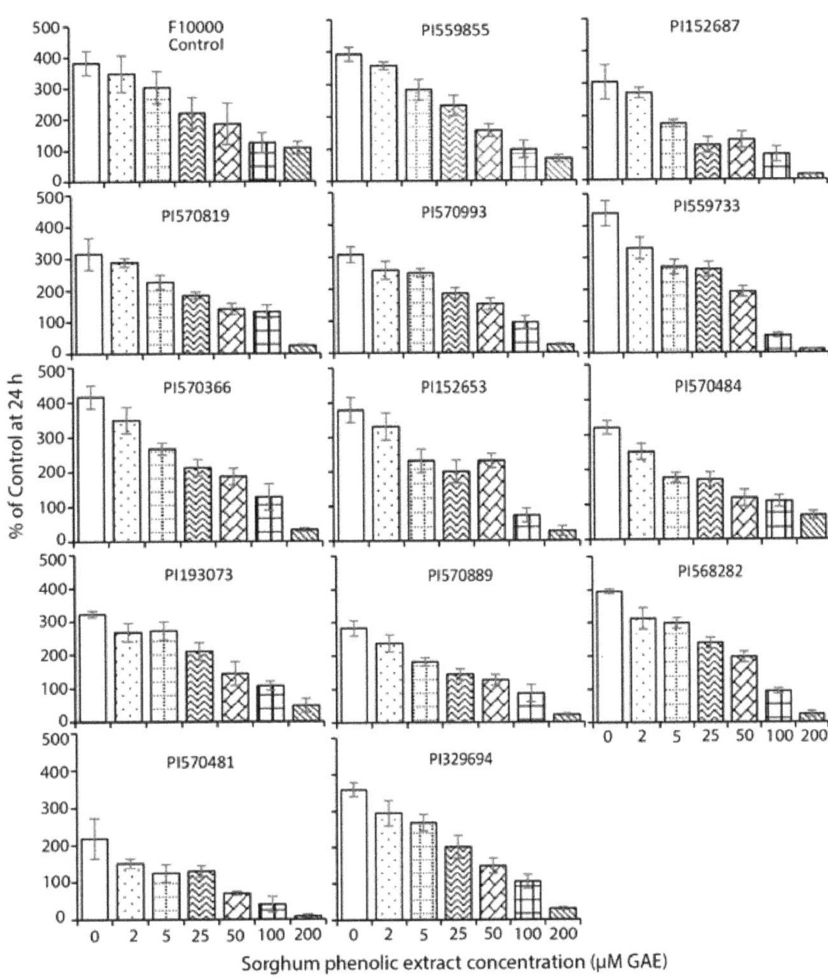

Figure 1. The effect of various sorghum phenolic extracts at 0–200 μM GAE for up to 72 h on cell growth in HepG2 cells. HepG2 cells were cultured with phenolic extracts at various concentrations (0–200 μM GAE) for up to 72 h in 6-well plates, then detached by trypsin-EDTA solution, and the cell number was counted by a hemacytometer. Values are expressed as Mean ± SD ($n = 3$). The significance of the trend for cell growth inhibition at various concentrations was analyzed with linear regression, $p < 0.05$.

Table 2. IC$_{50}$ Values of specialty sorghum phenolic extracts in HepG2 and Caco-2 cells.

Sorghum Accession No.	IC$_{50}$ [a] (mg DW/mL)	
	HepG2	Caco-2
F10000 (control)	1275.6	1131.3
PI 559855	221.8	
PI 152687	138.9	
PI 570819	192.1	
PI 570993	146.2	
PI 559733	90.8	
PI 570366	120.9	
PI 152653	127.9	
PI 570484	158.0	
PI 193073	177.3	
PI 570889	120.9	
PI 568282	113.3	
PI 570481	85.8	115.6
PI 329694	126.8	102.4

[a] Means $\pm$ SD, $n = 3$.

As PI329694 and PI570481 had the highest phenolic content and the lowest IC$_{50}$ value, respectively, extracts of these were selected for both HepG2 and Caco-2 treatments, resulting in a similar dose- and time-dependent reduction in cell number (Figure 2). Both IC$_{50}$ values were much lower than that of the control F10000 (Table 2).

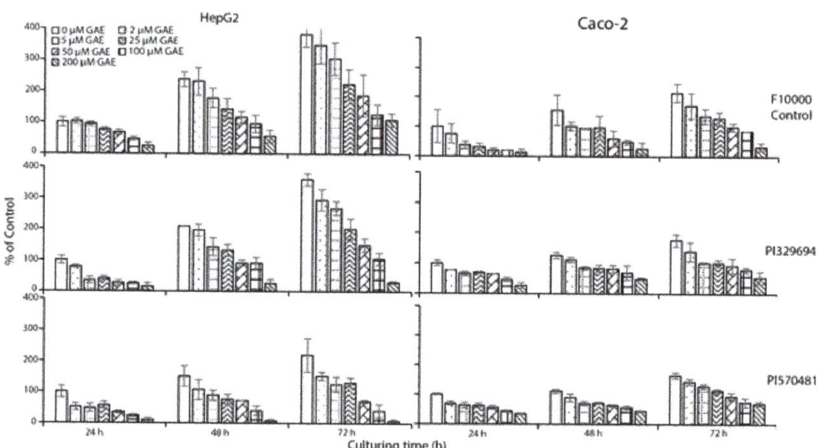

Figure 2. The effect of representative sorghum phenolic extracts at 0–200 µM GAE for up to 72 h on cell number in both HepG2 and Caco-2 cells. Cells were treated with sorghum phenolics extracted from the representative sorghum accessions PI329694 and PI570481 at various concentrations (0–200 µM GAE) for up to 72 h in 6-well plates, then detached by trypsin-EDTA solution, and the cell number was counted by hemacytometer at each timepoint. Values are expressed as a percentage of the untreated control at 24 h by Mean $\pm$ SD ($n = 3$). The significance of the trend for cell growth inhibition at various concentrations at each timepoint was analyzed with linear regression, $p < 0.05$.

3.3. Cytotoxicity Assay

Cell viability, as assessed by lactate dehydrogenase leakage, was generally greater than 85% in the adherent cells, and the treated cells did not differ significantly from the vehicle-treated control (data not shown).

3.4. Cell Cycle Arrest

The treatment of either HepG2 or Caco-2 cells with the high phenolic concentrations extracted from either PI329694 or PI570481 significantly induced cell cycle arrest at G2/M phase (Figure 3). The percentage of cells at G1 phase decreased correspondingly, while the proportion of cells at S phase was not significantly altered. When compared with the control F10000, which showed a similar induction, this was actually much less potent due to there being over 20-times less phenolic content.

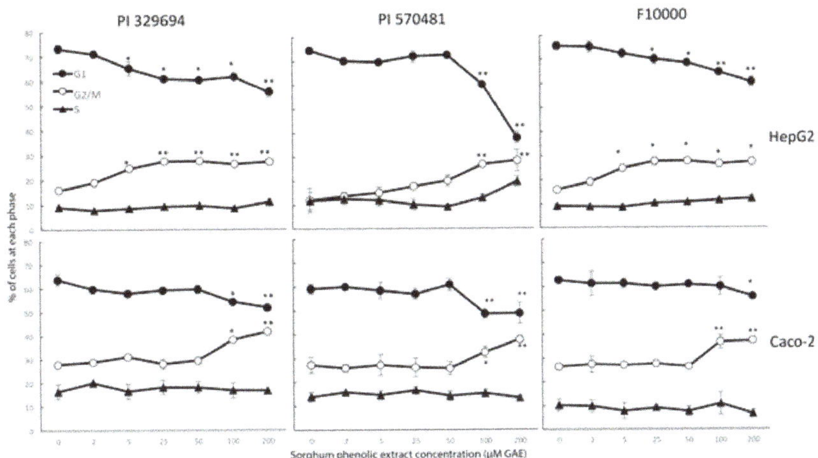

Figure 3. Cytostatic effect of representative sorghum phenolic extracts in HepG2 and Caco-2 cell lines. Cells were treated with sorghum phenolics extracted from the representative sorghum accessions PI329694 and PI570481 at 0–200 μM GAE for up to 72 h, and then cell cycle was monitored by a DNA flow cytometric analysis. Values are expressed as Mean ± SD ($n = 3$), * $p < 0.05$, ** $p < 0.01$ versus the vehicle control.

3.5. Apoptosis

As shown in Figure 4, the treatment of either HepG2 or Caco-2 cells with the phenolic extracts at high concentrations from PI329694 or positive control F10000 resulted in a significant increase in the percentage of apoptotic cells. A significant induction by the extract of PI329694 was also observed when compared with the control F10000.

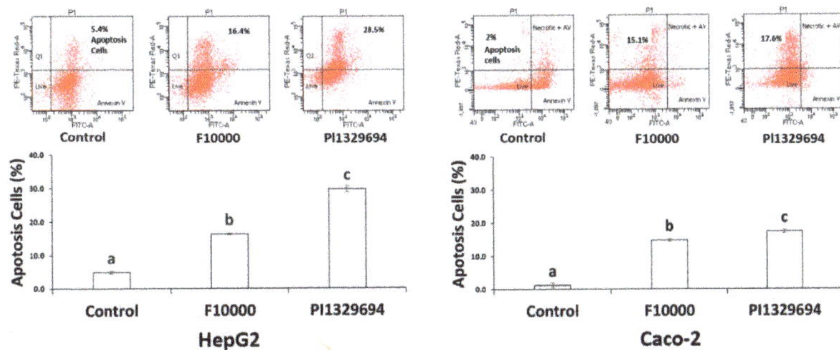

Figure 4. Apoptosis induced by representative sorghum phenolic extracts in HepG2 and Caco-2 cell lines. Cells were treated with sorghum phenolics extracted from the representative sorghum accessions PI329694 and F10000 at 0–200 μM GAE for up to 72 h, and then apoptosis was analyzed by FITC annexin V staining protocol. Values are expressed as Mean ± SD ($n = 3$), Means with different alphabetical letters differ significantly, $p \leq 0.05$.

3.6. Correlation Coefficient

As shown in Table 3, a significant inverse correlation was observed between total phenolic contents and IC_{50} values in the extracts of all accessions ($r = -0.6806$, $p < 0.05$), indicating that the cell growth inhibition by the sorghum phenolic extracts was significantly associated with the phenolic content. Meanwhile, significant inverse correlations were observed between the decrease in cell number and the increase in G2/M and apoptotic cells, suggesting that the decrease in cell number by sorghum phenolic extracts was associated with cell cycle arrest and apoptosis induction.

Table 3. Correlation coefficient (r) between phenolic contents and IC_{50} values from the phenolic extracts of 13 sorghum accessions in HepG2 cells, or between cell number and cell cycle arrest or apoptosis from control, PI329694 and PI570481 in both cell lines.

	IC_{50}		
Total phenolics	−0.6806 *		
		Cell Arrest at G2/M	**Apoptosis Cells**
HepG2			
F10000 control		−0.8608 *	−0.9377 **
PI329694	**Cell number**	−0.8281 *	−0.9764 **
PI570481		−0.9469 **	
Caco-2			
F10000 control		−0.7599 *	−0.9719 **
PI329694	**Cell number**	−0.7655 *	−0.8199 *
PI570481		−0.6840 *	

* $p < 0.05$, ** $p < 0.01$.

4. Discussion

Thirteen specialty sorghum accessions were selected based on their high levels of phenolics when compared with the control accession F10000, which contained the fewest phenolics. Two cell lines derived from liver and colorectal cancer were used as the liver is the major site for the metabolism of dietary compounds including phenolic compounds and the intestine is the major site for the absorption of phenolic compounds. To our knowledge, this is the first study to compare the cellular impact of thirteen different sorghum phenolic extracts on both liver and intestine cancer cells, which may provide a better understanding of the anti-cancer activities of sorghum phenolics with a long-term goal of promoting sorghum products as phenolic-sustainable functional foods for health benefits.

In the present study, total phenolic contents, determined by Folin–Ciocalteu assay, were 30–64 mg GAE/g DW in the selected 13 sorghum accessions (Table 1) while the control contained only 2.3 ± 0.2 mg GAE/g DW. This result is inconsistent with our previous study [13,23].

Treatment of sorghum phenolics at various concentrations for up to 72 h resulted in a dose- and time-dependent inhibition of cell growth in both HepG2 and Caco-2 cells (Figures 1 and 2). The values of IC_{50} varied from 85.8 to 221.8 mg DW/mL, with sorghum accession PI570481 being the lowest when compared with the control that showed the highest value (Table 2). Similar results were also found in Caco-2 cells when treated with various concentrations of phenolic extracts from PI329694 or PI570481 (Table 2 and Figure 2). These results indicate that, the more sorghum accessions are rich in phenolic compounds, the more they effectively suppress cancer cell growth. This result is in agreement with our previous study conducted in sorghum phenolics in HepG2 and Canc-2 cells [13,23].

To investigate the underlying mechanisms of cancer cell growth inhibition by sorghum phenolics, cytotoxicity was assessed for each concentration of phenolic extracts from various sorghum accessions in both cell lines, resulting in a difference in cytotoxicity of less than 15% between the treated and non-treated cells. It should be noted that the cytotoxicity detected did not differ significantly from the vehicle-treated control, suggesting sorghum phenolic extracts are non-toxic to the cells. In addition, the treatment of both cell lines by the phenolics extracted from PI329694 or PI570481 showed an induction of cell cycle arrest at G2/M phase (Figure 3) and apoptotic cells (Figure 4). These results indicate that the inhibition of cell growth by sorghum phenolic compounds appears to be cytostatic but not cytotoxic. The cell cycle arrest induced by sorghum phenolics may trigger the DNA repair machine, leading to apoptosis, as confirmed by our apoptosis analysis, suggesting that treatment of both cell lines with sorghum phenolic extracts significantly induced the apoptotic cells compared with the vehicle control.

Coefficient correlations were further calculated between total phenolic content and IC_{50} values, cell number and cell cycle arrest at G2/M phase, and cell number and proportion of apoptotic cells. Significant inverse correlations observed between total phenolic contents and IC_{50} values in thirteen sorghum accessions (Table 3) suggest that the cell growth inhibition by sorghum extracts is directly associated with their phenolic content. As strong significant inverse correlations were also observed between cell number and the accumulation of cell cycle arrest at G2/M phase and apoptosis in both cell lines (Table 3), the inhibitory effect of sorghum phenolic compounds on cell growth appears to be through a cytostatic mechanism.

5. Conclusions

In conclusion, the present study showed that phenolic extracts in various sorghum accessions effectively inhibited HepG2 or Caco-2 cancer cell growth in a dose- and time-dependent manner. The cell growth inhibition by the sorghum phenolic extracts was significantly associated with their phenolic content. Furthermore, the inhibition appeared to be mediated by cytostatic and apoptotic mechanisms rather than cytotoxicity. Taken together, this study is the first to investigate and compare the anti-cancer effect of sorghum phenolic compounds in both HepG2 and Caco-2 cell lines. The results suggest that sorghum is a valuable food and crop with health benefits to provide sustainable phenolics for potential cancer prevention.

Author Contributions: Conceptualization, T.H., D.S. and W.W.; methodology, X.C., J.S., J.X., D.S. and W.W.; validation, D.S. and W.W.; formal analysis, X.C. and J.S.; resources, R.P.; data curation, X.C., J.S., J.X. and W.W.; writing and original draft preparation, X.C.; writing, review and editing, T.H., D.S. and W.W.; supervision, T.H., D.S. and W.W.; project administration, T.H., D.S. and W.W.; funding acquisition, T.H., D.S. and W.W. All authors have read and agreed to the published version of the manuscript.

Funding: This research was supported in part by grants from USDA Cooperative Award 58-3020-5-016, KS511-1001903 (Contribution #17-359-J from the Kansas Agricultural Experiment Station), and a Doctoral Dissertation Research Award from College of Human Ecology, Kansas State University.

Data Availability Statement: The data is contained within the article.

Conflicts of Interest: The authors declare no conflict of interest.

Disclaimer: Mention of trade names or commercial products in this publication is solely for the purpose of providing specific information and does not imply recommendation or endorsement by the U.S. Department of Agriculture. USDA is an equal opportunity provider and employer.

References

1. Stefoska-Needham, A.; Beck, E.J.; Johnson, S.K.; Tapsell, L.C. Sorghum: An Underutilized Cereal Whole Grain with the Potential to Assist in the Prevention of Chronic Disease. *Food Rev. Int.* **2015**, *31*, 401–437. [CrossRef]
2. Hwang, K.; Weller, C.; Cuppett, S.L.; Hanna, M. Policosanol Contents and Composition of Grain Sorghum Kernels and Dried Distillers Grains. *Cereal Chem.* **2004**, *81*, 345–349. [CrossRef]
3. Su, X.; Rhodes, D.; Xu, J.; Chen, X.; Davis, H.; Wang, D.; Herald, T.J.; Wang, W. Phenotypic Diversity of Anthocyanins in Sorghum Accessions with Various Pericarp Pigments. *J. Nutr. Food Sci.* **2017**, *7*, 1000610. [CrossRef]
4. Shen, Y.; Su, X.; Rhodes, D.; Herald, T.; Xu, J.; Chen, X.; Smith, J.S.; Wang, W. The pigments of sorghum pericarp are associated with the contents of carotenoids and pro-vitamin A. *Int. J. Food Nutr. Sci.* **2017**, *6*, 48–56.

5. Davis, H.; Su, X.; Shen, Y.; Xu, J.; Wang, D.; Smith, J.S.; Aramouni, F.; Wang, W. Phenotypic diversity of colored phytochemicals in sorghum accessions with various pericarp pigments. In *Polyphenols in Plants*, 2nd ed.; Watson, R.R., Ed.; Academic Press: New York, NY, USA, 2019; pp. 123–131.

6. Doll, R.; Peto, R. The causes of cancer: Quantitative estimates of avoidable risks of cancer in the United States today. *J. Natl. Cancer Inst.* **1981**, *66*, 1191–1308. [CrossRef]

7. Ragaee, S.; Abdel-Aal, E.-S.M.; Noaman, M. Antioxidant activity and nutrient composition of selected cereals for food use. *Food Chem.* **2006**, *98*, 32–38. [CrossRef]

8. Awika, J.M.; Rooney, L.W. Sorghum phytochemicals and their potential impact on human health. *Phytochemistry* **2004**, *65*, 1199–1221. [CrossRef]

9. Wang, W.; Goodman, M.T. Antioxidant properties of dietary phenolic agents in a human LDL-oxidation ex vivo model: Interaction of protein binding activity. *Nutr. Res.* **1999**, *19*, 191–202. [CrossRef]

10. Van Rensburg, S.J. Epidemiologic and dietary evidence for a specific nutritional predisposition to esophageal cancer. *J. Natl. Cancer Inst.* **1981**, *67*, 243–251. [PubMed]

11. Chen, F.; Cole, P.; Mi, Z.; Xing, L.Y. Corn and wheat-flour consumption and mortality from esophageal cancer in shanxi, China. *Int. J. Cancer* **1993**, *53*, 902–906. [CrossRef]

12. Xu, S.; Shen, Y.; Xu, J.; Qi, G.; Chen, G.; Wang, W.; Sun, X.; Li, Y. Antioxidant and anticancer effects in human hepatocarcinoma (HepG2) cells of papain-hydrolyzed sorghum kafirin hydrolysates. *J. Funct. Foods* **2019**, *58*, 374–382. [CrossRef]

13. Cox, S.; Noronha, L.; Herald, T.; Bean, S.; Lee, S.H.; Perumal, R.; Wang, W.; Smolensky, D. Evaluation of ethanol-based extraction conditions of sorghum bran bioactive compounds with downstream anti-proliferative properties in human cancer cells. *Heliyon* **2019**, *5*, e01589. [CrossRef] [PubMed]

14. Isaacson, C. The change of the staple diet of black South Africans from sorghum to maize (corn) is the cause of the epidemic of squamous carcinoma of the oesophagus. *Med. Hypotheses* **2005**, *64*, 658–660. [CrossRef]

15. Birt, D.F.; Hendrich, S.; Wang, W. Dietary agents in cancer prevention: Flavonoids and isoflavonoids. *Pharmacol. Ther.* **2001**, *90*, 157–177. [CrossRef]

16. Park, J.H.; Darvin, P.; Lim, E.J.; Joung, Y.H.; Hong, D.Y.; Park, E.U.; Park, S.H.; Choi, S.K.; Moon, E.S.; Cho, B.W.; et al. Hwanggeumchal sorghum induces cell cycle arrest, and suppresses tumor growth and metastasis through Jak2/STAT pathways in breast cancer xenografts. *PLoS ONE* **2012**, *7*, e40531. [CrossRef] [PubMed]

17. Woo, H.J.; Oh, I.T.; Lee, J.Y.; Jun, D.Y.; Seu, M.C.; Woo, K.S.; Nam, M.H.; Kim, Y.H. Apigeninidin induces apoptosis through activation of Bak and Bax and subsequent mediation of mitochondrial damage in human promyelocytic leukemia HL-60 cells. *Process Biochem.* **2012**, *47*, 1861–1871. [CrossRef]

18. Devi, P.S.; Kumar, M.S.; Das, S.M. Evaluation of antiproliferative activity of red sorghum bran anthocyanin on a human breast cancer cell line (mcf-7). *Int. J. Breast Cancer* **2011**, *2011*, 891481. [CrossRef]

19. Devi, P.S.; Saravanakumar, M.; Mohandas, S. Identification of 3- deoxyanthocyanins from red sorghum (Sorghum bicolor) bran and its biological properties. *Afr. J. Pure Appl. Chem.* **2011**, *5*, 181–193.

20. Singleton, V.L.; Rossi, J.A. Colorimetry of total phenolics with phosphomolybdic-phosphotungstic acid reagents. *Am. J. Enol. Vitic.* **1965**, *16*, 144–158.

21. Ayella, A.; Lim, S.; Jiang, Y.; Iwamoto, T.; Lin, D.; Tomich, J.; Wang, W. Cytostatic inhibition of cancer cell growth by lignan secoisolariciresinol diglucoside. *Nutr. Res.* **2010**, *30*, 762–769. [CrossRef]

22. Qu, H.; Madl, R.L.; Takemoto, D.J.; Baybutt, R.C. Lignans Are Involved in the Antitumor Activity of Wheat Bran in Colon Cancer SW480 Cells. *J. Nutr.* **2005**, *135*, 598–602. [CrossRef] [PubMed]

23. Smolensky, D.; Rhodes, D.; McVey, D.S.; Fawver, Z.; Perumal, R.; Herald, T.; Noronha, L. High-Polyphenol Sorghum Bran Extract Inhibits Cancer Cell Growth Through ROS Induction, Cell Cycle Arrest, and Apoptosis. *J. Med. Food* **2018**, *21*, 990–998. [CrossRef] [PubMed]

 *foods*

Article

Allelic Variation at Glutenin Loci (*Glu-1*, *Glu-2* and *Glu-3*) in a Worldwide Durum Wheat Collection and Its Effect on Quality Attributes

Pablo F. Roncallo [1], Carlos Guzmán [2], Adelina O. Larsen [3], Ana L. Achilli [1], Susanne Dreisigacker [4], Elena Molfese [5], Valentina Astiz [6] and Viviana Echenique [1,*]

1   Centro de Recursos Naturales Renovables de la Zona Semiárida (CERZOS), Departamento de Agronomía, Universidad Nacional del Sur (UNS)-CONICET, Buenos Aires 8000, Argentina; roncallo@cerzos-conicet.gob.ar (P.F.R.); alachilli@cerzos-conicet.gob.ar (A.L.A.)
2   Departamento de Genética, Escuela Técnica Superior de Ingeniería Agronómica y de Montes, Edificio Gregor Mendel, Campus de Rabanales, Universidad de Córdoba, CeiA3, ES-14071 Córdoba, Spain; carlos.guzman@uco.es
3   Compañía Molinera del Sur, Buenos Aires 8000, Argentina; larsen.adelina@gmail.com
4   Global Wheat Program, International Maize and Wheat Improvement Center (CIMMYT), El Batán, Texcoco 56237, Edo. Mexico, Mexico; S.Dreisigacker@cgiar.org
5   CEI Barrow, Laboratorio de Calidad Industrial de Granos (Convenio INTA-Ministerio de Desarrollo Agrario), Buenos Aires 7500, Argentina; molfese.elenarosa@inta.gob.ar
6   EEA Cesareo Naredo, Instituto Nacional de Tecnología Agropecuaria (INTA), Buenos Aires 6417, Argentina; astiz.valentina@inta.gob.ar
*   Correspondence: echeniq@cerzos-conicet.gob.ar

Citation: Roncallo, P.F.; Guzmán, C.; Larsen, A.O.; Achilli, A.L.; Dreisigacker, S.; Molfese, E.; Astiz, V.; Echenique, V. Allelic Variation at Glutenin Loci (*Glu-1*, *Glu-2* and *Glu-3*) in a Worldwide Durum Wheat Collection and Its Effect on Quality Attributes. *Foods* **2021**, *10*, 2845. https://doi.org/10.3390/foods10112845

Academic Editor: Stefania Masci

Received: 21 September 2021
Accepted: 10 November 2021
Published: 18 November 2021

Abstract: Durum wheat grains (*Triticum turgidum* L. ssp. *durum*) are the main source for the production of pasta, bread and a variety of products consumed worldwide. The quality of pasta is mainly defined by the rheological properties of gluten, an elastic network in wheat endosperms formed of gliadins and glutenins. In this study, the allelic variation at five glutenin loci was analysed in 196 durum wheat genotypes. Two loci (*Glu-A1* and *Glu-B1*), encoding for high-molecular-weight glutenin subunits (HMW-GS), and three loci (*Glu-B2*, *Glu-A3* and *Glu-B3*), encoding for low molecular weight glutenin subunits (LMW-GS), were assessed by SDS-PAGE. The SDS-sedimentation test was used and the grain protein content was evaluated. A total of 32 glutenin subunits and 41 glutenin haplotypes were identified. Four novel alleles were detected. Fifteen haplotypes represented 85.7% of glutenin loci variability. Some haplotypes carrying the 7 + 15 and 7 + 22 banding patterns at *Glu-B1* showed a high gluten strength similar to those that carried the 7 + 8 or 6 + 8 alleles. A decreasing trend in grain protein content was observed over the last 85 years. Allelic frequencies at the three main loci (*Glu-B1*, *Glu-A3* and *Glu-B3*) changed over the 1915–2020 period. Gluten strength increased from 1970 to 2020 coinciding with the allelic changes observed. These results offer valuable information for glutenin haplotype-based selection for use in breeding programs.

Keywords: durum wheat; glutenins; gluten strength; grain protein content; haplotypes; SNPs

1. Introduction

Durum wheat grains (*Triticum turgidum* L. ssp. *durum* Desf. Husn) are widely consumed all around the world as an important part of the diet in several countries. Historically, it was used worldwide to make pasta and is also the main source for making different products consumed in the Mediterranean basin, in particular flat and leavened bread, couscous and bulgur, as well as freekeh in the WANA (West Asia and North Africa) region [1]. Pasta is also an important part of the food produced and consumed in Latin America. In particular, in Argentina, the consumption of pasta is 8.54 kg per capita per year, the seventh-highest in the world [2], with a durum wheat growing area of 129,255 ha during the crop season 2020/21, which is comparatively smaller than the area occupied by other cereals and oil crops, although it represents the greatest planted area in Latin America. Until the 1970s, Argentina exported high-quality durum wheat mainly to Italy [3]. However, grain quality decreased during the green revolution with the introduction of semi-dwarf wheat from CIMMYT, mainly due to an increase in *Fusarium* susceptibility, which caused a reduction in both the production area and exports.

The level and composition of protein in gluten is the reason why durum wheat is preferred for making pasta [4–6], since the gluten quality strongly affects the firmness of pasta after cooking [7,8]. Moreover, consumers prefer pasta products with a strong yellow colour and firmness, resulting in tasty and nutritionally superior food [9]. Gliadins and glutenins are the major storage proteins in wheat endosperm; they form an elastic network called gluten and can be distinguished based on their solubility or insolubility in aqueous alcohols, respectively [10]. The level [11] and composition [12,13] of protein are directly associated with the quality of wheat-derived products. Gliadins are responsible for gluten viscosity and extensibility and are encoded by two loci (*Gli-1* and *Gli-2*) located on the short arms of group 1 (-γ and -ω types) and 6 (-α and -β types) chromosomes [14,15]. Glutenins are associated with the viscoelastic properties of gluten [16]. Two groups of single monomeric glutenins were separated according to their mobility in SDS-PAGE, and were classified as high- (HMW-GS) and low-molecular-weight glutenin subunits (LMW-GS) controlled by orthologous genes located on the long and short arms of the group 1 chromosomes, respectively [17,18]. LMW-GS are encoded by *Glu-2* and *Glu-3* genes [18,19], whereas HMW-GS are encoded by *Glu-1* genes [20]. It was suggested that *Glu-A1* and *Glu-B1* in tetraploid wheat contained sequences encoding x- and y-type subunits as a result of an ancestral duplication [21]. In cultivated durum wheat the *Glu-A1* y subunit is always inactive.

The association between the γ-gliadin 45 and gluten strength was reported in durum wheat [22]. Later, it was demonstrated that linked gene/s coding for low-molecular-weight (LMW) glutenin subunits were mostly responsible for the changes in quality [23]. The LMW subunits are classified according to two models (1 and 2), associated with low and high quality, respectively [24]. The variability and role of high-molecular-weight (HMW) glutenin subunits were examined in wild emmer [25,26], durum [27] and hexaploid wheat subspecies [28,29].

In Argentina, the genetic variability in durum wheat storage proteins was previously evaluated [30,31] but only using a limited number of genotypes. However, a wider screening of modern and historical Argentinian genotypes, as well as worldwide accessions, could help to identify favourable alleles or allele combinations with greater effect on quality. For this purpose, we used a collection composed of South American and Mediterranean germplasms, also including CIMMYT/ICARDA genotypes or derivatives [32]. This collection was previously evaluated for genetic diversity, population structure and linkage disequilibrium patterns, exhibiting high genetic differences between subpopulations, making it possible to trace the origin of the South American germplasm.

The characterisation and availability of glutenin allelic composition could be beneficial to implement its use in marker-assisted selection (MAS) in breeding programs. In the present study, we assessed (i) the allelic variations in five glutenin genes using a worldwide durum wheat collection composed mainly of Argentinian genotypes [32], (ii) the effect of *Glu* haplotypes or individual allelic variants on gluten strength and protein content, (iii) the association of predictive traits and rheological parameters for different glutenin haplotypes using historical datasets, (iv) the changes in gluten composition due to breeding activities between periods or origins. In addition, a detailed description of LMW subunits, not previously available for Argentinian durum wheat, was undertaken.

2. Materials and Methods

2.1. Plant Material

The total evaluated plant material consisted of 196 durum wheat accessions (*Triticum turgidum* L. ssp. *durum*), mostly representative of the Argentinian breeding programs (85), but also including accessions from Italy (33), Chile (26), France (21), West Africa and North Asia (WANA) (17), CIMMYT (10) and the USA (4) (Supplementary Table S1).

2.2. Glutenin Characterisation and Haplotype Analysis

The genetic variability of high- (HMW) and low-molecular-weight (LMW) glutenin loci was evaluated in 196 durum wheat accessions using the sodium dodecyl sulphate (one dimensional SDS-PAGE) methodology, according to the protocol proposed by Peña et al. [33]. For this purpose, 20 mg of whole meal flour was incubated with 0.75 mL of 50% propanol (v/v) for 30 min in a Thermomixer (Eppendorf, Germany) at 1400 rpm and 65 °C to extract the gliadins, as described in Maryami et al. [34]. Specifically, tubes were centrifuged for 2 min at 10,000 rpm and the supernatant, including the gliadins fraction, was discarded. This step was repeated to remove any remaining gliadins. Subsequently, 100 uL of a solution with DTT at 1.5% (w/v) with 50 uL of propanol at 50% (v/v) and 50 uL of Tris-HCl 0.08 M pH 8.0 was added to the pellet. Tubes were mixed with vortex and incubated for 30 min in a Thermomixer at 1400 rpm and 65 °C. After that, the tubes were centrifuged for 2 min at 10,000 rpm with 100 uL of a solution of vinylpyridine at 1.4% (v/v), prepared with 50 uL of propanol at 50% (v/v), and 50 uL Tris-HCl 0.08 M pH 8.0, was added. Tubes were mixed again with vortex and incubated in a Thermomixer at 1400 rpm and 65 °C for 15 min. Then, the tubes were centrifuged for 2 min at 13,000 rpm. The supernatant was transferred to a new tube and 180 uL of a solution of Tris-HCl M pH 6.8, 2% SDS (w/v), 40% glycerol (w/v), and 0.02% (w/v) bromophenol blue was added. Tubes were mixed with vortex and incubated for 5 min in a Thermomixer at 1400 rpm at 90 °C, and then were centrifuged for 2 min at 13,000 rpm. From the supernatant, 6 uL was taken and used to run the gels. Separating gels with a concentration of 15% of acrylamide were prepared using 1 M Tris buffer at pH of 8.0 instead of the conventional 8.8. Gels were run at 12.5 mA per gel for 20 h and then stained using Coomasie blue. Five different loci, two encoding HMW glutenins (*Glu-A1*, *Glu-B1*), and three encoding LMW glutenins (*Glu-A3*, *Glu-B3* and *Glu-B2*) were characterised. Allele designations were made following previously proposed rules [35,36]. LMW-GS model classified the allelic combinations as 1 or 2 at *Glu-3* and *Glu-2* loci associated with low and intermediate/high gluten quality, respectively. Glutenin loci haplotypes were constructed considering all possible allele combinations for these five loci and named as Hap_[number].

2.3. KASP Markers

Additionally, two KASP markers for the *Gpc-B1* (6BS) locus affecting grain protein content (SNP marker: wMAS000017) and the glutenin locus *Glu-A1* (1AL) (SNP marker: Glu-Ax1/x2*_SNP) were assessed in the entire collection [37]. Allele-specific primers were designed using FAM and VIC tails and are shown in Supplementary Table S2. Three-week-old seedlings obtained from purified seeds were used for DNA extraction following a CTAB protocol [38]. For PCR, a touchdown protocol was used started with a 15 min hot enzyme activation at 94 °C followed by 11 cycles of 94° for 30 s, 65–55 °C for 60 s (-0.8 °C/cycle), 72 °C for 30 s. This was continued with 26 cycles of 94 °C for 30 s, 57 °C for 60 s, 72 °C for 30 s, and a final step at 10 °C. PCR was carried out using 5 µL of volume per well arrayed in a 384 PCR plate. DNA samples were briefly centrifuged and oven-dried at 60 °C for 1 h. SNP-specific KASP reagents, such as the assay mix and the 2X KASP Master mix, including the fluorescent dyes FAM and VIC, were added to dried DNA samples (150 ng/well). PCR-amplified products were subjected to an end-point fluorescence reading using the PHERAstar Plus plate reader from BMG LABTECH. Alleles were assigned based on the differential fluorescence reading using Excel software.

2.4. Field Experiments

Seven field trials were conducted during three growing seasons (three in 2011/12, three in 2014/15 and one in 2017/18) at three locations (Cabildo [39°36′ S, 61°64′ W], Barrow [38°20′ S, 60°13′ W] and Pieres [37°46′ S, 58°18′ W]) under rainfed conditions.

From the total genotyped accessions (196), 132 and 170 were phenotyped in successive years according to the population size availability in 2011 (132), 2014 (170) and 2017 (170). Plots were arranged using an alpha-lattice design [39] with two repetitions (incomplete blocks, 17 × 8 in 2011 and 10 × 17 in 2014 and 2017). Seed density was adjusted to achieve 300 viable plants per m^2. Plot size and growing conditions for each experiment are detailed in Supplementary Table S3. Trials were kept free of weeds and diseases and managed according to local practices.

Additional experiments from the National Durum Wheat Experimental Network (NDWEN) or from the Instituto Nacional de Tecnología Agropecuaria (INTA) breeding program, listed in Supplementary Table S4, were used. From these trials, Argentinian genotypes with contrasting glutenin haplotypes were selected. Unbalanced, historical quality datasets were used to assess the glutenin haplotype effect.

2.5. Phenotypic Evaluations

Whole-meal flour from each plot was obtained by milling 20 g samples of clean grains in a Cyclotec® 1093 mill (Foss Denmark). The grain protein content (GPC, %, adjusted to 13.5% H°) was determined by near-infrared spectroscopy (NIR Systems DS2500, Foss Denmark). Gluten strength was assessed using the sodium dodecyl sulfate (SDSS) micro-sedimentation test (mm) (adapted from AACC Method 56–70, 1984; for Argentinian strong gluten wheat). The GPC was obtained from each plot in all the trials, while SDSS was only assessed in the 2011 and 2014 field trials.

In the NDWEN or the INTA trials additional quality parameters, which were not evaluated in the collection, were considered. The percentage of gluten, wet and dry gluten, and the gluten index (GI) were obtained according to the IRAM 15864 (Part 2) method and the ICC standard 155 (International Association for Cereal Science and Technology, 1986) method using the Glutomatic system (Perten Instruments, Sweden). Farinograph parameters (Development time and Energy level) were assessed according to Irvine et al. [40] with modifications. A 50 g dough mixer, with constant water hydration (45%) and a fixed mixing time (8 min) were used. Development time (%) = peak height (min)-end height (min)/peak height (min) × 100. Energy level = (peak height (FU)/20) + Area (cm^2) [41].

2.6. Statistical Analyses

Genotype-adjusted means (LSMEANs) were obtained using a mixed linear model (PROC MIXED) with SAS 9.0® software [42], within a maximum restricted probability model (REML). The analysis was performed annually and by location for each trait. For ANOVA and LSMEAN, performed annually, the locations and incomplete blocks within repetitions were considered as random factors and the genotype and repetitions within environment as fixed factors. Analysis of variance (ANOVA) was performed for each trait considering the period, glutenin markers or haplotypes as fixed factors using the INFOSTAT software [43]. The Tukey–Kramer [44] test was implemented to evaluate allele ranking at each locus, while the BSS test [45,46] was applied to analyse the haplotype rank. This test proposed a recursive algorithm based on the combination of a clustering technique and a hierarchical analysis of variance. The combination of the analysis by year and by location was referred to as the environment. For the breeding period effect, the Duncan test ($p < 0.05$) was used.

3. Results

One hundred and ninety-six durum wheat genotypes from different origins and periods, including a large set of Argentinian genotypes, three landraces and historical genotypes, were characterised for their glutenin subunit banding patterns by using SDS-PAGE. A total of 32 glutenin subunits from five evaluated loci were identified in the collection (Table 1). The combination of these allelic variants formed a total of 41 glutenin haplotypes (Table 2).

Table 1. Allele frequency at five glutenin loci (HMW-GS and LMW-GS) detected in a worldwide collection of 196 durum wheat.

		Argentina (85)		Chile (26) 134		CIMMYT (10)		France (21)		Italy (33)		USA (4)		WANA (17)	
Locus/Banding Pattern	**Allele**	N	%	N	%	N	%	N	%	N	%	N	%	N	%

3.1. High-Molecular-Weight Glutenin Subunits (HMW-GS)

The highest variation between genes encoding for HMW-GS was exhibited by *Glu-B1*, contributing ten alleles to the collection, whereas, for *Glu-A1*, only two alleles were detected (2* (b) and null (c)). The 2* allele at *Glu-A1* was only found in one Argentinian breeding line (CBW 0105), also confirmed by using the KASP marker designed for the *Glu-A1* locus (Supplementary Table S2). At *Glu-B1*, the 7 + 8 (b), 6 + 8 (d) and 20x + 20y (e) alleles covered most of the variation (89.8%) with percentages of 38.3, 26.0 and 25.5, respectively. The frequency of the remaining seven alleles was lower than 4%, as the 7 + 15 (z) subunits detected in seven modern cultivars or breeding lines. The 13 + 16 (f) subunits were only observed in one old Argentinian (Bonaerense 202), one French (Neodur) and three Italians genotypes (Supplementary Table S1). Some specific allelic variants at *Glu-B1* were only detected in Argentinian (subunit 6), Italian (14 + 22* and 6 + 20y) and WANA (6* + 15*) germplasms (Figure 1). The combinations of glutenin subunits 14 + 22*, 6* + 15*, and 6+20y were not previously described and should be considered as novel alleles. We propose to name these alleles *Glu-B1cr*, *Glu-B1cs*, and *Glu-B1ct*, respectively, following the order of the Wheat Gene Catalogue and the latest publications on this topic [34].

Table 2. Frequency and allele composition of haplotypes for HMW-GS and LMW-GS detected in a worldwide collection of durum wheat.

	Loci							
Haplotype	*Glu-A1*	*Glu-B3*	*Glu-B2*	*Glu-B1*	*Glu-A3*	N	Frequency (%)	Quality Traits Assessed [1]
Hap_1	Ax2 *	2 + 4 + 15 + 19	12	6 + 8	6	1	0.5	yes
Hap_2	null	1 + 3 + 14 + 18	12 *	14 + 22 *	null	1	0.5	yes
Hap_3	null	2 + 4 + 15 + 16	12	20x + 20y	6 + 10	1	0.5	yes
Hap_4	null	2 + 4 + 15 + 16	12	6 * + 15 *	6	1	0.5	yes
Hap_5	null	2 + 4 + 15 + 16	12	20x + 20y	6	1	0.5	yes
Hap_6	null	2 + 4 + 15 + 16	null	6 + 8	6	1	0.5	no
Hap_7	null	2 + 4 + 15 + 18	12	20x + 20y	11	1	0.5	yes
Hap_8	null	2 + 4 + 15 + 18	12	7 + 15	null	1	0.5	yes
Hap_9	null	2 + 4 + 15 + 19	12	13 + 16	6	2	1.0	yes
Hap_10	null	2 + 4 + 15 + 19	12	13 + 16	6 + 10	1	0.5	yes
Hap_11	null	2 + 4 + 15 + 19	12	20x + 20y	6	25	12.8	yes
Hap_12	null	2 + 4 + 15 + 19	12	20x + 20y	6.1 + 10	3	1.5	yes
Hap_13	null	2 + 4 + 15 + 19	12	20x + 20y	6 + 10	11	5.6	yes
Hap_14	null	2 + 4 + 15 + 19	12	20x + 20y	null	3	1.5	yes
Hap_15	null	2 + 4 + 15 + 19	12	6 + 20y	6	1	0.5	no
Hap_16	null	2 + 4 + 15 + 19	12	6 + 8	6	25	12.8	yes
Hap_17	null	2 + 4 + 15 + 19	12	6 + 8	6 + 10	20	10.2	yes
Hap_18	null	2 + 4 + 15 + 19	12	7 + 15	11	1	0.5	yes
Hap_19	null	2 + 4 + 15 + 19	12	7 + 15	6	5	2.6	yes
Hap_20	null	2 + 4 + 15 + 19	12	7 + 22	6	2	1.0	yes
Hap_21	null	2 + 4 + 15 + 19	12	7 + 22	null	2	1.0	yes
Hap_22	null	2 + 4 + 15 + 19	12	7 + 8	6	44	22.4	yes
Hap_23	null	2 + 4 + 15 + 19	12	7 + 8	6 + 10	18	9.2	yes
Hap_24	null	2 + 4 + 15 + 19	12	7 + 8	6 + 11	4	2.0	yes
Hap_25	null	2 + 4 + 15 + 19	12	7 + 8	null	2	1.0	no
Hap_26	null	2 + 4 + 15 + 19	null	13 + 16	11	1	0.5	yes
Hap_27	null	2 + 4 + 15 + 19	null	20x + 20y	6	1	0.5	yes
Hap_28	null	2 + 4 + 15 + 19	null	20x + 20y	6 + 10	1	0.5	yes
Hap_29	null	2 + 4 + 15 + 19	null	20x + 20y	null	1	0.5	yes
Hap_30	null	2 + 4 + 15 + 19	null	6 + 8	6	1	0.5	no
Hap_31	null	2 + 4 + 15 + 19	null	7 + 8	10 + 11	1	0.5	yes
Hap_32	null	2 + 4 + 16	12	13 + 16	6	1	0.5	no
Hap_33	null	2 + 4 + 8 + 9 + 15 + 19	12	6	6	1	0.5	no
Hap_34	null	2 + 4 + 8 + 9 + 15 + 19	12	6 + 8	null	2	1.0	no
Hap_35	null	8 + 9 + 13 + 16	12	20x + 20y	5	1	0.5	yes
Hap_36	null	8 + 9 + 13 + 16	12	7 + 8	5	3	1.5	yes
Hap_37	null	8 + 9 + 13 + 16	12 *	7 + 8	5	1	0.5	yes
Hap_38	null	8 + 9 + 13 + 16	null	20x + 20y	5	1	0.5	yes
Hap_39	null	8 + 9 + 13 + 16	null	6 + 8	5	1	0.5	yes
Hap_40	null	8 + 9 + 13 + 16	null	6 + 8	5 + 10	1	0.5	yes
Hap_41	null	null	null	7 + 8	6	1	0.5	yes
Total (N or %)	2	8	3	10	9	196	100.0	

[1] Haplotypes reported without quality trait evaluations because they were not among the genotypes assessed in the 2011, 2014 and 2017 field trials. * corresponds with bands showing a slightly mobility difference in SDS-PAGE respect to the original band with the same number.

3.2. Low-Molecular-Weight Glutenin Subunits (LMW-GS)

The highest variation between LMW-GS coding genes was observed at the *Glu-A3* locus with nine alleles, followed by *Glu-B3* and *Glu-B2*, which presented eight and three alleles, respectively (Table 1). The most frequent glutenin subunits encoded by *Glu-A3* were 6 (a) and 6 + 10 (c), with frequencies of 57.7 and 26.5 percent, respectively, followed by null (h) and 5 (b) subunits (Table 1). The *Glu-A3ax* (6.1 subunit) allele was detected privatively in genotypes from Argentina (Buck Cristal and two breeding lines, VF0154 and VF042), but our study conformed a new banding pattern (6.1 + 10). In addition, 10 + 11 and 5 + 10 banding patterns were only detected in the Biensur and Langdon (Dic-3A)-10 cultivars, respectively (Supplementary Table S1). At the *Glu-B3* locus, the *Glu-B3a* (2 + 4 + 15 + 19) allele was present in approximately 90% of the genotypes, followed by 8 + 9 + 13 + 16 and 2 + 4 + 15 + 16 subunit combinations. A new banding pattern (2 + 4 + 8 + 9 + 15 + 19) was detected in three Argentinian breeding lines (Figure 1) and a rare variant (2 + 4 + 16) was obtained in the old cultivar Bonaerense 202 (Supplementary Table S1). The analysis at *Glu-B2* resulted in the detection of three alleles (a, b and c), subunit 12 (a) being present in 93.4% of the genotypes. The 12* allele (c) was the least frequent, only detected in two genotypes from Italy and the WANA region, while the null (b) allele was carried by 11 genotypes and it was more widely distributed among the origins than *Glu-B2c* (Table 1).

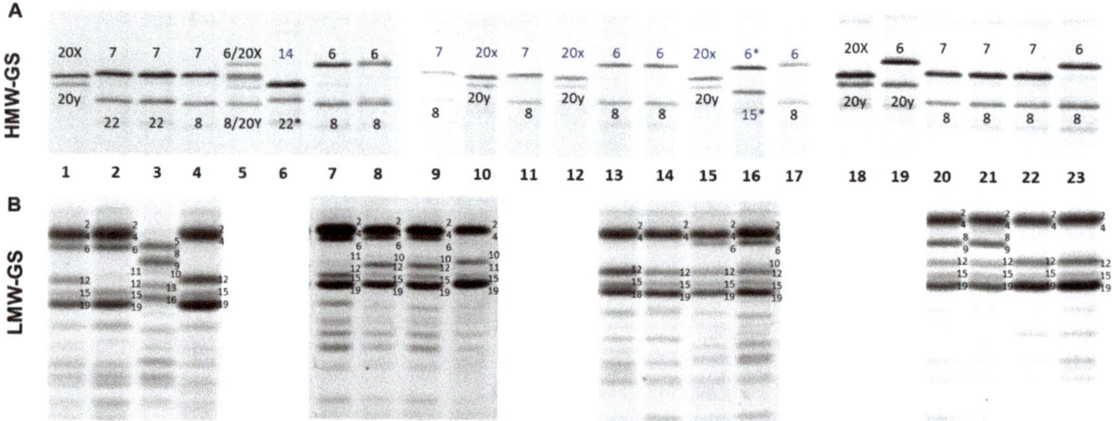

Figure 1. Representative variability of HMW-GS (**A**) and LMW-GS (**B**) found in the evaluated durum wheat collection. Genotypes are as follows: (**A**), 1, Candeal Durumbuck; 2, CRZ-1.12; 3, GAB 125; 4, Gerardo 574; 5, Maristella; 6, Polesine; 7, Buck#33 (33.1123.16-3-4-3); 8, Langdon; 9, Coccorit; 10, Gan; 11, Cham 1 = Waha; 12, Korifla = Cham 3; 13, Focha; 14, Bha; 15, Buck No6; 16, Haurani; 17, Langdon; 18, Cappelli; 19, Capeiti8; 20, Chagual INIA; 21, ACA 5284.06; 22, ACA 3571.13; 23, ACA 3576.13. (**B**), 1, Quc 3462-2009; 2, Quc 3763-2008; 3, Langdon (Dic-3A)-10; 4, CBW 05024; 7, Dupri; 8, Durobonus; 9, Joyau; 10, Biensur; 13, Ardente; 14, Appullo; 15, Ixos; 16, Buck Granate; 20, CBW 0416; 21, CBW 09161; 22, CBW 09270; 23, CBW 09280.

3.3. Effect of HMW-GS and LMW-GS on Gluten Strength and Protein Content

The relevance of glutenin alleles for improving gluten strength (measured by the SDSS test) and grain protein content (GPC) was assessed in samples from a total of seven field trials (Tables 3 and 4). ANOVA was conducted, taking into account the characterised alleles, in 132 (2011) and 170 (2014–2017) genotypes phenotypically evaluated for quality traits. Data for the genotypes numbered from 171 to 196 are not available since they were added to the collection after the field trials were conducted. For HMW-GS, only the alleles at the *Glu-B1* locus were associated with highly significant differences ($p < 0.001$) in the SDSS test and significant differences in GPC (in four out of seven trials). This locus explained the 36–41% and 18–21% of the variation in the SDSS test during 2011 and 2014, respectively

(Supplementary Table S5). The Glu-A1 locus was practically fixed with only one genotype carrying a differential allele, and it became difficult to test for differences.

Considering LMW-GS, all three loci (*Glu-A3*, *Glu-B3* and *Glu-B2*) showed significant differences between alleles in the SDSS test values in all the evaluated experiments. According to the ANOVA test, the SDSS variance was explained in the following order *Glu-B3* > *Glu-A3* > *Glu-B2* in both years. In addition, the *Glu-A3* locus also significantly affected the grain protein content in BW 2011, PS 2014 and BW 2017, whereas the *Glu-B3* showed a significant effect on the GPC in BW 2011, CA 2014, PS 2014 and BW 2017.

For all loci, the SDSS variance was explained significantly in the following order: *Glu-B1* > *Glu-B3* > *Glu-A3* > *Glu-B2*, in 2011, and *Glu-B3* > *Glu-A3* > *Glu-B1* > *Glu-B2*, in 2014 (Supplementary Table S5). After ANOVA, the rank of alleles was established using the Tukey–Kramer test ($p < 0.01$). Banding patterns 7 + 8, 6 + 8, 7 + 15, 7 + 22 and 6* + 15* at the *Glu-B1* locus were associated with the highest mean values of the SDSS test (Tables 3 and 4). Additionally, the 13 + 16 allele, carried mainly by European genotypes, reached high SDSS test values in several environments, whereas the 20x + 20y banding pattern showed intermediate values in both evaluated years. In addition, the 14 + 22* banding pattern was associated with a detrimental effect on the SDSS test value, but a high GPC value. On the other hand, the 6, 11, 6 + 11 and 6 + 10 banding patterns at the *Glu-A3* locus were associated with high SDSS test values, their effects being significantly different to other alleles in most of the environments. The worst performance was exhibited by genotypes carrying the subunit 5, followed by the 5 + 10 and 6.1 + 10 banding patterns.

At *Glu-B3*, the 2 + 4 + 15 + 19 (allele a) banding pattern was responsible for significantly increased SDSS test values. On the contrary, genotypes carrying the 8 + 9 + 13 + 16 and 1 + 3 + 14 + 18 banding patterns showed the lowest SDSS test values. A null allele (j) at the *Glu-B3* locus, not previously described in durum wheat, was identified in the Chilean breeding line Quc 3506-2009. This allele was associated with a high SDSS test value. Regarding the effect of the *Glu-B2* alleles on the SDSS test, it was 12 (a) > null (b) > 12* (c), although significant differences were only detected between 12* and null, or 12* and 12 subunits.

Table 3. The effect of glutenin alleles on grain protein content and the SDSS test in 132 durum wheat genotypes grown in Argentina (2011).

		2011		GPC [1]				SDSS Test			
Locus/Banding Pattern	Allele	N	%	CA 2011	BW 2011	PS 2011	LSmean	CA 2011	BW 2011	PS 2011	LSmean
HMW glutenin subunits											
Glu-A1 (2)											
2 *	b	1	0.8	13.92 a	12.29 a	11.5 a	12.58 a	84.50 a	86.66 a	68.51 a	80.17 a
null	c	131	99.2	14.62 a	13.39 a	11.95 a	13.32 a	62.11 a	66.51 a	54.16 a	60.92 a
Glu-B1 (10)											
7 + 8	b	42	31.8	14.49 a	13.28 a	11.93 a	13.23 a	67.6 b	71.18 b	61.94 b	66.34 b
6 + 8	d	33	25.0	14.81 a	13.32 ab	11.85 a	13.32 a	71.97 b	77.19 b	62.58 b	70.63 b
20x + 20y	e	43	32.6	14.59 a	13.43 ab	11.96 a	13.33 a	48.69 ab	52.66 ab	38.87 ab	46.71 ab
13 + 16	f	4	3.0	14.48 a	13.45 ab	11.78 a	13.23 a	61.0 b	74.37 b	54.24 ab	63.34 b
7 + 15	z	4	3.0	14.57 a	13.49 ab	12.15 a	13.41 a	73.13 b	77.85 b	69.14 b	73.38 b
6	an	na	na	na	na	na	na	na	na	na	na
7 + 22	ch	4	3.0	14.54 a	13.84 ab	12.35 a	13.59 a	69.0 b	71.62 b	60.43 b	66.92 b
14 + 22 *	new (cr)	1	0.8	14.99 a	14.78 b	13.18 a	14.3 a	24.5 a	19.61 a	20.6 a	21.83 a
6 * + 15 *	new (cs)	1	0.8	14.99 a	14.17 ab	12.47 a	13.9 a	80.5 b	88.57 b	70.82 b	79.83 b
6 + 20y	new (ct)	na	na	na	na	na	na	na	na	na	na
LMW glutenin subunits											
Glu-A3 (9)											
6	a	74	56.1	14.65 a	13.41 ab	11.99 a	13.35 a	64.72 b	68.53 b	56.55 ab	63.28 b
5	b	6	4.5	14.84 a	13.85 ab	12.35 a	13.67 a	22.0 a	22.36 a	18.25 a	20.94 a
6 + 10	c	37	28.0	14.46 a	13.21 ab	11.8 a	13.15 a	64.31 b	70.92 b	56.27 ab	63.83 b
6 + 11	d	1	0.8	15.09 a	13.25 ab	11.94 a	13.4 a	56.5 ab	75.98 b	52.84 ab	61.67 b
11	e	3	2.3	14.6 a	13.23 ab	11.88 a	13.24 a	69.8 b	71.37 b	52.37 ab	64.61 b
null	h	7	5.3	14.94 a	14.14 b	12.4 a	13.84 a	57.0 ab	60.11 ab	48.49 ab	55.0 ab
6.1 + 10	ax	3	2.3	14.09 a	12.64 a	11.0 a	12.57 a	59.2 ab	57.3 ab	54.02 ab	56.83 ab
5 + 10		na	na	na	na	na	na	na	na	na	na
10 + 11		1	0.8	14.34 a	13.01 ab	11.9 a	13.12 a	78.0 b	86.87 b	75.79 b	80.17 b
Glu-B3 (8)											
2 + 4 + 15 + 19	a	120	90.9	14.59 a	13.34 a	11.9 a	13.28 a	64.9 c	69.6 b	56.79 a	63.78 b
8 + 9 + 13 + 16	b	6	4.5	14.84 a	13.85 ab	12.35 a	13.67 a	22.0 a	22.36 a	18.25 a	20.94 a
2 + 4 + 15 + 16	g	3	2.3	14.41 a	13.69 ab	12.47 a	13.53 a	51.8 abc	52.75 ab	45.42 a	49.78 ab
1 + 3 + 14 + 18	h	1	0.8	14.99 a	14.78 b	13.18 a	14.3 a	24.5 ab	15.5 a	20.6 a	21.83 a
2 + 4 + 15 + 18	ax	2	1.5	15.34 a	13.54 a	12.35 a	13.75 a	58.3 bc	67.07 b	41.03 a	55.25 ab
2 + 4 + 16	ag	na	na	na	na	na	na	na	na	na	na
2 + 4 + 8 + 9 + 15 + 19	new (aw)	na	na	na	na	na	na	na	na	na	na
Null	j	na	na	na	na	na	na	na	na	na	na
Glu-B2 (3)											
12	a	125	94.7	14.60 a	13.37 a	11.92 a	13.3 a	63.76 b	68.35 b	55.48 b	62.52 b
null	b	5	3.8	14.86 a	13.51 a	12.23 a	13.55 a	42.8 ab	44.22 ab	39.5 ab	42.23 ab
12 *	c	2	1.5	14.47 a	14.08 a	12.84 a	13.78 a	18.75 a	16.02 a	15.50 a	17.17 b

Table 4. The effect of glutenin alleles on grain protein content and the SDSS test in 170 genotypes grown in Argentina (2014 and 2017).

Locus/Banding Pattern	2014/17			GPC [1]					SDSS Test			
	Allele	N	%	CA 2014	BW 2014	PS 2014	BW 2017	LSmean	CA 2014	BW 2014	PS 2014	LSmean
HMW glutenin subunits												
Glu-A1 (2)												
2 *	b	1	0.6	13.2 a	10.61 a	11.54 a	10.51 a	11.64 a	72.08 a	68.99 a	59.08 a	66.84 a
null	c	161	94.7	12.98 a	12.2 a	12.37 a	11.14 a	12.16 a	58.81 a	66.38 a	66.28 a	63.83 a
Glu-B1 (10)												
7 + 8	b	65	38.2	12.49 ab	12.31 a	12.31 a	10.93 a	12.0 a	61.99 b	73.03 b	71.16 b	68.69 b
6 + 8	d	44	25.9	13.68 ab	11.94 a	12.34 a	11.23 ab	12.29 a	64.82 b	66.39 ab	67.48 b	66.31 b
20x + 20y	e	47	27.6	13.08 ab	12.26 a	12.41 a	11.3 ab	12.27 a	48.12 ab	55.19 ab	56.46 ab	53.27 ab
13 + 16	f	4	2.4	11.73 ab	11.6 a	11.98 a	11.23 ab	11.59 a	59.34 ab	77.87 b	78.02 b	71.61 b
7 + 15	z	4	2.4	12.32 ab	11.81 a	12.42 a	10.94 a	11.9 a	61.59 b	69.14 b	78.82 b	69.63 b
6	an	na	na									
7 + 22	ch	4	2.4	14.04 ab	12.58 a	12.33 a	11.26 ab	12.48 ab	74.9 b	80.61 b	72.01 b	76.15 b
14 + 22 *	new (cr)	1	0.6	15.38 b	12.5 a	14.82 b	12.7 b	14.02 b	18.41 a	21.97 a	21.28 a	20.31 a
6 * + 15 *	new (cs)	1	0.6	11.43 a	13.31 a	13.67 b	11.82 ab	12.35 a	68.67 b	87.02 b	72.0 b	75.48 b
6 + 20y	new (ct)	na	na									
LMW glutenin subunits												
Glu-A3 (9)												
6	a	98	57.6	12.98 a	12.23 a	12.39 ab	11.11 ab	12.07 a	61.92 abc	69.18 bc	70.11 b	67.10 c
5	b	6	3.5	13.06 a	13.12 a	13.05 ab	11.69 ab	12.77 a	18.72 a	19.9 a	23.6 a	20.66 a
6 + 10	c	48	28.2	12.89 a	12.08 a	12.23 ab	11.08 ab	12.07 a	58.11 abc	68.68 bc	64.69 ab	63.81 bc
6 + 11	d	2	1.2	13.80 a	11.97 a	12.44 ab	10.73 ab	12.24 a	67.58 bc	73.09 c	60.76 ab	67.15 c
11	e	3	1.8	11.92 a	11.77 a	12.03 ab	11.42 ab	11.78 a	61.25 abc	67.0 abc	72.51 b	66.55 c
null	h	8	4.7	13.35 a	12.15 a	12.65 ab	11.44 ab	12.43 a	57.36 abc	64.48 abc	64.27 ab	62.03 bc
6.1 + 10	ax	3	1.8	13.86 a	11.29 a	11.15 a	10.8 ab	11.72 a	51.01 abc	42.33 abc	58.57 ab	50.7 abc
5 + 10		1	0.6	14.24 a	11.99 a	11.96 ab	12.3 ab	12.64 a	31.45 ab	21.96 ab	23.78 a	25.97 ab
10 + 11		1	0.6	11.79 a	12.33 a	13.29 b	10.42 a	11.69 a	78.4 c	80.56 c	90.74 b	83.22 c
Glu-B3 (8)												
2 + 4 + 15 + 19	a	156	117.3	12.99 a	12.15 a	12.32 a	11.10 ab	12.13 a	61.17 ab	68.92 bc	68.42 b	66.18 b
8 + 9 + 13 + 16	b	7	5.3	13.23 a	12.96 a	12.98 ab	11.78 ab	12.75 ab	20.54 a	20.20 a	23.63 a	21.42 a
2 + 4 + 15 + 16	g	3	2.3	11.96 a	12.32 a	12.26 a	11.32 ab	11.95 a	49 ab	67.52 abc	58.82 ab	58.59 ab
1 + 3 + 14 + 18	h	1	0.8	15.38 b	12.5 a	14.82 b	12.70 b	14.02 b	18.41 a	21.97 ab	21.88 a	20.31 a
2 + 4 + 15 + 18	ax	2	1.5	10.92 a	12.54 a	12.81 ab	10.88 ab	11.96 a	45.83 ab	44.0 abc	70.91 b	53.48 ab
2 + 4 + 16	ag	na	na									
2 + 4 + 8 + 9 + 15 + 19	new (aw)	na	na									
Null	j	1	0.8	14.61 ab	10.97 a	12.11 a	10.60 a	12.06 a	66.5 b	82.04 c	81.69 b	76.94 b
Glu-B2 (3)												
12	a	159	93.5	12.96 a	12.19 a	12.33 a	11.11 a	12.14 a	60.06 b	68.02 b	67.54 b	65.22 b
null	b	9	5.3	13.16 a	11.96 a	12.44 a	11.36 a	12.25 a	48.00 b	48.89 b	54.30 b	50.31 b
12 *	c	2	1.2	14.08 a	12.88 a	14.48 b	12.07 a	13.43 b	14.65 a	16.43 a	16.66 a	15.52 a
LMW-GS Model												
1		7	4.1	13.23 a	12.96 b	12.98 a	11.78 a	12.74 a	20.54 a	20.20 a	23.63 a	21.42 a
2		162	95.3	12.96 a	12.16 ab	12.34 a	11.11 a	12.14 a	60.5 b	68.3 b	67.98 b	65.6 b
none		1	0.588	14.61 a	10.88 a	12.11 a	10.60 a	12.06 a	66.5 b	82.04 b	81.68 b	76.94 b

[1] LSMEAN estimated using 132 genotypes in each environment and over the environments. GPC: grain protein content; SDSS: sodium dodecyl sulfate micro-sedimentation. The environmental references are CA: Cabildo; BW: Barrow; PS: Pieres, followed by the year of sowing. Values in the same column for each locus followed by a different letter are significantly different using the Tukey–Kramer test ($p < 0.01$). * corresponds with bands showing a slightly mobility difference in SDS-PAGE respect to the original band with the same number.

3.4. HMW-GS and LMW-GS Distribution along Different Periods and Origins

Six breeding periods were considered for the analysis of the allele distribution at the three main loci that showed high variability (*Glu-B1*, *Glu-A3* and *Glu-B3*). The results are summarised in Figure 2. The effect of breeding on gluten strength showed a positive selection of alleles associated with high values of the sedimentation test (SDSS). At the *Glu-B1* locus, the allele 20x + 20y (e) was associated with low–intermediate gluten quality decreased its frequency over time, whereas the alleles 6 + 8 (d) and 7 + 8 (b) were favoured by selection and their proportion increased, mostly in the 7 + 8 allele. At the *Glu-A3* locus, the proportion of genotypes carrying the allele 6 (a) did not change over time, whereas the genotypes with 6 + 10 (c) and 6 + 11 (d) increased from the old to the modern germplasm. In addition, the null allele (h) and the subunit 5 (b) were progressively discarded from the 1915–1979 to the 2000–2020 periods. On the other hand, for *Glu-B3* the greatest variability was observed from 1915 to 1979 with a predominance of the 2 + 4 + 15 + 19 (a) allele, which progressively increased to reach 100% in modern genotypes. The allelic distribution by origin is shown in Supplementary Figure S1. At *Glu-B1*, the 6 + 8 (d) was the main allele detected in the Argentinian genotypes, while the 7 + 8 (b) subunits were more frequent in

Chilean and CIMMYT germplasm. The 20x + 20y (e) banding pattern was mostly detected in genotypes from France, WANA and Italy. At *Glu-A3*, the subunit 6 was mostly detected in the Argentinian, Italian and Chilean genotypes, whereas 6 + 10 (c) was the main allele in germplasm from WANA and France.

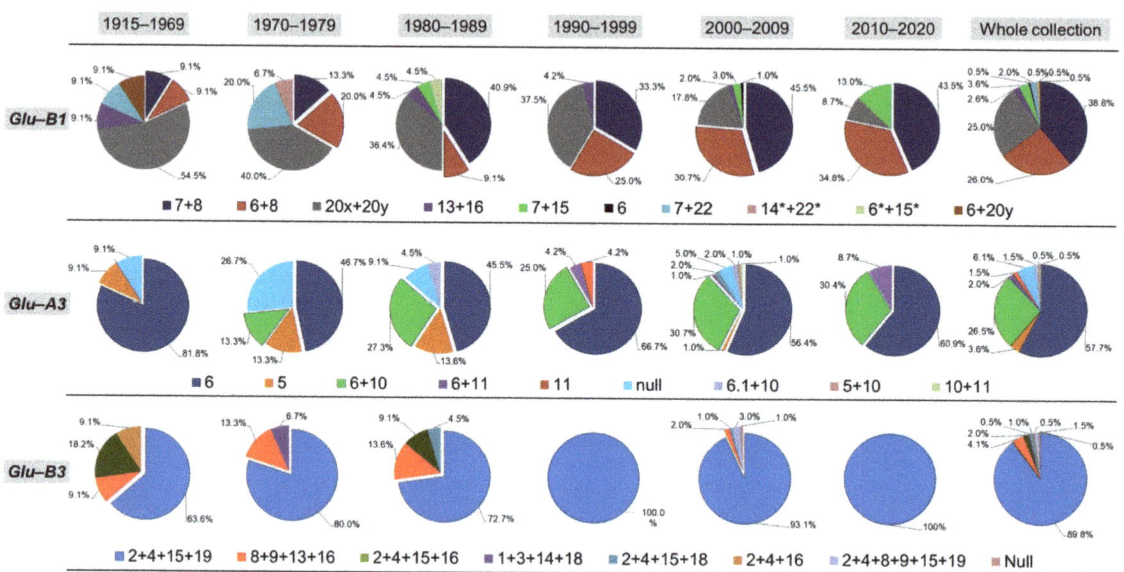

Figure 2. Allelic distribution of the three main glutenin loci (*Glu-B1*, *Glu-A3* and *Glu-B3*) in a worldwide collection of durum wheat over six breeding periods.

3.5. Quality Traits Variation over Different Breeding Periods

The variation in GPC and the SDSS test over the six breeding periods previously mentioned was assessed by ANOVA and Duncan tests ($p < 0.05$) for each environment and year. A decreasing trend in GPC was observed from the 1915–1969 to the 2000–2009 periods in six out of seven experiments, with a slight recovery also observed during the 2010–2020 period in some environments (Supplementary Figure S2). This decrease in GPC represented an average of 5.32% (in relative value) from 1934 to 2020 (-0.07% year^{-1} in GPC). On the other hand, the SDSS test values also decreased from 1915–1969 to 1970–1979, but this was later followed by a growth trend in the subsequent periods (Supplementary Figure S2). The overall estimated genetic gain for the gluten strength (SDSS test) was 10.8% from 1934–1969 to 2010–2020 (0.14% year^{-1} in SDSS). However, when the first breeding interval was removed, the overall genetic gain from 1970–1979 to 2010–2020 was 37.8%, representing an increase of 0.48% year^{-1} caused by breeding in our collection.

3.6. Glutenin Haplotype Frequency, Distribution and Effect

A total of 41 haplotypes could be conformed based on the alleles detected at the five glutenin loci (Table 2). Most of the haplotypes (30 of total) were poorly represented, showing frequencies of up to 1%. The most frequent haplotype, carried by 22.4% of the genotypes, was Hap_22, followed by Hap_11 and Hap_16 with 12.8% each, Hap 17 (10.2%) and Hap_23 (9.2%). The distribution of haplotypes by origin is shown in Supplementary Table S6. The most frequently observed haplotypes among the genotypes that ranked in the top 20% for the best SDSS test values were Hap_22 (for ex. BonINTA Cumenay), Hap_23 (CBW 09034), Hap_17 (CBW 0111) and Hap_16 (Buck#33). Additionally, some of the less frequent haplotypes obtained high SDSS values in both years (Hap_9, Hap_19,

Hap_31, Hap_1, Hap_4 and Hap_18). Significant differences between haplotypes in the SDSS test were observed using the Bautista multiple comparison test $p < 0.05$ (BSS, [45]). A significant haplotype effect on GPC was only observed in three environments (BW 2011, PS 2014 and BW 2017) and the LSMEAN in the 2014/2017 trials (Figure 3). The ANOVA test considering the haplotype effect explained 70–77% of the variance in the SDSS test during 2011, and 48–53% during 2014, whereas, for GPC, it explained 33% of the phenotypic variation (Supplementary Table S5a,b).

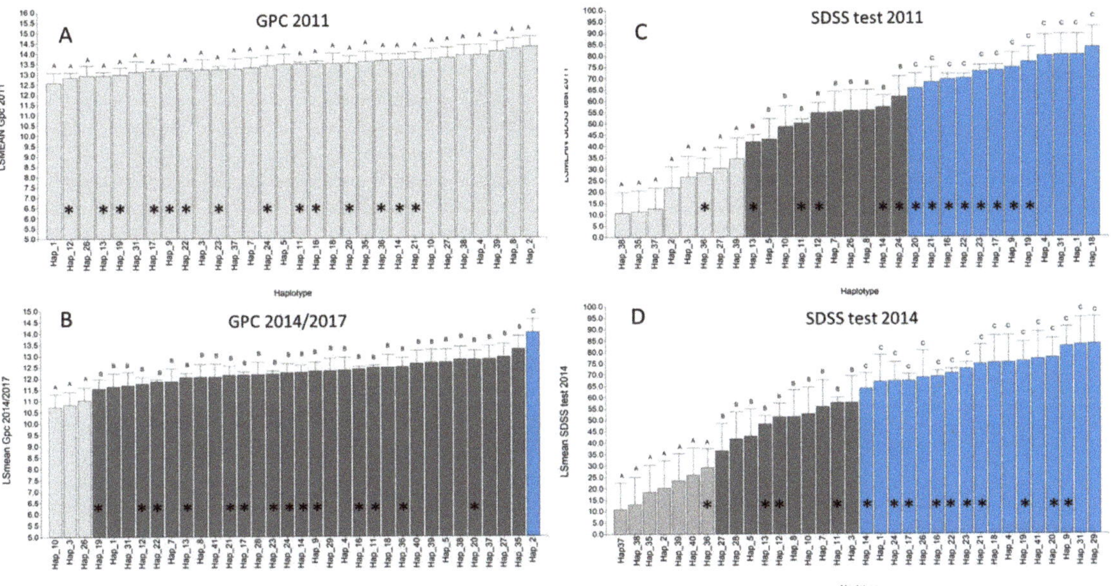

Figure 3. Multiple comparison test (BSS) between haplotypes considering the LSMEANs of grain protein content (GPC) and sodium dodecyl sulfate micro-sedimentation (SDSS) test by year (2011 (**A**,**C**) and 2014 (**D**) or grouping 2014/2017 (**B**)). Bars with a symbol (*) represent haplotypes detected in at least two genotypes. Bars with different letters at the top and colours indicate significant differences according to the BSS test. Hap_: Haplotype_(number).

3.7. Haplotype Effect on Additional Quality Parameters in Argentinian Genotypes

The contrasting haplotype effects in additional quality parameters, such as % gluten, wet gluten, dry gluten, gluten index and farinograph measurements (Development time and Energy level) as well as protein content, were analysed using historical datasets of Argentinian genotypes. For these analyses, 13 genotypes grown in up to 11 environments (11–175) from 1995/96 to 2017/18, representing a total of eight haplotypes (detailed in Supplementary Table S4), were used. These haplotypes varied at the *Glu-B1*, *Glu-B2* and *Glu-A3* loci. The ANOVA test showed a significant effect of the haplotypes on the Gluten Index, Grain protein, Energy level and Development time (Supplementary Figure S3). The haplotypes Hap_23 and Hap_22 exhibited the highest gluten index and Energy level, but obtained the lowest Development time from the farinograph. Additionally, the Hap_29 carrying 20x + 20y subunits at *Glu-B1* but a triple null at *Glu-A1*, *Glu-B2* and *Glu-A3* and *Glu-B3a* also showed a high gluten index and Energy level.

4. Discussion

The importance of durum wheat prolamins (glutenins and gliadins) and their effect on gluten strength for pasta and breadmaking was mainly studied in different genetic backgrounds [4,7,47–51]. Our study focused on the characterisation of HMW-GS and LMW-GS variation in a worldwide durum wheat collection. The effect of these variants on quality traits was analysed gene-by-gene or as haplotypes, highlighting the importance of common and rare allelic variants to improve gluten quality. A description was also made of the glutenin profile of landraces, as well as old and modern germplasms, mainly from Argentina and other countries.

4.1. Glutenin Allelic and Haplotype Variation and Its Effect on GPC and the SDSS Test

In this study, a high number of alleles were detected at the glutenin loci, the *Glu-B1*, *Glu-A3* and *Glu-B3* genes being the main sources of variation in modern durum wheat germplasms exploited for gluten strength improvement. Most of the SDSS variance was explained alternatively by *Glu-B1* or *Glu-B3* in the two years evaluated, indicating that the expression of these genes was affected environmentally. However, these genes exerted a small effect on GPC in one year (2014), suggesting a weak association with this trait. Previous studies yielded controversial results regarding the relative importance of *Glu-B1* and *Glu-B3* and their association with gluten quality. According to Martínez et al. [52], the alleles at *Glu-B3* strongly affect the gluten quality measured by the SDSS, mixograph and alveograph tests. On the contrary, *Glu-B1* was also reported to play a major role in the end-use quality of durum wheat, affecting W and P/L ratio values [53]. However, other authors did not find any effect of *Glu-B1* on gluten quality [54].

A better association between the glutenin loci and quality traits was obtained based on haplotypes, which could explain most of the variation in SDSS, most likely because the haplotype analysis also considers the possible effect of the interactions between loci.

Considering individual loci, *Glu-A1* (HMW-GS) and *Glu-B2* (LMW-GS) showed the lowest variation in our collection. The low variation at *Glu-A1* in modern germplasms was also mentioned by several authors [27,50]. The null allele was practically fixed in our collection as well as in modern germplasms worldwide [51,53,55]. In our analysis, the 2* (b) allele was the unique differential subunit at *Glu-A1* and was also detected at a very low frequency in durum wheat cultivars by other authors [27]. However, frequencies of 18.4% [56], 23.3% [57] and 41.1% [34] were observed in Iranian and Spanish durum wheats and Iranian bread wheat landraces, respectively. We found the 2* subunit to be associated with high gluten strength (higher SDSS), as was previously reported [56,58]. Nonetheless, the contribution of the non-null *Glu-A1* alleles to improve durum wheat quality is still not well defined [55].

In our study, the highest variation was observed at the *Glu-B1* locus with ten alleles, which coincided with previous reports on landraces and modern cultivars of durum wheat [50]. *Glu-B1b* (7 + 8) followed by *Glu-B1d* (6 + 8) and *Glu-B1e* (20x + 20y) were the most frequent alleles. Similar results were obtained in durum wheat, but with different frequencies, with 20x + 20y being > 6 + 8 > 7 + 8 [27]. In this study, the *Glu-B1d* (6 + 8) or *Glu-B1b* (7 + 8) alleles were among the four alleles with higher SDSS values. However, this effect was environmentally impacted, showing *Glu-B1d* with higher SDSS values in 2011 and *Glu-B1b* in 2014. The lower precipitation and mean temperature values obtained at all locations in 2011, suggested that an environmental effect could affect the allelic performance. *Glu-B1b* was widely associated with strong gluten and a good pasta-making quality [53]. Conversely, subunits 6 + 8 were associated with higher SDSS values than 7 + 8 and 20x + 20y [52]. According to Ammar et al. [59] *Glu-B1d* exhibited a better overall breadmaking quality compared to the 7 + 8 or 20 banding patterns. Based on our results, we cannot establish a clear order of importance at *Glu-B1* between the 7 + 8 and 6 + 8 banding patterns or the haplotypes carrying these alleles, but a small difference of *Glu-B1b* (7 + 8) increasing SDSS was observed.

Previously, the 20x + 20y banding pattern was associated with inferior quality [48,59]. In our study, the *Glu-B1e* (20x + 20y) allele was associated with intermediate values of SDSS. This result was coincident with previous studies [31]. However, genotypes with Hap_14 and Hap_29, where *Glu-B1e* was combined with the *Glu-A3h* (null) allele, resulted in similar mean SDSS values to the genotypes with the 6 + 8 banding pattern. The breeding line CBW

05024 (Hap_29) carried *Glu-B1e*/*Glu-A3h*, but also a null allele at *Glu-A1* and *Glu-B2*, which could be an interesting alternative haplotype, showing a high gluten strength.

Two additional *Glu-B1* banding patterns involving the Bx7 subunit (7 + 15 and 7 + 22) were identified among old, intermediate and modern genotypes from Argentina, France, Italy and CIMMYT. *Glu-B1z* (7 + 15) was reported in CIMMYT-derived germplasms and in one Iranian landrace [56,60]. In addition, the *Glu-B1ch* (7 + 22) allele was reported in one Mediterranean landrace [50] and three Iranian landraces [56]. The latter authors also associated *Glu-B1ch* with low gluten quality. Conversely, in the present study the haplotypes (Hap_19, Hap_20 and Hap_21) carrying 7 + 15 or 7 + 22 banding patterns ranked among the best glutenin profiles increasing the SDSS values, suggesting that they should be more carefully evaluated for use in durum wheat breeding programs. The *Glu-B1f* (13 + 16) allele was detected at a low frequency, which was also reported in other cultivars and durum wheat landraces [53,61,62]. However, the frequency of *Glu-B1f* was high (21.38%) in an Algerian durum wheat collection [63] and it was suggested that this allele was associated with low gluten quality [62]. However, our results showed that this allele in Hap_9 was associated with high gluten quality. Hap_9, ranked among the best haplotypes in both years.

A unique banding pattern was carried by the Italian cultivar, Polesine (14 + 22*), associated with low gluten quality (as SDSS), but showing a high GPC value. Two other different banding patterns (13 + 16 and 23 + 18) were previously reported for this genotype [27]. Although the genetics and morphology observed in Polesine, such as plant height and a light green colour (typically only observed in old Italian germplasms) the differences in banding pattern could suggest a misclassification of this material in our seed stock. As far as we know, this is the first report of the 14 + 22* subunit combination in durum wheat. Subunit 22* was previously mentioned as a rare variant [56]. Another unique banding pattern (6* + 15*) was obtained in the landrace, Haurani. A previous study described the 6 + 16 banding pattern in this landrace [60]. Subunit 6* has a higher mobility than subunit 6 in SDS-PAGE. In our study, Haurani showed high SDSS values in both years evaluated, suggesting that the 6* + 15* subunits could be a useful resource for breeding. Additionally, the 6 + 20y banding pattern resulted in a new combination of subunits at the *Glu-B1* locus, only observed in the Italian cultivar, Capeiti8. Nevertheless, different patterns for this cultivar were reported in previous studies as 20x + 20y and 7 + 8 [60,64]. The 6 + 20y banding pattern was not evaluated phenotypically. According to the Catalogue of Gene Symbols for Wheat [65] and the 2020 Supplement (https://wheat.pw.usda.gov/GG3/wgc, accessed date on 2 September 2021), the 14 + 22*, 6 + 20y and 6* + 15* banding patterns were not annotated, and we tentatively propose the nomenclature of *Glu-B1cq*, *Glu-B1cr* and *Glu-B1cs* for these subunit combinations.

Significant interactions between *Glu-B1* and *Glu-B3* or *Glu-A3* were indicated as playing an important effect on quality parameters [52]. Supporting this, the presence of the *Glu-B3b* (8 + 9 + 13 + 16) or *Glu-B3h* (1 + 3 + 14 + 18) allele causes a detrimental effect on SDSS, regardless of which allele is present at *Glu-B1*. Since *Glu-B3a* (2 + 4 + 15 + 19) represented about 90% of the total variation and was gradually fixed in modern germplasms, most of the differences in quality traits were due to the allelic effect from *Glu-B1* and *Glu-A3* and eventually at *Glu-B2*. A low level of variation was observed at *Glu-B3* in comparison with other studies conducted on landraces [50,55,56] or durum wheat germplasm from North Africa [60] and Spain [66]. Contrary to our study, the 2 + 4 + 15 + 18 banding pattern was recorded with a relatively high frequency in Mediterranean landraces [67] and was also reported by other authors [35,56,66].

The rare allelic variant *Glu-B3ag* (2 + 4 + 16) [56] was only detected in the old Argentinian cultivar Bonaerense 202 (1966). In addition, the 2 + 4 + 8 + 9 + 15 + 19 banding pattern was not previously reported, and we tentatively suggest that this new allele be known as *Glu-B3aw*, following the order of the Catalogue of Gene Symbols for wheat [65] and its 2020 Supplement (https://wheat.pw.usda.gov/GG3/wgc, accessed date on 2 September

2021). Additionally, the subunit combination 2 + 4 + 15 + 18 was not previously named and we propose the name of *Glu-B3ax* for this allele.

Another rare variant at *Glu-B3j* (null) was previously mentioned as a result of the wheat–rye translocation 1BL/1RS [68,69] that caused the loss of *Glu-B3* LMW-GS and possibly the linked gliadins (*Gli-B1* locus), and incorporated the secalins (*Sec-1* locus) into the wheat, causing dough stickiness and a reduced gluten quality [70–72]. However, the unique genotype which carried *Glu-B3j* (Quc 3506-2009) showed high gluten strength. This Chilean breeding line (Hap_41) is a triple null genotype (*Glu-A1*, *Glu-B3* and *Glu-B2*), also carrying the *Glu-B1b* and *Glu-A3a* alleles, which showed bread wheat ancestry as being the possible origin of the rye translocation. To our knowledge this allele was not previously reported in durum wheat.

Among LMW-GS the *Glu-A3* locus exhibited a slightly higher variability but explained less of the SDSS test variance than *Glu-B3*. The level of polymorphism at *Glu-A3* observed in our study was comparable to previous reports [55,57] in Moroccan and Spanish durum wheat germplasms. The most common variants at *Glu-A3*, 6 and 6 + 10, associated with high gluten strength, are also widely distributed worldwide [36,50,55–57,62,66]. However, we could not find a clear significant difference favouring these two alleles over the others. Among the *Glu-A3* alleles, the unique banding patterns 5 + 10 (Langdon (Dic-3A)-10) and 10 + 11 (Biensur) were only previously described in Mediterranean durum wheat germplasms [50]. Only the presence of the subunit 5 (*Glu-A3b*), or combined as 5 + 10, was clearly associated with low SDSS values (Hap_35 to Hap_40). The absence of subunit 5 at *Glu-A3* in old, but also modern, Argentinian genotypes could be one of the reasons for this, since Argentinian durum wheat was considered to be of high quality in the past.

In our study, the 6.1 + 10 banding pattern was only detected in three Argentinian genotypes. The *Glu-A3ax* (6.1) allele was described previously and named in the Argentinian cultivar Buck Cristal [31]. However, we observed the 6.1 + 10 banding pattern in Buck Cristal. This could be attributed to different resolution levels in the methodology. The 6.1 subunit [66] is equivalent to the previously designated 7* subunit [73]. *Glu-A3e* (11) was described in durum wheat by several authors [36,55,62] but only detected in the French and Italian germplasm in our collection.

As previously mentioned, the interaction between HMW-GS and LMW-GS, in particular for the *Glu-1* and *Glu-3* loci, has an important influence on technological properties [74–76]. According to He et al., interactions such as *Glu-B1* × *Glu-B3* and *Glu-D1* × *Glu-A3*, strongly affect the SDS sedimentation value, farinograph stability and loaf volume [77]. Similarly, both additive and epistatic effects between glutenin loci significantly affect the dough characteristics [78]. To address this problem, we analysed the effects of these genes on quality traits using haplotypes as in previous reports [33,62,66]. The analysis of allelic combinations instead of individual loci better explained the SDSS and partially explained the GPC variations. Some haplotypes were only detected in the Argentinian (10), Italian (8), French (5), WANA (4), USA (2) and Chilean (1) genotypes, showing the benefits of germplasm exchange. Some rare haplotypes in genotypes from the USA, Italy and Chile were likely a consequence of the allele's introgression through wide crosses during the breeding process.

4.2. Variability in Quality Traits between Breeding Periods and Its Relationship with Allelic Variation at Glutenin Loci

Our results showed a decreasing trend in GPC over six breeding periods in most of the environments evaluated. The negative relationship between GPC and grain yield under most environmental conditions is well known [79,80]. This result agrees well with the increase in grain yield over time, reported by our group [81] using a subset of our collection, as well as from other authors [51,53,82,83]. The estimated reduction in GPC based on our data over time (-0.07% year^{-1}) was about half of the value reported in previous studies [53,62,84].

Moreover, the SDSS test showed an initial decrease from the first period (1915–1969) to the second (1970–1979), followed by a consistent improvement in gluten quality until the 2010–2020 period. The positive growth rate period was similar to results obtained for Italian and Spanish cultivars [53,62]. In the same way, an increased glutenin content without any changes in the albumin/globulin rate from 1891 to 2010 was reported in bread wheat [85]. Nevertheless, the improvement in gluten strength over the last 50 years partially recovered the loss observed during the first and second breeding periods. This finding supports the idea that a reduction in grain quality occurred after the introduction of semi-dwarfism during the 1970s. This agrees well with the fact that all genotypes from the 1934–1969 period were tall and carried the Rht-B1a allele [86]. The SDSS genetic gain from 1970–1979 to 2010–2020 (37.8%) was slightly higher than that reported for Spanish and Italian germplasms [62].

Our study shows that the increases in gluten strength over time are strongly associated with changes in allele composition in the HMW and LMW glutenin subunits and were associated with a progressive replacement of the 20x + 20y banding pattern by the 7 + 8 and 6 + 8 subunits at *Glu-B1* as also reported previously [62]. Negative effect alleles, such as *Glu-A3h* and *Glu-A3b*, were also progressively replaced by more advantageous ones (6 + 10 and 6 + 11). The *Glu-A1*, *Glu-B3* and *Glu-B2* loci were progressively fixed (null(c), 2 + 4 + 15 + 19 (a), 12 (a), respectively) from old to modern accessions, indicating that most of the variation observed today is due to allelic variations at the *Glu-B1* and *Glu-A3* loci. The most frequent haplotypes (22, 23, 17 and 16) carried this conserved allelic combination. The *Glu-B3a/Glu-B2a* combination was also present in 75% of all the intermediate and in 100% of the modern Italian cultivars [51].

4.3. Haplotype Effect on Additional Quality Parameters

The gluten strength in durum wheat is mainly responsible for pasta quality, and several methodologies were proposed as predictors or as direct tests for rheological properties (micromixograph, viscoelastograph, farinograph, alveograph, gluten index and SDSS) [87]. The SDSS test is widely used as a predictor of gluten strength [24,48,88,89]. Although the SDSS test and Gluten Index are highly correlated, it is noted that the SDSS is clearly influenced by protein content [90]. We evaluated the effect of the five most frequent haplotypes and some contrasting haplotypes in selected Argentinian genotypes by using historical datasets for quality traits, including the gluten index. Our results confirmed that Hap_23 and Hap_22, both carrying 7 + 8 subunits at *Glu-B1*, showed a superior performance based both on the gluten index or farinograph parameters. This analysis provides additional evidence for the high gluten strength associated with Hap_29 by the SDSS test. These results confirm the suitability of SDSS as a gluten strength predictor and its association with allelic variants at glutenin loci and haplotypes.

5. Conclusions

In the present study, the allelic variations at five durum wheat glutenin loci were characterised and four new alleles were detected. Additionally, the contribution of individual alleles to improve gluten quality, and their influence on grain protein content, was highlighted, and the haplotype analysis offered valuable information for use in durum wheat breeding programs. Our results showed a decreasing trend in grain protein content over the last 85 years, which could be attributed to a dilution effect due to grain yield improvements. The changes in gluten strength measured by the SDSS test over the same breeding period were associated with the variation in the allele frequency at glutenin loci. Furthermore, some haplotypes poorly represented in modern germplasms were shown to be associated with a high SDSS and quality performance and should be considered for use in breeding programs.

Supplementary Materials: The following are available online at https://www.mdpi.com/2304-815 8/10/11/2845/s1, Figure S1: Allelic distribution of the three main glutenin loci (*Glu-B1*, *Glu-A3* and *Glu-B3*) considering the origin of accession in the whole collection. Figure S2: Multiple comparison tests (Duncan test, $p < 0.05$) performed on the GPC and SDSS means considering six breeding periods in each environment and by year. The environmental references are CA: Cabildo; BW: Barrow; PS:

Pieres, followed by the year of sowing. Bars in the same environment with a different letter on the top are significantly different. Figure S3: Bar plot showing additional quality traits mean values according to the corresponding haplotype. Means with a different letter on the top indicate significant differences according to the Tukey–Kramer test ($p < 0.05$). Table S1: Allelic characterisation of HMW-GS and LMW-GS glutenin loci in a worldwide durum wheat collection of 196 genotypes. Table S2: List of KASP markers for *Glu-A1* and *Gpc-B1* used in this study. Table S3: Precipitation, mean temperature and adopted management in the seven field trials. Table S4: Summary of genotypes/haplotypes and trials considered with additional quality parameters. Table S5: Percentage of variance explained by each locus, model and haplotypes for GPC and SDS sedimentation test. Table S6: Frequency and distribution of haplotypes according to the origin of accessions.

Author Contributions: Conceptualization, P.F.R. and V.E.; Methodology, P.F.R., A.O.L., C.G., S.D., E.M., V.A.; Validation, P.F.R., A.O.L., C.G.; Formal Analysis, P.F.R., A.L.A.; Investigation, P.F.R., C.G., A.O.L., A.L.A., E.M. and V.A.; Resources, P.F.R., V.E., C.G., S.D., E.M.; Data Curation, P.F.R., A.O.L., C.G., E.M.; Visualization P.F.R., C.G.; Writing—Original Draft Preparation, P.F.R.; Writing—Review and Editing, P.F.R., V.E., A.L.A., C.G.; Funding Acquisition, V.E., P.F.R., C.G., S.D.; Supervision, P.F.R., V.E. and C.G.; Project Administration, V.E., P.F.R. All authors have read and agreed to the published version of the manuscript.

Funding: This research was funded by CONICET, CERZOS-CCT CONICET Bahía Blanca, Universidad Nacional del Sur/ME (NUTRIPASTA PCESU9-UNS993), ANPCYT: PICT 2015-1401. We also acknowledge the support of the H2020-MSCA-RISE-2015 project EXPOSEED (grant no. 691109) and CGIAR Research Program WHEAT. Carlos Guzman gratefully acknowledges the European Social Fund and the Spanish State Research Agency (Ministry of Science and Innovation) for financial funding through the Ramon y Cajal Program (RYC-2017-21891).

Data Availability Statement: Data in this study are available in the article and Supplementary Materials. Plant material and raw data are available upon request from the first author.

Acknowledgments: Thanks are due to all seed donors for building the collection analysed in this study. Additionally, we would like to thank ACA Coop LTDA, BUCK Semillas Companies and INTA Barrow for facilitating their experimental fields. We especially thank CIMMYT for expertise support in genetic markers.

Conflicts of Interest: The authors declare no conflict of interest. The funders had no role in decision in this manuscript.

References

1. Matsuo, R.R. Durum wheat: Its unique pasta-making properties. In *Wheat: Production, Properties and Quality*; Bushuk, W., Rasper, V.F., Eds.; Springer: Boston, MA, USA, 1994; pp. 169–178. [CrossRef]
2. Unión de Industriales Fideeros de la República Argentina (UIFRA). Pastas Secas, Estadísticas Sectoriales. Unión De Industriales Fideeros De La República Argentina. 2020, pp. 1–12. Available online: https://uifra.org.ar/wp-content/uploads/2020/04/dossier2020v2.pdf (accessed on 20 September 2021).
3. Motzo, R.; Giunta, F.; Fois, S.; Deidda, M. *Evoluzione Varietale e Qualità in Frumento Duro (Triticum turgidum subsp. durum): Dalle Vecchie Popolazioni Alle Attuali Cultivar*; Dipartimento di Scienze Agronomiche e Genetica Vegetale Agraria, Facoltà di Agraria, Università degli Studi di Sassari: Sassari, Italy, 2001; p. 30.
4. Matsuo, R.R.; Bradley, J.W.; Irvine, G.M. Effect of protein content on the cooking quality of spaghetti. *Cereal Chem.* **1972**, *49*, 707.
5. Dexter, J.E.; Matsuo, R.R. Relationship between Durum Wheat Protein Properties and Pasta Dough Rheology and Spaghetti Cooking Quality. *J. Agric. Food Chem.* **1980**, *8*, 899–902. [CrossRef]
6. Boggini, G.; Pogna, N.E. The Breadmaking Quality and Storage Protein Composition of Italian Durum Wheat. *J. Cereal Sci.* **1989**, *9*, 131–138. [CrossRef]
7. Matsuo, R.R.; Irvine, G.M. Effect of gluten on the cooking quality of spaghetti. *Cereal Chem.* **1970**, *47*, 173–180.
8. Impiglia, A.; Nachit, M.; Lafiandra, D.; Porceddu, E. Effect of gliadin and glutenin components on gluten strength in durum wheat. In *Durum Wheat Quality in the Mediterranean Region*; Di Fonzo, N., Kaan, F., Nachit, M., Eds.; Options Méditerranéennes: Série, A. Séminaires Méditerranéens n 22; CIHEAM: Zaragoza, Spain, 1995; pp. 167–172.
9. Troccoli, A.; Borrelli, G.M.; De Vita, P.; Fares, C.; Di Fonzo, N. Durum Wheat Quality: A Multidisciplinary Concept. Mini Review. *J. Cereal Sci.* **2000**, *32*, 99–113. [CrossRef]
10. Osborne, T.B. *The Proteins of the Wheat Kernel*; Carnegie Inst Washington Publication: Washington, WA, USA, 1907; Volume 84, pp. 1–119.
11. Dexter, J.E.; Matsuo, R.R. Influence of protein content on some durum wheat quality parameters. *Can. J. Plant Sci.* **1977**, *57*, 717–727. [CrossRef]

12. Payne, P.I.; Holt, L.M.; Lawrence, G.J.; Law, C.N. The genetics of gliadin and glutenin, the major storage proteins of the wheat endosperm. *Plant Foods Hum. Nutr.* **1982**, *31*, 229–241. [CrossRef]
13. Porceddu, E. Durum wheat quality in the Mediterranean countries. In *Durum Wheat Quality in the Mediterranean Region*; Di Fonzo, N., Kaan, F., Nachit, M., Eds.; Options Méditerranéennes: Série, A. Séminaires Méditerranéens n 22; CIHEAM: Zaragoza, Spain, 1995; pp. 11–21.
14. Shepherd, K.W. Chromosomal control of endosperm proteins in wheat and rye. In Proceedings of the 3rd International Wheat Genetic Symptoms, Australian Academy of Sciences, Canberra, Australia, 5–9 August 1968; pp. 86–96.
15. Joppa, L.R.; Khan, K.; Williams, N.D. Chromosomal location of genes for gliadin polypeptides in durum wheat *Triticum turgidum*. *Theor. Appl. Genet.* **1983**, *64*, 289–293. [CrossRef]
16. Pogna, N.; Lafiandra, D.; Feillet, P.; Autran, J.C. Evidence for direct casual effect of low molecular weight subunits of glutenins on gluten viscoelasticity in durum wheats. *J. Cereal Sci.* **1988**, *7*, 211–214. [CrossRef]
17. Lawrence, G.J.; Shepherd, K.W. Variation in Glutenin Protein Subunits of Wheat. *Aust. J. Bioi. Sci.* **1980**, *33*, 221–234. [CrossRef]
18. Payne, P.I. Genetics of wheat storage proteins and the effect of allelic variation on bread-making quality. *Ann. Rev. Plant Physiol.* **1987**, *38*, 141–153. [CrossRef]
19. Jackson, E.A.; Holt, L.M.; Payne, P.I. *Glu-B2*, a storage protein locus controlling the D group of LMW glutenin subunits in bread wheat (*Triticum aestivum*). *Genet. Res.* **1985**, *46*, 11–17. [CrossRef]
20. Shewry, P.; Tatham, A.S.; Forde, J.; Kreis, M.; Miflin, B.J. The Classification and Nomenclature of Wheat Gluten Proteins: A Reassessment. *J. Cereal Sci.* **1986**, *4*, 97–106. [CrossRef]
21. Harberd, N.P.; Bartels, D.; Thompson, R.D. DNA restriction-fragment variation in the gene family encoding high molecular weight (HMW) glutenin subunits of wheat. *Biochem. Genet.* **1986**, *24*, 579–596. [CrossRef] [PubMed]
22. Damidaux, R.; Grignac, P.; Feillet, P. Relation applicable en sélection entre l'électrophoregramme des gliadines et les propriétés viscoélastiques du gluten de *Triticum durum* Desf. *C. R. Acad. Sci. Paris* **1978**, *287*, 701–704.
23. Payne, P.I.; Jackson, E.A.; Holt, L.M. The association between γ-gliadin 45 and gluten strength in durum wheat varieties: A direct causal effect or the result of genetic linkage? *J. Cereal Sci.* **1984**, *2*, 73–81. [CrossRef]
24. Autran, J.-C.; Laignelet, B.; Morel, M.-H. Characterization and quantification of low molecular weight glutenins in durum wheats. *Biochimie* **1987**, *69*, 699–711. [CrossRef]
25. Levy, A.A.; Galili, G.; Feldman, M. Polymorphism and genetic control of high molecular weight glutenin subunits in wild tetraploid wheat *Triticum turgidum* var. *dicoccoides. Heredity* **1988**, *61*, 63–72. [CrossRef]
26. Xu, S.S.; Khan, K.; Klindworth, D.L.; Faris, J.D.; Nygard, G. Chromosomal location of genes for novel glutenin subunits and gliadins in wild emmer wheat (*Triticum turgidum* L. var. *dicoccoides*). *Theor. Appl. Genet.* **2004**, *108*, 1221–1228. [CrossRef]
27. Branlard, G.; Autran, C.; Monneveux, P. High molecular weight glutenin subunit in durum wheat (*T. durum*). *Theor. Appl. Genet.* **1989**, *78*, 353–358. [CrossRef]
28. Caballero, L.; Martín, M.A.; Alvarez, J.B. Allelic variation for the high- and low-molecular-weight glutenin subunits in wild diploid wheat (*Triticum urartu*) and its comparison with durum wheats. *Aust. J. Agric. Res.* **2008**, *59*, 906–910. [CrossRef]
29. Alvarez, J.B.; Guzmán, C. Recovery of Wheat Heritage for Traditional Food: Genetic Variation for High Molecular Weight Glutenin Subunits in Neglected/Underutilized Wheat. *Agronomy* **2019**, *9*, 755. [CrossRef]
30. Wallace, J.; Bainotti, C.; Nisi, M.M.; Formica, B.; Seghezzo, M.L.; Molfese, E.; Jensen, C.; Nisi, J.; Helguera, M. Variabilidad genética de proteínas de reserva en trigos candeales argentinos y su interacción con la calidad industrial. *AgriScientia* **2003**, *20*, 19–27. [CrossRef]
31. Lerner, S.E.; Cogliatti, M.; Ponzio, N.R.; Seghezzo, M.L.; Molfese, E.R.; Rogers, W.J. Genetic variation for grain protein components and industrial quality of durum wheat cultivars sown in Argentina. *J. Cereal Sci.* **2004**, *40*, 161–166. [CrossRef]
32. Roncallo, P.F.; Larsen, A.O.; Achilli, A.L.; Saint Pierre, C.; Gallo, C.A.; Dreisigacker, S.; Echenique, V. Linkage disequilibrium patterns, population structure and diversity analysis in a worldwide durum wheat collection including Argentinian genotypes. *BMC Genom.* **2021**, *22*, 233. [CrossRef] [PubMed]
33. Peña, R.J.; Gonzalez-Santoyo, J.; Cervantes, F. Relationship between *Glu-D1/Glu-B3* allelic combinations and bread-making quality-related parameters commonly used in wheat breeding. In *The Gluten Proteins, Proceedings of the 8th Gluten Workshop, Viterbo, Italy, 8–10 September 2003*; Lafiandra, D., Masci, S., D'Ovidio, R., Eds.; Royal Society of Chemistry: Cambridge, UK, 2003; pp. 156–157. [CrossRef]
34. Maryami, Z.; Huertas-García, A.B.; Azimi, M.R.; Hernández-Espinosa, N.; Payne, T.; Cervantes, F.; Govindan, V.; Ibba, M.I.; Guzman, C. Variability for Glutenins, Gluten Quality, Iron, Zinc and Phytic Acid in a Set of One Hundred and Fifty-Eight Common Wheat Landraces from Iran. *Agronomy* **2020**, *10*, 1797. [CrossRef]
35. Payne, P.I.; Lawrence, G.J. Catalogue of alleles for the complex gene loci, *Glu-A1*, *Glu-B1*, and *Glu-D1* which code for high-molecular-weight subunits of glutenin in hexaploid wheat. *Cereal Res. Commun.* **1983**, *11*, 29–35.
36. Nieto-Taladriz, M.T.; Ruiz, M.; Martínez, M.C.; Vazquez, J.F.; Carrillo, J.M. Variation and classification of B low-molecular-weight glutenin subunit alleles in durum wheat. *Theor. Appl. Genet.* **1997**, *95*, 1155–1160. [CrossRef]
37. Rasheed, A.; Wen, W.; Gao, F.; Zhai, S.; Jin, H.; Liu, J.; Guo, Q.; Zhang, Y.; Dreisigacker, S.; Xia, X.; et al. Development and validation of KASP assays for genes underpinning key economic traits in bread wheat. *Theor. Appl. Genet.* **2016**, *129*, 1843–1860. [CrossRef]

38. Saghai-Maroof, M.A.; Soliman, K.M.; Jorgensen, R.A.; Allard, R.W. Ribosomal DNA spacer length polymorphism in barley: Mendelian inheritance, chromosomal location and population dynamics. *Proc. Nat. Acad. Sci. USA* **1984**, *81*, 8014–8019. [CrossRef]

39. Patterson, H.D.; Williams, E.R. A new class of resolvable incomplete block designs. *Biometrika* **1976**, *63*, 83–92. [CrossRef]

40. Irvine, G.; Bradley, J.; Martin, G. A farinograph technique for macaroni doughs. *Cereal Chem.* **1961**, *38*, 153–164.

41. Molfese, E.; Astiz, V.; Seghezzo, M.L. Evaluación de la calidad del trigo candeal (*Triticum turgidum* L. subsp. *durum*) en los programas de mejoramiento de Argentina. *RIA* **2017**, *43*, 303–311.

42. SAS Institute INC. *Base SAS 9.2 Procedures Guide*; SAS Institute INC: Cary, NC, USA, 2010.

43. Di Rienzo, J.A.; Casanoves, F.; Balzarini, M.G.; Gonzalez, L.; Tablada, M.; Robledo, C.W. *InfoStat Versión 2016*; Grupo InfoStat (FCA, Universidad Nacional de Córdoba): Cordoba, Argentina, 2016; Available online: http://www.infostat.com.ar (accessed on 2 September 2021).

44. Miller, R.G. *Simultaneous Statistical Inference*, 2nd ed.; Springer: New York, NY, USA, 1981. [CrossRef]

45. Bautista, M.G.; Smith, D.W.; Steiner, R.L. A Cluster-Based approach to Means Separation. *J. Agric. Biol. Environ. Stat.* **1997**, *2*, 179–197. [CrossRef]

46. Di Rienzo, J.A.; Guzman, A.W.; Casanoves, F. A multiple-comparisons method based on the distribution of the root node distance of a binary tree. *JABES* **2002**, *7*, 129–142. [CrossRef]

47. Kosmolak, F.G.; Dexter, J.E.; Matsuo, R.R.; Leisle, D.; Marchylo, B.A. A relationship between durum wheat quality and gliadin electrophoregrams. *Can. J. Plant Sci.* **1980**, *60*, 427–432. [CrossRef]

48. Carrillo, J.M.; Vazquez, J.F.; Orellana, J. Relationship between gluten strength and glutenin proteins in durum wheat cultivars. *Plant Breed* **1990**, *104*, 325–333. [CrossRef]

49. Carrillo, J.M.; Martínez, M.C.; Brites, M.C.; Nieto-Taladriz, M.T.; Vázquez, J.F. Proteins and quality in durum wheat (*Triticum turgidum* L. var. *durum*). In *Durum Wheat Quality in the Mediterranean Region: New Challenges*; Royo, C., Nachit, M., Di Fonzo, N., Araus, J.L., Eds.; Options Méditerranéennes: Série, A. Séminaires Méditerranéens; n. 40; CIHEAM: Zaragoza, Spain, 2000; pp. 463–467.

50. Nazco, R.; Peña, R.J.; Ammar, K.; Villegas, D.; Crossa, J.; Moragues, M.; Royo, C. Variability in glutenin subunit composition of Mediterranean durum wheat germplasm and its relationship with gluten strength. *J. Agric. Sci.* **2014**, *152*, 379–393. [CrossRef]

51. De Santis, M.A.; Giuliani, M.M.; Giuzio, L.; De Vita, P.; Lovegrove, A.; Shewry, P.R.; Flagella, Z. Differences in gluten protein composition between old and modern durum wheat genotypes in relation to 20th century breeding in Italy. *Eur. J. Agron.* **2017**, *87*, 19–29. [CrossRef]

52. Martínez, M.C.; Ruiz, M.; Carrillo, J.M. Effect of different prolamin alleles on durum wheat quality properties. *J. Cereal Sci.* **2005**, *41*, 123–131. [CrossRef]

53. De Vita, P.; Nicosia, O.L.D.; Nigro, F.; Platani, C.; Riefolo, C.; Di Fonzo, N.; Cattivelli, L. Breeding progress in morpho-physiological, agronomical and qualitative traits of durum wheat cultivars released in Italy during the 20th century. *Eur. J. Agron.* **2007**, *26*, 39–53. [CrossRef]

54. Vázquez, J.F.; Ruiz, M.; Nieto-Taladriz, M.T.; Albuquerque, M.M. Effects on gluten strength of low Mr glutenin subunits coded by alleles at *Glu-A3* and *Glu-B3* loci in durum wheat. *J. Cereal Sci.* **1996**, *24*, 125–130. [CrossRef]

55. Chegdali, Y.; Ouabbou, H.; Essamadi, A.; Cervantes, F.; Ibba, M.I.; Guzmán, C. Assessment of the glutenin subunits diversity in a durum wheat (*T. turgidum* ssp. *durum*) collection from Morocco. *Agronomy* **2020**, *10*, 957. [CrossRef]

56. Hernández-Espinosa, N.; Payne, T.; Huerta-Espino, J.; Cervantes, F.; Gonzalez-Santoyo, H.; Ammar, K.; Guzmán, C. Preliminary characterization for grain quality traits and high and low molecular weight glutenins subunits composition of durum wheat landraces from Iran and Mexico. *J. Cereal Sci.* **2019**, *88*, 47–56. [CrossRef]

57. Chacón, E.A.; Vázquez, F.J.; Giraldo, P.; Carrillo, J.M.; Benavente, E.; Rodríguez-Quijano, M. Allelic Variation for Prolamins in Spanish Durum Wheat Landraces and Its Relationship with Quality Traits. *Agronomy* **2020**, *10*, 136. [CrossRef]

58. Pasqualone, A.; Piarulli, L.; Mangini, G.; Gadaleta, A.; Blanco, A.; Simeone, R. Quality characteristics of parental lines of wheat mapping populations. *Agric. Food Sci.* **2015**, *24*, 118–127. [CrossRef]

59. Ammar, K.; Kronstad, W.E.; Morris, C.F. Breadmaking Quality of Selected Durum Wheat Genotypes and Its Relationship with High Molecular Weight Glutenin Subunits Allelic Variation and Gluten Protein Polymeric Composition. *Cereal Chem.* **2000**, *77*, 230–236. [CrossRef]

60. Bellil, I.; Hamdi, O.; Khelifi, D. Diversity of five glutenin loci within durum wheat (*Triticum turgidum* L. ssp. *durum* (Desf.) Husn.) germplasm grown in Algeria. *Plant Breed* **2014**, *133*, 179–183. [CrossRef]

61. Carrillo, J.M. Variability for glutenin proteins in Spanis h durum wheat landraces. In *Durum Wheat Quality in the Mediterranean Region*; Di Fonzo, N., Kaan, F., Nachit, M., Eds.; Options Méditerranéennes: Série, A. Séminaires Méditerranéens n. 22; CIHEAM: Zaragoza, Spain, 1995; pp. 143–147.

62. Subira, J.; Peña, R.J.; Álvaro, F.; Ammar, K.; Ramdani, A.; Royo, C. Breeding progress in the pasta-making quality of durum wheat cultivars released in Italy and Spain during the 20th Century. *Crop. Pasture Sci.* **2014**, *65*, 16–26. [CrossRef]

63. Hamdi, W.; Bellil, I.; Branlard, G.; Douadi, K. Genetic Variation and Geographical Diversity for Seed Storage Proteins of Seventeen Durum Wheat Populations Collected in Algeria. *Not. Bot. Horti Agrobot. Cluj-Napoca* **2010**, *38*, 22–32. [CrossRef]

64. Vallega, V. High-molecular-weight glutenin subunit composition of Italian *Triticum durum* cultivars and spaghetti cooking quality. *Cer. Res. Comm.* **1986**, *14*, 251–257.

65. Mcintosh, R.A.; Yamazaki, Y.; Dubcovsky, J.; Rogers, W.J.; Morris, C.; Appels, R.; Xia, X.C. Catalogue of Gene Symbols for Wheat. 2013. Available online: https://shigen.nig.ac.jp/wheat/komugi/genes/symbolClassList.jsp (accessed on 31 August 2021).

66. Ruiz, M.; Bernal, G.; Giraldo, P. An update of low molecular weight glutenin subunits in durum wheat relevant to breeding for quality. *J. Cereal Sci.* **2018**, *83*, 236–244. [CrossRef]

67. Nazco, R.; Peña, R.J.; Ammar, K.; Villegas, D.; Crossa, J.; Royo, C. Durum wheat (*Triticum durum* Desf.) Mediterranean landraces as sources of variability for allelic combinations at *Glu-1/Glu-3* loci affecting gluten strength and pasta cooking quality. *Genet. Resour. Crop. Evol.* **2014**, *61*, 1219–1236. [CrossRef]

68. Martínez-Cruz, E.; Espitia-Rangel, E.; Villaseñor-Mir, H.E.; Molina-Galán, J.D.; Benítez-Riquelme, I.; Santacruz-Varela, A.; Peña-Bautista, R.J. Diversidad genética de gluteninas y gliadinas en trigos harineros (*Triticum aestivum* L.) mexicanos. *Agrociencia* **2010**, *44*, 187–195.

69. Howell, T.; Hale, I.; Jankuloski, L.; Bonafede, M.; Gilbert, M.; Dubcovsky, J. Mapping a region within the 1RS.1BL translocation in common wheat affecting grain yield and canopy water status. *Theor. Appl. Genet.* **2014**, *127*, 2695–2709. [CrossRef] [PubMed]

70. Rajaram, S.R.; Braun, H.J. Wheat yield potential. In *International Symposium on Wheat Yield Potential: Challenges to International Wheat Breeding*; Reynolds, M., Pietragalla, P.J., Braun, H.J., Eds.; CIMMYT: Mexico City, México, 2008; pp. 103–107.

71. Fenn, D.; Lukow, O.M.; Bushuk, W.; Depauw, R.M. Milling and baking quality of 1BL/1RS translocation wheats. I. Effects of genotype and environment. *Cereal Chem.* **1994**, *71*, 189–195.

72. Liu, L.; He, H.Z.; Yan, Y.; Xia, X.C.; Peña, R.J. Allelic variations at the *Glu-1* and *Glu-3* loci, presence of the 1B.1R translocation, and their effects on mixographic properties in Chinese bread wheats. *Euphytica* **2005**, *142*, 197–204. [CrossRef]

73. Aguiriano, E.; Ruiz, M.; Fité, R.; Carrillo, J.M. Genetic variation for glutenin and gliadins associated with quality in durum wheat (*Triticum turgidum* L. ssp. *turgidum*) landraces from Spain. *Span. J. Agric. Res.* **2008**, *6*, 599–609. [CrossRef]

74. Gupta, R.B.; Paul, J.G.; Cornish, B.B.; Palmer, G.A.; Bekes, F.; Rathjen, A.J. Allelic variation at glutenin subunit and gliadin loci, *Glu-1*, *Glu-3* and *Gli-1*, of common wheats. I. Its additive and interaction effects on dough properties. *J. Cereal Sci.* **1994**, *19*, 9–17. [CrossRef]

75. Ito, M.; Fushie, S.; Maruyama-Funatsuki, W.; Ikeda, T.M.; Nishio, Z.; Nagasawa, K.; Tabiki, T.; Yamauchi, H. Effect of allelic variation in three glutenin loci on dough properties and breadmaking qualities of winter wheat. *Breed Sci.* **2011**, *61*, 281–287. [CrossRef]

76. Krystkowiak, K.; Langner, M.; Adamski, T.; Salmanowicz, B.P.; Kaczmarek, Z.; Krajewski, P.; Surma, M. Interactions between *Glu-1* and *Glu-3* loci and associations of selected molecular markers with quality traits in winter wheat (*Triticum aestivum* L.) DH lines. *J. Appl. Genet.* **2017**, *58*, 37–48. [CrossRef]

77. He, Z.H.; Liu, L.; Xia, X.C.; Liu, J.J.; Peña, R.J. Composition of HMW and LMW Glutenin Subunits and Their Effects on Dough Properties, Pan Bread, and Noodle Quality of Chinese Bread Wheats. *Cereal Chem.* **2005**, *82*, 345–350. [CrossRef]

78. Nieto-Taladriz, M.T.; Perretant, M.R.; Rousset, M. Effect of gliadins and HMW and LMW subunits of glutenin on dough properties in the F6 recombinant inbred lines from a bread wheat cross. *Theor. Appl. Genet.* **1994**, *88*, 81–88. [CrossRef] [PubMed]

79. Laidig, F.; Piepho, H.P.; Rentel, D.; Drobek, T.; Meyer, U.; Huesken, A. Breeding progress, environmental variation and correlation of winter wheat yield and quality traits in german official variety trials and on-farm during 1983–2014. *Theor. Appl. Genet.* **2017**, *130*, 223–245. [CrossRef] [PubMed]

80. Eichi, V.R.; Okamoto, M.; Garnett, T.; Eckermann, P.; Darrier, B.; Riboni, M.; Langridge, P. Strengths and Weaknesses of National Variety Trial Data for Multi-Environment Analysis: A Case Study on Grain Yield and Protein Content. *Agronomy* **2020**, *10*, 753. [CrossRef]

81. Achilli, A.L.; Roncallo, P.F.; Echenique, V. Genetic gains in grain yield and agronomic traits of Argentinian durum wheat from 1934 to 2015. CERZOS-CONICET, Bahía Blanca, Buenos Aires. *Argentina* **2021**. manuscript being submitted.

82. Mefleh, M.; Conte, P.; Fadda, C.; Giunta, F.; Piga, A.; Hassoun, G.; Motzo, R. From ancient to old and modern durum wheat varieties: Interaction among cultivar traits, management, and technological quality. *J. Sci. Food Agric.* **2019**, *99*, 2059–2067. [CrossRef]

83. Kibite, S.; Evans, L.E. Causes of negative correlations between grain yield and grain protein concentration in common wheat. *Euphytica* **1984**, *33*, 801–810. [CrossRef]

84. Taghouti, M.; Nsarellah, N.; Rhrib, K.; Benbrahim, N.; Amallah, L.; Rochdi, A. Evolution from durum wheat landraces to recent improved varieties in Morocco in terms of productivity increase to the detriment of grain quality. *Rev. Mar. Sci. Agron. Vét.* **2017**, *5*, 351–358.

85. Pronin, D.; Börner, A.; Weber, H.; Scherf, K.A. Wheat (*Triticum aestivum* L.) Breeding from 1891 to 2010 Contributed to Increasing Yield and Glutenin Contents but Decreasing Protein and Gliadin Contents. *J. Agric. Food Chem.* **2020**, *68*, 13247–13256. [CrossRef]

86. Roncallo, P.F.; Larsen, A.O.; Achilli, A.L.; Ammar, K.; Huerta Espino, J.; Gonzalez, L.; Campos, P.; Dreisigacker, S.; Beker, M.; Echenique, V. Effect of major genes and loci identified by association mapping on yield, plant height and head date in durum wheat. In Proceedings of the RAFV Conference 202, XXXIII Argentinian Meeting of Plant Physiology, Santa Fe, Argentina, 13–17 September 2021; Available online: https://rafv2020.wixsite.com/santa-fe (accessed on 2 September 2021).

87. Cubadda, R.; Carcea, M.; Pasqui, L.A. Suitability of the gluten index method for assessing gluten strength in durum wheat and semolina. *Cereal Foods World* **1992**, *37*, 866–869.

88. Dexter, J.E.; Matsuo, R.R.; Kosmolak, F.G.; Leisle, D.; Marchylo, B.A. The suitability of the SDS-sedimentation test for assessing gluten strength in durum wheat. *Can. J. Plant Sci.* **1980**, *60*, 427. [CrossRef]

89. Dick, J.W.; Quick, J.S. A modified screening test for the rapid estimation of gluten strength in early-generation durum wheat breeding lines. *Cereal Chem.* **1983**, *60*, 315.
90. Clarke, F.R.; Clarke, J.M.; Ames, N.A.; Knox, R.E.; Ross, J.R. Gluten index compared with SDS sedimentation volume for early generation selection for gluten strength in durum wheat. *Can. J. Plant Sci.* **2010**, *90*, 1–11. [CrossRef]

*foods*

Article

Identifying Quality Protein Maize Inbred Lines for Improved Nutritional Value of Maize in Southern Africa

Isaac Amegbor [1,2], Angeline van Biljon [1], Nemera Shargie [3], Amsal Tarekegne [4] and Maryke Labuschagne [1,*]

1   Department of Plant Sciences, University of the Free State, Bloemfontein 9300, South Africa; isaacamegbor@gmail.com (I.A.); avbiljon@ufs.ac.za (A.v.B.)
2   CSIR—Savanna Agricultural Research Institute, Tamale 00233, Ghana
3   Agricultural Research Council—Grain Crops, Potchefstroom 2520, South Africa; shargien@arc.agric.za
4   Zamseeds, 6483 Mugoti Road, Lusaka 10101, Zambia; a.tarekegne63@gmail.com
*   Correspondence: labuscm@ufs.ac.za; Tel.: +27-51-4012-715

Abstract: Malnutrition, as a result of deficiency in essential nutrients in cereal food products and consumption of a poorly balanced diet, is a major challenge facing millions of people in developing countries. However, developing maize inbred lines that are high yielding with enhanced nutritional traits for hybrid development remains a challenge. This study evaluated 40 inbred lines: 26 quality protein maize (QPM) lines, nine non-QPM lines, and five checks (three QPM lines and two non-QPM lines) in four optimum environments in Zimbabwe and South Africa. The objective of the study was to identify good-quality QPM inbred lines for future hybrid breeding efforts in order to increase the nutritional value of maize. The QPM lines had a lower protein content (7% lower) than that of the non-QPM lines but had 1.9 times more tryptophan and double the quality index. The lysine- and tryptophan-poor α-zein protein fraction was 41% lower in QPM than in non-QPM, with a subsequent increase in γ-zein. There was significant variation within the QPM inbred lines for all measured quality characteristics, indicating that the best lines can be selected from this material without a yield penalty. QPM lines that had both high protein and tryptophan levels, which can be used as parents for highly nutritious hybrids, were identified.

Keywords: maize inbred lines; nutritional value; protein quality

Citation: Amegbor, I.; van Biljon, A.; Shargie, N.; Tarekegne, A.; Labuschagne, M. Identifying Quality Protein Maize Inbred Lines for Improved Nutritional Value of Maize in Southern Africa. *Foods* **2022**, *11*, 898. https://doi.org/10.3390/foods11070898

Academic Editors: Barbara Laddomada and Weiqun Wang

Received: 25 February 2022
Accepted: 17 March 2022
Published: 22 March 2022

Publisher's Note: MDPI stays neutral with regard to jurisdictional claims in published maps and institutional affiliations.

1. Introduction

Malnutrition, due to lack of a balanced diet, has become a chronic disease in under-developed and developing countries, affecting about two billion people [1] and leading to about 45% child mortality among infants under the age of five years [2]. Consequently, malnutrition has become an international problem and has caused an 11% loss of annual gross domestic product (GDP) in Africa and in other developing countries [1]. Maize is the third most important staple cereal food crop in the world after wheat and rice; it contributes about 30% of food-calorie intake and is a source of protein for more than four billion people in 94 developing countries [3].

Maize kernels consist of 61 to 78% starch, 6 to 12% protein, and 3 to 6% fat [4,5]. Maize protein exists largely in the form of zein proteins, subdivided into β-, γ-, and α-zein fractions based on amino acid sequences [5,6], with α-zein being the most abundant fraction. The zein storage proteins are deficient in lysine and tryptophan, which are essential amino acids for human and other monogastric animals, thus negatively affecting the crop's nutritional value [7,8]. This led to the development of quality protein maize (QPM) in the 1960s, by conventional breeding through the introgression of the *opaque2* gene, and later modifier genes to harden the endosperm, to improve the nutritional content of traditional maize varieties to combat malnutrition. QPM has double the amount of lysine and tryptophan of normal maize and can supply 70 to 80% of human protein requirements, while non-QPM genotypes can only supply a maximum of 46% [9,10]. Several studies have been conducted on QPM to determine the lysine and tryptophan content of maize grains [11,12]. It is difficult to combine high grain yield with high-quality protein content in elite maize varieties because these two characteristics are often negatively correlated [13], either because the grain yield is directly involved in the process of seed modification or because the modifier gene(s) could be tightly linked to those responsible for protein synthesis [14].

Maize parental lines are the fundamental genetic materials essential for understanding the mechanism and principles of breeding [15,16]. In recent years, a large number of QPM inbred lines have been developed and successfully used as parents for generating hybrids and synthetic varieties [17,18]. Parental materials form the basis on which the development of stable and high yielding maize hybrids revolve [19]. As such, inbred lines should be extensively evaluated at different locations for consistency in the performance of traits of interest. Examining genotypes by environment interaction (GEI) is very important in crop development because several factors contribute to the performance of a genotype, since the production environment is variable [16]. The aim of this study is to assess the seed composition, the nutritional value and zein protein composition, and the yield performance of the new QPM inbred lines developed by the International Maize and Wheat Improvement Centre (CIMMYT) in order to identify the best lines to use as parents in developing high yielding and nutritious hybrids.

2. Materials and Methods

2.1. Genetic Material and Field Evaluation

The genetic material used consisted of 26 QPM inbred lines and nine non-QPM lines, as well as five commercial checks (three QPM and two non-QPM) (Table 1), developed by CIMMYT, Zimbabwe. The experiment was carried out under optimum conditions at Cedara (latitude $-29.54°$, longitude $30.26°$, with an elevation of 1066 m above sea level, with reddish brown clay soils) and Potchefstroom (latitude $-26.73°$, longitude $27.08°$, with an elevation of 1349 m above sea level, with brown sandy loam soils) in South Africa. All the trial locations are in summer rainfall regions, with December, January, and February receiving the highest rainfall. Average annual rainfall is 824 mm (Cedara) and 600 mm (Potchefstroom). The maximum temperatures during the maize growing season (October to April) vary between 22–25 °C (Cedara) and 24–29 °C (Potchefstroom). Trials in Zimbabwe were grown at CIMMYT, Harare (latitude $17°46'$, longitude $31°02'$, with an elevation of 1406 m above sea level), and Gwebi Agricultural College (latitude $17°13'$, longitude $31°$ E, with an altitude of 1406 m above sea level). Gwebi is 27 km from Harare and has a similar climate. The annual rainfall of Harare is 845 mm with maximum temperatures of 25–31 °C in the growing season. All trials were conducted under natural rainfall without supplementary irrigation.

The experimental design used was a 5×8 randomised complete block design with two replications at each location. The experimental unit consisted of one-row plots, each 4 m long with inter-row spacing of 0.75 m and spacing within rows of 0.25 m. Two seeds were planted per hill, and seedlings were thinned to one plant per hill at four weeks after emergence to give a final plant population density of about 53,333 plants ha^{-1}.

At Potchefstroom, the fertiliser regime consisted of compound fertiliser 3:2:1 (25) + Zn, applied as a basal application at planting, at a rate of 200 kg NPK (nitrogen, phosphorus, potassium) ha^{-1}. LAN (Limestone Ammonium Nitrate) with 28% N was used for top-dressing in two equal splits at 28 and 56 days after emergence and at a rate of 100 kg ha^{-1} each. At Cedara, MAP (Monoammonium Phosphate) at 250 kg ha^{-1} was applied at planting, and LAN was given at 150 kg ha^{-1} in two equal splits of 75 kg ha^{-1} at 28 and 56 days after emergence. For the Zimbabwe trials, fertilisers were applied at the recommended rates of 250 kg ha^{-1} N, 83 kg ha^{-1} P, and 111 kg ha^{-1} K. Basal fertiliser application was conducted in the form of NPK, and an additional N application was conducted four weeks after seed emergence.

Table 1. Description of the QPM and non-QPM inbred lines and checks used in the study.

Code	Name	Donor
L1	CZL1330	QPM progeny
L2	CZL15041	QPM progeny
L3	CZL15055	QPM progeny
L4	CZL15073	QPM progeny
L5	CZL1471	QPM progeny
L6	TL135470	QPM progeny
L7	VL06378	QPM progeny
L8	TL155805	QPM progeny
L9	TL147078	QPM progeny
L10	TL147070	QPM progeny
L11	TL13609	QPM progeny
L12	TL145743	QPM progeny
L13	TL156614	QPM progeny
L14	CZL1477	QPM progeny
L15	CZL15074	QPM donor
L16	CZL0616	QPM progeny
L17	CZL083	QPM progeny
L18	CML572	Non-QPM parent
L19	EBL167787	Non-QPM check
L20	CZL0520	Non-QPM parent
L21	CZL99005	Non-QPM parent
L22	CML502	QPM donor
L23	CZL0920	QPM donor
L24	CML144	QPM donor
L25	CML159	QPM donor
L26	CML181	QPM donor
L27	CML197	Non-QPM parent
L28	CML312SR	Non-QPM parent
L29	CML488	Non-QPM parent
L30	CML491	QPM donor
L31	LH51	Non-QPM parent
L32	CZL00025	Non-QPM parent
L33	CZL15049	QPM tester
L34	CZL059	QPM tester
L35	CML444	Non-QPM tester
L36	CML395	Non-QPM tester
L37	CZL01005	QPM check
L38	CML511	QPM check
L39	CML312	Non-QPM check
L40	CZL1470	QPM check

Yield was measured on a plot basis, but as the difference between the average QPM and non-QPM inbred lines was not significant, it was decided to focus only on quality characteristics in this study.

2.2. Seed Samples

The seed samples were obtained by self-pollinating two cobs from each entry in the two replications from all sites in order to prevent cross-pollination with unknown pollen, which could affect the quality characteristics. All the self-pollinated cobs were harvested and shelled manually and then bulked for each plot. The seed was placed in cold storage (5 °C) until laboratory analyses, which was carried out within a few weeks of harvesting.

2.3. Sample Preparation

One hundred kernels of uniform size were randomly selected from the bulked seed samples of each entry and replication; milled into flour using an IKA, A10 Yellowline grinder (Merck Chemicals Pty Ltd., Darmstadt, Germany); and then sieved with a 1 mm screen mesh.

2.4. Zein

Zein analysis was conducted using reverse-phase high-performance liquid chromatography (RP-HPLC). Zein extraction was adapted from the extraction method described previously [20]. Maize flour (0.20 g) was placed in 2 mL reaction tubes. An aqueous solution, containing 70% ethanol (Merck; 96% v/v), 24.5% filtered double distilled water, 5% beta-mercaptoethanol (Sigma; $\geq$99% v/v), and 0.5% sodium acetate (Saarchem AR; w/v), was prepared as a stock solution [21,22]. One millilitre of this solution was added to each tube. The mixture was agitated continuously on a vortex for $\pm$16 h at ambient temperature, and the suspension was centrifuged for 15 min at 6000 revolutions per minute. The obtained supernatant was filtered through a 0.45 μm membrane filter using a syringe into glass vials, and it was then kept in a refrigerator at 4 °C until injection into the HPLC system.

RP-HPLC was performed on a Shimadzu Prominence LC System using a Jupiter C18 column (Phenomenex$^{®}$) of 250 × 4.6 mm, with a 5 μm particle size and 300 Å pore size. Samples of 50 μL were injected and eluted with the solvent at 1 mL per minute flow rate with a column temperature of 55 °C. The two solvents used were labelled A and B. Solvent A was made up of LiChrosolv Acetonitrile (Merck) containing 0.1% (v/v) HiPersolv trifluoroacetic acid (TFA) (VWR chemicals), and Solvent B was filtered with deionised water comprising 0.1% (v/v) TFA and was set to run for 75 min per sample. Zein fractions were determined in a chromatograph using Shimadzu Class-VP 6.14 SP1 software. Based on the distinct peak retention time, β-, γ-, and α-zeins were determined as a percentage of the total zein content.

2.5. Amylose Content

Maize flour was used to determine the amylose content [23]. Absolute ethanol (Merck; v/v) was diluted with distilled water to 95% concentration. A 40 g sodium hydroxide (NaOH, Merck; w/v) was dissolved in double distilled water to prepare a 1 L solution. Other solutions were acetic acid, prepared by adding 57.2 mL of glacial acetic acid (Merck, Mr = 60.05; density = 1.05 g per mL) to 1 L distilled water, and iodine solution, prepared by dissolving 0.2 g iodine and 2.0 g potassium iodide in 100 mL distilled water.

A solution containing 1 ml of 95% ethanol (v/v) and 9 mL 1 M NaOH was added to each flour sample (0.1 g) in a 100 mL volumetric flask and vortexed. The samples were incubated and placed in boiling water to gelatinise the starch for 30 min before being allowed to cool for 1 h at room temperature, and they were then centrifuged for 5 min at 3000 rpm. A 100 μL sample was transferred into each 15 mL test tube, and 20 μL 1 M acetic acid and the prepared 200 μL iodine solutions were added. The final volume was made up to 10 mL using double distilled water, thoroughly mixed, and left to stand for 20 min for colour development to take place, before reading the absorbance at 620 nm on a UV–Vis spectrophotometer (Jenway Spectrophotometer Model 7315, UK, ST15 OSA).

The amylose percentage was calculated as follows:

$$\text{Amylose (\%)} = \frac{\text{Total amylose in sample (mg)}}{\text{Sample mass (mg)}} \times 100$$

2.6. Tryptophan and Starch Content

Lysine and tryptophan levels are generally highly positively correlated in maize kernels [7,24], and, for this reason, only tryptophan was measured in this study. Flour samples (2 g) were defatted and analysed for tryptophan content, using a colorimetric method based on a glyoxylic acid reaction with the tryptophan present in the flour, in the presence of ferric chloride and sulphuric acid (H_2SO_4), according to the protocol of Nurit [24]. The optical density of tryptophan was read at 560 nm using a Jenway Spectrophotometer, Model 7315. The tryptophan percentage was then calculated [24]:

$$\text{Tryptophan (\%)} = \frac{\text{OD}_{60mm}}{\text{slope}} \times \frac{\text{hydrolysis volume}}{\text{sample weight}} \times 100$$

The total starch was determined using a polarimetric method [25]. Flour samples of 2.5 g were weighed into 100 mL Erlenmeyer flasks. Then, 50 ml of a 32% (v/v) GR HCL (Merck) solution was added, and the flask was placed in boiling water for 15 min, with the samples stirred every 5 min. The samples were taken out and allowed to cool to approximately 20 °C in a cool water bath. The mixture was then quantitatively transferred to a 100 mL volumetric flask. Then, 10 ml of a 4% (w/v) GR tungstophosphoric acid (Merck) was added to the solution, made up to 100 mL using double distilled water, and inverted several times. Double filtration with Whatman no. 4 filter paper followed until approximately 70 mL filtrate was collected. The filtrate was then read on an automatic polarimeter (ATAGO® AP-300) at 589 nm.

The percentage of starch in the sample was computed as follows:

$$\text{Starch (\%)} = \frac{10,000 \times P}{L \times [a]_D^{20°} \times S}$$

where P = the measured angle of the optical rotation in degrees; L = length (dm) of the sample tube, $[a]_D^{20°}$ = specific rotation of the pure starch, and S = exact mass of the sample.

2.7. Protein and Oil Content, and Quality Index

A 500 g sample of the self-pollinated seeds was used to determine the protein and oil contents (percentage weight basis, %wt) for each sample, using a near-infrared transmission spectroscopy (NIR) Perten Grain Analyzer (Model DA 7250, Perten, Instruments AB, Stockholm, Sweden), with three subsamples for each sample. The NIR calibration was confirmed with results from the wet chemistry of 50 samples before use. The quality index was calculated as the proportion of tryptophan to protein content.

2.8. Data Analysis

Analysis of variance was performed on the measured characteristics with Genstat (21st edition) [26] for the separate locations and across locations.

3. Results

Analysis of Variance for Individual and Combined Locations

The effect of the inbreds was significant for all measured characteristics at all four locations and across locations, with the exceptions of protein content and β-zein content at Cedara and Gwebi (Table 2). Across all locations, the effect of the environment was highly significant ($p > 0.01$) for all characteristics. Genotype-by-environment interaction was highly significant ($p > 0.01$) for all characteristics except for β-zein, indicating that the inbreds did not rank the same across the different environments for the measured characteristics (Table 2).

Table 2. Analysis of variance of quality traits analysed for 40 inbred lines at four environments in Zimbabwe and South Africa during the 2017/2018 cropping season.

Source	Protein %	Oil %	Starch %	Tryptophan %	Amylose %	β-Zein	γ-Zein	α-Zein	QI
Cedara									
Rep	0.61	0.03	23.62 *	0.000003	15.77	54.90	14.30	110.80	0.03
Entry	0.86	1.44 *	9.59 *	0.000700 **	224.68 **	29.47	814.77 **	596.97 *	0.11 **
Error	0.59	0.47	4.81	0.000026	3.99	43.81	185.02	277.94	0.01
Potch									

Table 2. *Cont.*

Source	Protein %	Oil %	Starch %	Tryptophan %	Amylose %	β-Zein	γ-Zein	α-Zein	QI
Rep	0.01	0.82	5.61	0.00012 *	3.43	2.89	506.42	184.40	0.01
Entry	1.77 **	1.77 **	11.59 **	0.00007 **	127.90 **	2.56 *	626.25 **	749.91 **	0.11 **
Error	0.41	0.44	3.38	0.00001	5.62	1.09	173.98	102.07	0.01
Harare									
Rep	0.01	0.01	0.44	0.00005	25.97	51.47	265.39	1124.40 *	0.05
Entry	2.12 *	1.05 *	3.16 **	0.00048 **	2459.23	159.55 **	455.05	491.71 **	0.16 **
Error	0.45	0.08	0.13	0.00004	1861.53	31.43	279.22	132.02	0.01
Gwebi									
Rep	0.23	0.30	0.04	0.00005	23.99	34.95	8228.20 *	52.26	0.01
Entry	2.33	0.97 *	4.70	0.00050 **	166.79	11.12	390.22	591.20	0.11 **
Error	1.00	0.13	1.41	0.00002	7.54	23.68	454.75	806.43	0.01
Combined analysis									
Entry (G)	7.42 **	4.41 **	0.003 **	7.42 **	249.07 **	9.80 **	1490.51 **	1577.48 **	0.99 **
Environment (E)	592.77 **	4.14 **	0.008 **	592.77 **	823.45 **	67.80 **	976.73 **	606.62 **	20.71 **
GxE	1.03 **	0.88 **	0.000 **	1.03 **	232.70 **	2.21	213.41 **	209.52 **	0.21 **
Error	0.32	0.25	0.000	0.32	10.01	1.67	96.01	98.11	0.04

* $p < 0.05$, ** $p < 0.01$; QI = quality index.

On average, across the locations (Table 3), the protein content was significantly higher for non-QPM than for QPM, and the values ranged from 6.38% (QPM L6) to 9.78% (non-QPM L27). The oil content was similar for QPM and non-QPM, with the values ranging between 3.18% (QPM L30) and 5.65% (QPM L33). The starch content of QPM and non-QPM was similar, ranging from 60.20% (QPM L14) to 67.49% (non-QPM L28). As expected, the tryptophan content of QPM was much higher than that of non-QPM (1.9 times), with values ranging between 0.034% (non-QPM L36) and 0.091% (QPM L5 and L16). The quality index of QPM was double that of non-QPM, varying from 0.42 (non-QPM L39) to 1.73 (QPM L33). Amylose content was similar for QPM and non-QPM but varied extensively from 31.88% (non-QPM L39) to 53.18% (QPM L25). β-zein content was similar for QPM and non-QPM and varied from 2.35% (non-QPM L31) to 6.75% (QPM L31). There was a very large difference (41%) between QPM and non-QPM for γ-zein content, with the QPM values being the highest. The differences between the highest and lowest values were also very high (22.71% for non-QPM L20 and 70.62% for QPM L215). Likewise, the difference between QPM and non-QPM for α-zein content was 37% with non-QPM having the highest value and the values overall ranging from 26.68% (QPM L5) to 68.4% (non-QPM L20).

Table 3. Means of quality traits for 40 inbred lines analysed across four environments in Zimbabwe and South Africa during the 2017/2018 cropping season.

Line	Line Status	Protein %	Oil %	Starch %	Tryptophan %	QI	Amylose %	β-Zein %	γ-Zein %	α-Zein %
L1	QPM	7.21	4.27	65.69	0.079	1.21	41.23	4.98	62.70	33.06
L2	QPM	8.44	3.94	66.35	0.086	1.11	44.24	4.33	46.49	49.26
L3	QPM	7.11	4.94	65.17	0.076	1.19	42.04	5.15	56.96	37.88
L4	QPM	7.54	3.78	66.84	0.068	1.01	45.92	6.00	35.27	58.66
L5	QPM	7.80	5.50	63.11	0.091	1.28	38.94	5.21	68.03	26.68
L6	QPM	6.38	3.83	66.72	0.079	1.68	46.56	4.79	44.18	38.19
L7	QPM	8.54	3.54	61.51	0.089	1.01	38.50	5.41	65.21	29.30
L8	QPM	9.43	5.09	63.25	0.076	0.88	46.10	5.92	40.08	54.05
L9	QPM	8.21	4.97	64.36	0.089	1.18	50.65	5.11	57.19	37.79
L10	QPM	9.64	6.31	61.14	0.076	0.83	37.30	5.81	37.52	56.67
L11	QPM	7.22	5.27	65.83	0.078	1.30	35.38	4.44	43.70	51.93
L12	QPM	7.13	3.62	66.91	0.078	1.12	36.07	6.75	62.61	30.56
L13	QPM	6.75	4.26	66.56	0.081	1.21	34.67	5.36	56.16	38.51
L14	QPM	8.63	4.15	60.20	0.083	1.01	33.75	5.48	63.42	31.02
L15	QPM	6.96	4.33	66.32	0.083	1.29	44.08	6.35	64.34	29.39
L16	QPM	8.98	5.06	63.61	0.091	1.07	44.12	4.71	49.60	45.69
L17	QPM	6.69	5.30	65.11	0.085	1.40	48.00	4.81	60.26	34.93

Table 3. *Cont.*

Line	Line Status	Protein %	Oil %	Starch %	Tryptophan %	QI	Amylose %	β-Zein %	γ-Zein %	α-Zein %
L18	non-QPM	7.81	3.47	66.20	0.035	0.51	39.82	6.18	44.82	48.92
L19	non-QPM	8.46	3.65	67.01	0.039	0.70	48.10	6.46	30.40	62.84
L20	non-QPM	8.17	4.50	65.06	0.044	0.71	49.42	4.55	22.71	72.83
L21	QPM	8.57	3.75	65.03	0.046	0.67	44.46	2.59	40.02	57.32
L22	QPM	8.00	4.57	63.45	0.081	1.56	37.23	5.38	51.73	42.90
L23	QPM	8.63	3.92	65.10	0.075	0.96	39.01	4.95	43.68	51.29
L24	QPM	8.24	4.12	65.50	0.085	1.15	43.61	5.71	64.69	29.60
L25	QPM	6.38	4.96	65.30	0.083	1.48	53.18	6.58	70.62	22.89
L26	QPM	8.53	3.86	65.64	0.075	1.06	39.37	3.06	43.13	53.81
L27	non-QPM	9.78	5.75	61.46	0.045	0.45	32.22	3.72	34.26	62.15
L28	non-QPM	7.09	4.07	67.49	0.038	0.65	36.04	5.64	32.26	62.10
L29	non-QPM	7.79	5.34	66.32	0.046	0.69	39.48	3.75	34.95	61.22
L30	QPM	7.95	3.18	66.14	0.043	0.58	44.74	6.62	57.88	35.43
L31	non-QPM	8.06	3.90	65.71	0.045	0.72	39.40	2.35	29.17	68.40
L32	non-QPM	9.30	4.09	63.85	0.040	0.46	31.32	3.69	30.40	65.99
L33	QPM	6.63	5.65	65.38	0.086	1.73	50.20	4.27	57.41	38.32
L34	QPM	6.53	5.64	65.01	0.075	1.30	45.43	4.58	64.01	31.49
L35	non-QPM	8.17	4.70	64.68	0.045	0.56	46.37	3.43	31.41	65.17
L36	non-QPM	7.16	4.77	66.42	0.034	0.59	31.88	6.85	29.98	63.16
L37	QPM	6.60	4.06	67.10	0.066	1.29	37.68	4.20	68.67	27.04
L38	QPM	7.53	5.04	66.19	0.079	1.38	39.50	5.41	47.62	44.80
L39	non-QPM	9.49	4.97	61.85	0.041	0.42	45.60	5.40	33.52	61.20
L40	QPM	6.77	4.86	66.10	0.076	1.36	44.49	4.93	55.98	39.08
QPM mean		7.69	4.54	64.99	0.078	1.18	42.29	5.13	54.45	39.92
Non-QPM mean		8.30	4.47	65.10	0.041	0.59	40.00	4.73	32.17	63.09
Total mean		7.86	4.52	65.02	0.068	1.02	41.68	5.02	48.33	46.29
LSD (0.05)		0.47	0.42	1.08	0.0051	0.16	2.62	1.07	8.11	8.20

QPM = quality protein maize; non-QPM = non quality protein maize; QI = quality index.

The values of significant correlations were low (Table 4), except for those for α- and γ-zeins, which were directly negatively correlated (r = −0.94). For the rest, only correlations of higher than 0.3 and $p < 0.01$ are mentioned. α-Zein content was negatively correlated with tryptophan content, while β-zein content was significantly negatively correlated with protein content. γ-Zein content was positively correlated with tryptophan content.

Table 4. Correlations between measured characteristics across all locations.

Characteristic 1	Characteristic 2	Correlation
α-Zein	γ-Zein	−0.94 **
	Tryptophan	−0.47 **
Amylose	Starch	0.18 *
β-Zein	Protein	−0.31 **
	Starch	0.15 *
Oil	Protein	0.22 **
	Tryptophan	0.19 *
γ-Zein	Tryptophan	0.41 **
Tryptophan	Starch	0.20 *

* $p < 0.01$, ** $p < 0.001$.

4. Discussion

With the development of QPM, it was reported that the protein content of QPM and that of non-QPM were similar [27], but in the inbred lines in this study, the protein content of the non-QPM was significantly higher (by 7%) than that of the QPM. Oil, starch, amylose, and β-zein were similar for the two groups of material, but there was a wide range of values for these characteristics, indicating that there are QPM lines with high protein content.

QPM contains the mutant gene *opaque* 2 (*o2*), which is responsible for the reduction in α-zeins, which contain no lysine or tryptophan, with an increase in non-zein proteins. Originally, this gene caused soft endosperm, which was overcome by using *o2* modifiers that ensured hard endosperm without the loss of the high lysine trait [28,29]. In this study, the tryptophan content of the QPM lines was 1.9 times that of non-QPM; this was also reflected in the QI, which was double in QPM compared to non-QPM. For a maize grain to be classified as a QPM, it must have a QI equal to or above 0.8 [30,31]. In *o2* maize, the zein fraction was found to be reduced by 50% [14], but it had higher contents of non-zein protein (albumin, globulin, and glutelin fractions), which are rich in lysine and tryptophan [32]. Therefore, the presence of the *o2* gene in the QPM lines would explain the large decrease in α-zeins in the QPM lines compared to non-QPM and the simultaneous increase in γ-zein in the QPM, where γ-zein was highly significantly correlated with tryptophan content. The highly significant negative correlation between α- and γ-zeins supports this finding. γ-Zeins are subdivided into three classes, namely, 16 kDa, 27 kDa, and 50 kDa. QPM lines were reported to have 2–3-fold higher levels of the 27 kDa γ-zein than non-QPM lines [32]. This may have caused the increase in the γ-zein fraction in the QPM lines. It was reported [33] that maize kernels with hard endosperm produced more α-zein (22 kDa and 19 kDa), while the intensity of these bands declined in QPM genotypes [19,33]. In the current study, the α-zein concentration was also much lower in QPM lines.

The significant genotype-by-environment interaction, observed for all the quality characteristics analysed except for β-zein, indicates that the environment plays an important role in the expression of these characteristics and protein fractions, changing the ranking of lines in different environments. Differences in amino acid contents, as a result of variations in soil nitrogen contents in different soybean varieties and different planting dates of buckwheat, were reported [34,35], demonstrating the environmental influences on protein composition, which was also the case in this study. This indicates that not only the line but also the environment influence the nutritional value of the grain.

Generally, the range of starch content values in the present study for the inbred lines is comparable to the values reported previously [36,37], with ample variation in all lines. Similarly, amylose, which is a major constituent of starch, varied significantly for the lines evaluated.

Some QPM lines, such as L16, L7, and L2, had a tryptophan content higher than 0.086 and could still maintain a protein content of 8.98, 8.54 and 8.44%, respectively. These lines can be used as parents in hybrid crosses in order to enhance the protein and tryptophan content in the seed.

5. Conclusions

The yield potential of the QPM and non-QPM inbred lines was similar (data not shown) but the QPM lines had much higher tryptophan and QI, and a much lower amount of the lysine- and tryptophan-poor α-zein fraction. Within the QPM lines, there was significant variation in protein content and α- and γ-zeins, as well as in starch and amylose content. The QPM lines had much better protein quality than that of the non-QPM lines, although the protein content was 7% lower on average. Some QPM lines were identified to have acceptable yield and to have both high protein quantity and quality; these can be used for future hybridisation to produce highly nutritious hybrids, which can be distributed to farmers in the region to address malnutrition caused by a maize-based diet.

Author Contributions: Conceptualisation, A.T. and M.L.; methodology, M.L., N.S., A.v.B. and I.A.; formal analysis, I.A. and M.L.; investigation, I.A.; resources, A.T., N.S. and M.L.; data curation, M.L. and I.A.; writing—I.A.; writing—review and editing, M.L., N.S., A.T., A.v.B. and I.A.; supervision, M.L., A.v.B., N.S. and A.T.; funding acquisition, M.L., N.S. and A.T. All authors have read and agreed to the published version of the manuscript.

Funding: This research was funded by the National Research Foundation through the South African Research Chairs Initiative (UID 84647). Field trials were funded by the Agricultural Research Council and CIMMYT (Zimbabwe).

Institutional Review Board Statement: Not applicable.

Informed Consent Statement: Not applicable.

Data Availability Statement: Data are available from the corresponding author.

Acknowledgments: We thank the Agricultural Research Council and CIMMYT (Zimbabwe) for their support in carrying out field trials.

References

1. Sarika, K.; Hossain, F.; Muthusamy, F.; Zunjare, R.U.; Baveja, A.; Goswami, R.; Bhat, J.S.; Saha, S.; Gupta, H.S. Marker-assisted pyramiding of *opaque2* and novel *opaque16* genes for further enrichment of lysine and tryptophan in sub-tropical maize. *Plant Sci.* **2018**, *272*, 142–152. [CrossRef]
2. Black, R.E.; Victora, C.G.; Walker, S.P.; Bhutta, Z.A.; Christian, P.; De Onis, M.; Ezzati, M.; Grantham-Mcgregor, S.; Katz, J.; Martorell, R.; et al. Maternal and child undernutrition and overweight in low-income and middle-income countries. *Lancet* **2013**, *382*, 427–451. [CrossRef]
3. Shiferaw, B.; Prasanna, B.M.; Hellin, J.; Bänziger, M. Crops that feed the world 6. Past successes and future challenges to the role played by maize in global food security. *Food Sec.* **2011**, *3*, 307–327. [CrossRef]
4. Ai, Y.; Jane, J.L. Macronutrients in corn and human nutrition. *Comp. Rev. Food Sci. Food Safety* **2016**, *15*, 581–598. [CrossRef]
5. Singh, A.A.; Agrawal, S.B.; Shahi, J.P.; Agrawal, M. Yield and kernel nutritional quality in normal maize and quality protein maize cultivars exposed to ozone. *J. Sci. Food Agric.* **2019**, *99*, 2205–2214. [CrossRef]
6. Salleh, S.N.; Fairus, A.A.H.; Mohd Nizam Zahary, M.N.; Raj, N.B.; Jalil, A.M.M. Unravelling the effects of soluble dietary fibre supplementation on energy intake and perceived satiety in healthy adults: Evidence from systematic review and meta-analysis of randomised-controlled trials. *Foods* **2019**, *8*, 15. [CrossRef] [PubMed]
7. Krivanek, A.F.; De Groote, H.; Gunaratna, N.; Diallo, A.; Friesen, D. Breeding and disseminating quality protein maize (QPM) for Africa. *Afr. J. Biotech.* **2007**, *6*, 312–324.
8. Sofi, P.A.; Wani, S.A.; Rather, A.G.; Wani, S.H. Quality protein maize (QPM): Genetic manipulation for nutritional fortification of maize. *Crop Sci.* **2009**, *1*, 244–253.
9. Babu, R.; Prasanna, B.M. Molecular Breeding for Quality Protein Maize (QPM). In *Genomics of Plant Genetic Resources*; Tuberosa, R., Graner, A., Frison, E., Eds.; Springer: Berlin/Heidelberg, Germany, 2014; pp. 490–505.
10. Teklewold, A.; Wegary, D.; Tadesse, A.; Tadesse, B.; Bantte, K.; Friesen, D.; Prasanna, B.M. *Quality Protein (QPM) Maize. A Guide to the Technology and Its Promotion in Ethiopia*; Nutritious Maize for Ethiopia (NuME) Project; CIMMYT: Addis Ababa, Ethiopia, 2015.
11. Mbuya, K.; Nkongolo, K.K.; Kalonji-Mbuyi, A. Nutritional analysis of quality protein maize varieties selected for agronomic characteristics in a breeding program. *Int. J. Plant Breed. Gen.* **2011**, *5*, 317–327. [CrossRef]
12. Wegary, D.; Labuschagne, M.T.; Vivek, B.S. Protein quality and endosperm modification of quality protein maize (*Zea mays* L.) under two contrasting soil nitrogen environments. *Field Crops Res.* **2011**, *121*, 408–415. [CrossRef]
13. Tandzi, L.N.; Mutengwa, C.S.; Ngonkew, E.L.M.; Woïn, N.; Gracen, V. Breeding for quality protein maize (QPM) varieties: A review. *Agronomy* **2017**, *7*, 80. [CrossRef]
14. Prasanna, B.M.; Vasal, S.K.; Kasshun, B.; Singh, N.N. Quality protein maize. *Curr. Sci.* **2001**, *81*, 1308–1319.
15. Semagn, K.; Magorokosho, C.; Vivek, B.S.; Makumbi, D.; Beyene, Y.; Mugo, S.; Prasanna, B.M.; Warburton, M.L. Molecular characterization of diverse CIMMYT maize inbred lines from eastern and southern Africa using single nucleotide polymorphic markers. *BMC Genom.* **2012**, *13*, 113. [CrossRef] [PubMed]
16. Wegary, D.; Vivek, B.S.; Labuschagne, M.T. Genetic relationships and heterotic structure of quality protein maize (*Zea mays* L.) inbred lines adapted to eastern and southern Africa. *Euphytica* **2018**, *214*, 1–11. [CrossRef]
17. Njeri, S.G.; Makumbi, D.; Warburton, M.L.; Diallo, A.; Jumbo, M.B.; Chemining'wa, G. Genetic analysis of tropical quality protein maize (*Zea mays* L.) germplasm. *Euphytica* **2017**, *213*, 1–19. [CrossRef] [PubMed]
18. Sood, A.; Thakur, K.; Sharma, P.N.; Gupta, D.; Singode, A.; Rana, R.; Lata, S. A comprehensive study of variation in selected QPM and Non-QPM maize inbred lines. *Agric. Res.* **2017**, *6*, 103–113. [CrossRef]
19. Chandel, U.; Mankotia, S.B.; Thakur, K. Evaluation of CIMMYT maize (*Zea mays* L.) germplasm by tropical inbred testers. *Bangl. J. Bot.* **2014**, *43*, 131–139. [CrossRef]
20. Dombrink-Kurtzman, M.A. Examination of opaque mutants of maize by reversed-phase high-performance liquid chromatography and scanning electron microscopy. *J. Cer. Sci.* **1994**, *19*, 57–64. [CrossRef]
21. Gupta, J.; Wilson, B.W.; Vadlan, P.V. Evaluation of green solvents for a sustainable zein extraction from ethanol industry DDGS. *Biomass Bioeng.* **2016**, *85*, 313–319. [CrossRef]
22. O'Kennedy, K.; Fox, G. Zein characterisation of South African maize hybrids and their respective parental lines using MALDI-TOF MS. *Food Anal. Meth.* **2017**, *10*, 1661–1668. [CrossRef]
23. Deja Cruz, N.; Khush, G.S. Rice grain quality evaluation procedures. In *Aromatic Rice*; Singh, R.K., Singh, U.S., Khush, G.S., Eds.; Oxford and IBA Publishing Co: New Dheli, Calcuta, 2000.
24. Nurit, E.; Tiessen, A.; Pixley, K.; Palacios-Rojas, N. Reliable and inexpensive colorimetric method for determining protein-bound tryptophan in maize kernels. *J. Agric. Food Chem.* **2009**, *57*, 7233–7238. [CrossRef]

25. Caprita, R.; Caprita, A.; Cretescu, I. Effective extraction of soluble non-starch polysaccharides and viscosity determination of aqueous extracts from wheat and barley. In *Proceedings of the World Congress on Engineering and Computer Science II, San Francisco, CA, USA, 19–21 October 2011*; Newswood Academic Publishing: Hong Kong, China, 2011; Volume 2, pp. 1–3.

26. VSN International. Genstat for Windows 17th Edition. 2021. Available online: vsni.co.uk/software/genstat (accessed on 10 January 2022).

27. Kaur, N.; Singh, B.; Sharma, S. Comparison of quality protein maize (QPM) and normal maize with respect to properties of instant porridge. *LWT Food Sci. Tech.* **2019**, *99*, 291–298. [CrossRef]

28. Ngaboyisonga, C.; Njoroge, K.; Kirubi, D.; Githiri, S.M. Quality protein maize under low nitrogen and drought environments: Endosperm modification, protein and tryptophan concentration in grain. *Agric. J.* **2012**, *7*, 327–338. [CrossRef]

29. Bello, O.B.; Olawuyi, O.J.; Ige, S.A.; Mahamood, J.; Afalobi, M.S.; Azeez, M.A.; Abdulmaliq, S.Y. Agro-nutritional variations of quality protein maize (*Zea mays* L.) in Nigeria. *J. Agric. Sci.* **2014**, *59*, 101–116. [CrossRef]

30. Twumasi-Afriye, S.; Rojas, P.N.; Friesen, D.; Teklewold, A.; Gissa, D.W.; De Groote, H.; Prasanna, B.M. *Guidelines for the Quality Control of Quality Protein Maize (QPM) Seed and Grain*; CIMMYT: Addis Ababa, Ethiopia, 2016.

31. Wu, Y.; Holding, D.R.; Messing, J. γ-Zeins are essential for endosperm modification in quality protein maize. *Proc. Natl. Acad. Sci. USA* **2010**, *107*, 12810–12815. [CrossRef]

32. Vivek, B.S.; Krivanek, A.F.; Palacios-Rojas, N.; Twumasi-Afriyie, S.; Diallo, A.O. *Breeding Quality Protein Maize (QPM): Protocols for Developing QPM Cultivars*; CIMMYT: Addis Ababa, Ethiopia, 2008.

33. Mansilla, P.S.; Nazar, M.C.; Pérez, G.T. Evaluation and comparison of protein composition and quality in half-sib families of *opaque-2* maize (*Zea mays* L.) from Argentina. *Argentina* **2019**, *36*, 39–53. [CrossRef]

34. Zimmer, S.; Messmer, M.; Haase, T.; Piephoe, H.P.; Mindermann, A.; Schulz, H.; Habekuss, A.; Ordon, F.; Wilboisg, K.P.; Heß, J. Effects of soybean variety and Bradyrhizobium strains on yield, protein content and biological nitrogen fixation under cool growing conditions in Germany. *Eur. J. Agron.* **2016**, *72*, 38–46. [CrossRef]

35. Siracusa, L.; Gresta, F.; Sperlinga, E.; Ruberto, G. Effect of sowing time and soil water content on grain yield and phenolic profile of four buckwheat (*Fagopyrum esculentum* Moench.) varieties in a Mediterranean environment. *J. Food Comp. Anal.* **2017**, *62*, 1–7. [CrossRef]

36. Sharma, S.; Carena, M.J. Grain quality in maize (*Zea mays* L.): Breeding implications for short-season drought environments. *Euphytica* **2016**, *212*, 247–260. [CrossRef]

37. Lu, X.; Chen, J.; Zheng, M.; Guo, J.; Qi, J.; Chen, Y.; Miao, S.; Zheng, B. Effect of high-intensity ultrasound irradiation on the stability and structural features of coconut-grain milk composite systems utilizing maize kernels and starch with different amylose contents. *Ultrason Sonochem.* **2019**, *55*, 135–148. [CrossRef]

 *foods*

Review

Wheat/Gluten-Related Disorders and Gluten-Free Diet Misconceptions: A Review

Carolina Sabença [1,2,3], Miguel Ribeiro [1,2,3], Telma de Sousa [1,2,3], Patrícia Poeta [3,4], Ana Sofia Bagulho [5] and Gilberto Igrejas [1,2,3,*]

1. Department of Genetics and Biotechnology, University of Trás-os-Montes and Alto Douro, 5000-801 Vila Real, Portugal; carolinasabenca@hotmail.com (C.S.); jmribeiro@utad.pt (M.R.); telmaslsousa@hotmail.com (T.d.S.)
2. Functional Genomics and Proteomics Unity, University of Trás-os-Montes and Alto Douro, 5000-801 Vila Real, Portugal
3. LAQV-REQUIMTE, Faculty of Science and Technology, University Nova of Lisbon, 2829-546 Lisbon, Portugal; ppoeta@utad.pt
4. Microbiology and Antibiotic Resistance Team (MicroART), Department of Veterinary Sciences, University of Trás-os-Montes and Alto Douro, 5000-801 Vila Real, Portugal
5. National Institute for Agrarian and Veterinarian Research (INIAV), Estrada Gil Vaz, Ap. 6, 7350-901 Elvas, Portugal; ana.bagulho@iniav.pt
* Correspondence: gigrejas@utad.pt

Abstract: In the last 10,000 years, wheat has become one of the most important cereals in the human diet and today, it is widely consumed in many processed food products. Mostly considered a source of energy, wheat also contains other essential nutrients, including fiber, proteins, and minor components, such as phytochemicals, vitamins, lipids, and minerals, that together promote a healthy diet. Apart from its nutritional properties, wheat has a set of proteins, the gluten, which confer key technical properties, but also trigger severe immune-mediated diseases, such as celiac disease. We are currently witnessing a rise in the number of people adhering to gluten-free diets unwarranted by any medical need. In this dynamic context, this review aims to critically discuss the nutritional components of wheat, highlighting both the health benefits and wheat/gluten-related disorders, in order to address common misconceptions associated with wheat consumption.

Keywords: wheat; nutrients; celiac disease; wheat allergy; non-celiac wheat/gluten sensitivity

Citation: Sabença, C.; Ribeiro, M.; Sousa, T.d.; Poeta, P.; Bagulho, A.S.; Igrejas, G. Wheat/Gluten-Related Disorders and Gluten-Free Diet Misconceptions: A Review. *Foods* **2021**, *10*, 1765. https://doi.org/10.3390/foods10081765

Academic Editors: Barbara Laddomada, Antonella Pasqualone and Weiqun Wang

Received: 19 May 2021
Accepted: 28 July 2021
Published: 30 July 2021

Publisher's Note: MDPI stays neutral with regard to jurisdictional claims in published maps and institutional affiliations.

1. Introduction

The domestication of wheat revolutionized the human diet as this cereal provided a significant source of energy. Globally, wheat accounts for the largest harvested area of any crop [1] and provides more protein and calories than any other cereal crop [2]. Wheat is nutritious, simple to transport and store, and can be transformed into several types of food. The most valuable modern wheat species are hexaploid bread wheat (*Triticum aestivum* L.) and tetraploid durum wheat (*T. turgidum* L. var. *durum*), which have distinct genomes, grain composition, and end-use quality attributes. Wheat adapts to all climatic conditions common in agricultural fields (except for the hot tropics), so globally, it is harvested all year round [3].

Wheat is a valuable source of essential nutrients, providing carbohydrate-based energy and fiber, protein, B vitamins, calcium, magnesium, phosphorus, potassium, zinc, and iron [4]. In low and medium-income countries, grain-based foods still make up the central part of the diet. The wheat seed can be ground into flour or semolina, for example, which form the essential ingredients of bread, pasta, noodles, and other food products, essentially the primary source of nutrients for most of the world population [5]. Conversely, the lack of grains too often signifies hunger and malnutrition. The characteristic that has given wheat an advantage over other temperate crops is the unique viscoelastic properties of dough formed from wheat flours, which allow it to be processed into such an array of forms [6]. Dough viscoelasticity depends on the structures and interactions that occur between grain storage proteins that form the gluten protein complex [7].

Gluten, which is now an almost ubiquitous ingredient in the food industry, is implicated in several immune-mediated disorders, such as celiac disease (CD). Both CD and other intolerances are of increasing concern [8,9], and the prevalence of CD is predicted to rise [10]. These disorders demand a gluten-free diet (GFD), but a GFD can itself be associated with digestive problems due to insufficient intake of dietary fiber and other nutrients [11].

This review focuses on wheat from a human health perspective. We will present the positive impacts of wheat, referring to the benefits of the different components of the wheat grain on human health, and juxtapose this with the negative impacts on the health of sensitive and genetically susceptible individuals caused by wheat components. At the same time, we draw attention to common gluten-related misconceptions and try to demystify them.

2. The Health Benefits of Wheat

Wheat grain is composed of the germ (2–3%), the bran (13–17%), and the endosperm (80–85%) [5] (Figure 1). Wheat germ is the embryo of the wheat kernel and is relatively rich in protein, lipids, and several of the B-vitamins [5,12]. Whole-wheat flour includes the bran, which contains a limited amount of protein, larger quantities of the B-complex vitamins, trace minerals, and indigestible cellulose material called dietary fiber [5,12]. White flour originates from the endosperm. The endosperm contains most of the protein in the whole kernel, iron, carbohydrates, and many B-complex vitamins, such as riboflavin, thiamine, and niacin [5,12].

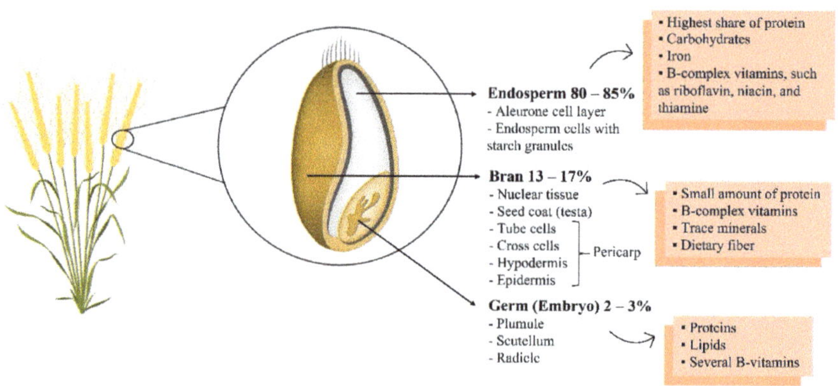

Figure 1. Wheat grain constitution.

The consumption of wheat brings many health benefits. In the European Prospective Investigation into Cancer and Nutrition (EPIC) study populations, 27% of total carbohydrate intake was from bread [13]. Epidemiological studies show that cereal dietary fiber and wholegrain consumption protects against the fast increasing chronic diseases related to a sedentary lifestyle, such as type 2 diabetes and cardiovascular disease [14–17].

2.1. Proteins

Protein is an essential nutrient for humans and animals [5]. Protein content is used to classify wheat. Breeders target this feature by regularly selecting for protein content traits in breeding programs; wheat with a low protein content is suitable for animal feed other uses, while wheat with a high protein content is necessary for breadmaking [4,7]. Protein content differs depending on the growing conditions, type or class of wheat, and fertilizer inputs, especially nitrogen [4,18]. Thus, there is no such thing as typical protein content, but on average, it can vary between 9–18% of the grain weight [4,5,19–22]. Protein is unequally distributed in the grain. A percentage of 5.1% of protein was reported in the pericarp, 5.7% in the testa, 22.8% in the aleurone, and 34.1% in the germ [21]. T.B. Osborne demonstrated that wheat proteins could be classified according to their extractability and solubility in distinct solvents [23]. Globulins are insoluble in pure water, and high NaCl concentrations but soluble in dilute NaCl solutions; albumins are soluble in water; glutenins are soluble in dilute acid or sodium hydroxide solutions, and gliadins are soluble in 70% ethyl alcohol [5].

Of the 20 amino acids commonly present in proteins, lysine, leucine, isoleucine, phenylalanine, threonine, tyrosine, tryptophan, histidine, valine, and methionine (and potentially cysteine since it can only be synthesized from methionine) are considered essential because they must be provided in the diet as animals cannot synthesize them [7]. Average contents of essential amino acids reported for whole wheat, wholemeal, and white flour are compared with the physiological requirements for older children, adolescents, and adults in Table 1. All cereals have a low content of lysine. In wheat, barley, rye, corn, and oats, methionine content is also low. Both amino acids are substantially lower in flour than in meat, milk, or egg proteins [24]. The data in Table 1 support the notion that lysine is the most limiting amino acid in wheat grains, with other essential amino acids being present in adequate amounts for older children, adolescents, and adults, making this cereal an excellent food for nutrition for any age group. Wheat breeders have attempted to improve the grain's essential amino acid content. The approach has been effective in producing high-lysine barley and numerous maize cultivars [25]. Additional efforts were made using genetic engineering approaches to increase the synthesis and reduce the catabolism of these essential amino acids and express enriched recombinant proteins [26].

Table 1. Essential amino acid levels recommended for older children, adolescents, and adults compared with those in whole wheat, wholemeal wheat, and white flour (expressed as g/g protein).

Amino Acids	FAO Recommended Intake Levels			Amino Acids Content		
	Older Children (Age 11–14)	Adolescents (Age 15–18)	Adults (Age > 18)	Whole Wheat	Wholemeal Wheat	White Flour
Histidine	0.016	0.016	0.015	0.022	0.0266	0.0269
Isoleucine	0.030	0.030	0.030	0.038	0.0314	0.0309
Leucine	0.061	0.060	0.059	0.067	0.0594	0.0565
Lysine	0.048	0.047	0.045	0.027	0.0288	0.0222
Methionine + cysteine (sulfur amino acids)	0.023	0.023	0.022	0.039	0.0363	0.033
Phenylalanine + tyrosine (aromatic amino acids)	0.041	0.040	0.038	0.077	0.0544	0.0514
Threonine	0.025	0.024	0.023	0.029	0.0254	0.0224
Tryptophan	0.0066	0.0063	0.006	0.012	-	0.0085
Valine	0.040	0.040	0.039	0.047	0.0388	0.0354
Authors (Reference)	Food and Agriculture Organization of the United Nations [27]			Khan et al. [4]	Shewry et al. [21]	

2.2. Carbohydrates

The wheat grain consists of 85% carbohydrate at maturity, 80% of which is the starchy endosperm. The non-starch carbohydrate is constituted of approximately 7% mono-, di-, and oligosaccharides and fructans, along with about 12% of cell wall polysaccharides [28,29]. In addition to being an essential energy source in the human nutrition and animal feed [30], wheat starch is the substrate for the production of alcoholic beverages and fuel ethanol by fermentation [31] and is the raw material for several other industries [32]. Polysaccharides are the main structural elements of the protoplasts walls present in all cells of the grain tissues. The cell wall polysaccharides are essential in human diet as sources of dietary fiber and have an impact on end-use quality and grain consumption [32].

Carbohydrates are recognized by WHO/FAO [33] as the macronutrient humans need to consume the most. Many countries have nutritional guidelines that emphasize the importance of cereals and cereal carbohydrates as the foundation of a healthy diet [34], mainly because the primary benefit of carbohydrates is as a source fuel, glucose. All body tissues, including brain tissue, require glucose. While the brain consist of only 2% of body mass, it uses 20% of the fuel [35]. Dietary carbohydrate is also vital in ensuring gastrointestinal integrity and function and glycemic homeostasis. Unlike protein and fat, high levels of complex carbohydrate are not associated with adverse health consequences to the extent that diets high in complex carbohydrates are less likely to lead to obesity and its morbid consequences than diets high in fat [36]. In an ideal diet, at least 55% of total energy should come from carbohydrates obtained from various food sources [37].

2.3. Lipids

Lipids are a minor constituent of wheat, mostly in the germ, making up 3–4% of the whole grain weight and 1–2.5% of directly milled flour [29]. Lipids have a critical role in baking processes, dough mixing, and the acceptance of the finished products by consumers. Their ability to associate with gluten proteins and form complexes contributes to stabilizing the gas-cell structure, significantly influencing the final texture of baked products and loaf volume [38].

Wheat grain lipids can be classified into polar and non-polar lipids. In all membranes we can find polar lipids such as phospholipids and glycolipids. Half of the total non-polar lipids in wheat are triglycerides that are deposited in spherosomes surrounded by monolayer membranes. The remaining non-polar lipids are mono- and diglycerides, sterol esters, and fatty acids. Wheat lipids can also be divided into saponifiable and non-saponifiable lipids. Glycolipids, acylglycerols, fatty acids, sterols, and phospholipids are saponifiable lipids. Tocopherols and carotenoids are non-saponifiable lipids [24,29].

Palatability, including aroma, texture or juiciness, and taste are improved by lipids, since they carry fat-soluble flavor molecules. The satiety value of foods is increased by the slow movement of lipids in the gastrointestinal tract [29,39]. Lipids play an essential role in our diet through vital biochemical and physiological processes. They are a source of high energy and form the structure of cell membranes. Fats, oils, specific fat-soluble vitamins, hormones, and most non-protein membrane components are lipids.

2.4. Minerals

Over two billion people suffer from micronutrient deficiency according to the World Health Organisation (WHO). The most predominant micronutrient deficiencies are iron, zinc, iodine, and vitamin A [40]. Proximally a third of the world's population is affected by iron and zinc deficiencies [41]. The determining factor in mineral malnutrition is insufficient dietary intake, so nutrition is the most potent environmental factor that can be targeted to reduce the problem over the course of an individual's lifetime [42,43].

Iron is concentrated in the aleurone and zinc in the embryo [44]. Deficiencies in iron and zinc micronutrients are common in populations that consume wheat as a staple because wheat products are usually low in bioavailable forms of these micronutrients. In wheat, two features contribute largely to the low content in bioavailable iron and zinc: the most consumed form is white flour, which contains low concentrations of these minerals, and the existence of phytates in mineral-rich bran fractions that retain minerals in a form that is not bioavailable.

Interest in improving the mineral and vitamin contents of cereal crops has been growing in the last 20 years [45]. Genetic modification and conventional breeding are the two main biofortification strategies. Conventional breeding in conjunction with foliar application of $ZnSO_4$ was used to develop high-zinc types of wheat [46]. This approach has not worked for iron. Transgenic approaches have increased iron and zinc contents in white flour by transforming the starchy endosperm tissue into a "sink" for minerals [47]. The overexpression of metal transporter genes increased single mineral content in starchy endosperm cells according to the respective highly specific metal transporters targeted. For example, expression of a wheat vacuolar iron transporter (TaVIT2) under the control of an endosperm-specific promoter more than doubled the iron content of the white flour fraction [48], while expression of the barley metal tolerance protein 1 (HvMTP1) using an identical promoter considerably increased the zinc content in the endosperm of barley grains [49].

Moreover, selenium (Se) is an essential micronutrient for regular cell metabolism in animals and humans, being present as selenocysteine in several enzymes [50], but no function is known in plants. For over a century, Se was known only as a toxin [51], but in the late 1950s, it was first recognized as an essential micronutrient for animals [52]. Many people worldwide have a low dietary intake of Se. This is due to the low bioavailability of Se in some soils and, therefore, low Se concentrations in plant tissues. In livestock, Se deficiency is common, causing diseases such as white-muscle disease in cattle and sheep. In humans, severe Se deficiency has been associated with two conditions: Keshan disease, a cardiomyopathy occurring in people living in a geographic area stretching from north-east to south-west China [53–55], and Kaschin-Beck disease, an osteoarthropathy occurring in China and less widely in south-east Siberia [53,56].

In most diets, meats, fish, and cereals are the primary sources of Se [54]. Wheat grain Se concentration can vary from about 3 µg/kg in places like the Keshan disease area in China where the Se concentrations in staple crops and Se availability in soils are very low, to over 2000 µg/kg in North and South Dakota in the USA [54]. The minimum nutritional level for humans and animals is about 50–100 µg Se/kg in dry food, and intake below this range may cause Se deficiency [57].

Biofortification strategies to improve wheat Se content have been employed to diminished Se deficiency and those public-related health issues. Using a meta-analysis approach, Ros et al. [58] showed that fertilizers based on selenate could increase Se uptake by crops and consequently Se intake in humans and animals. In Finland, Se biofortification approaches have been practiced commercially in regions deficient in Se by adding to soils Se-amended inorganic fertilizers [59,60]. A solution containing Se is dosed onto the crops leaf surface, enriching the Se content in agricultural products [61]. Ros et al. [58] estimated that the most efficient fertilizer method to increase crop Se uptake in most arable crops is the selenate fertilization of the foliar. Genetically-modified plants have also been developed to improve the uptake of Se from the soil. Possible genetic targets for strengthening the Se content of wheat are in the acquisition and distribution processes, usually catalyzed by transporters. Overexpressing genes encoding transporters for selenite, selenate, or seleno-amino acids in the plasma membrane of particular cells can improve Se uptake and transport capacity inside the plant [62].

2.5. B Vitamins

The B vitamin complex, which at first was thought to be a single compound, comprises eight water-soluble components, which often co-occur in the same foods. They are unequally spread in the wheat kernel and are primarily found in wheat bran and the germ; hence they are present in reduced quantities in refined flours [29]. Cereals are dietary sources of several B vitamins, particularly riboflavin (B2), folates (B9), thiamine (B1), pyridoxine (B6), and niacin (B3) [21]. These molecules play an essential role in metabolism, particularly thiamine in the metabolism of carbohydrates, and riboflavin and pyridoxine in the metabolism of proteins and fats [63]. Consumption of wholemeal products provides 40% of the recommended daily allowance of thiamine, 10% of riboflavin, 22% of niacin, 33% of vitamin B6, and 13% of folate recommended [64]. Niacin is of particular concern as only a proportion of the total present in cereals is bioavailable in a chemically bound form, nicotinic acid [65]. Multiple studies have focused on this issue [66], with some reports of increasing niacin bioavailability by treatment with alkali [67].

Dietary vitamins are required to prevent deficiency disorders [68]. Many of these deficiency diseases, such as beriberi (B1 deficiency), and pellagra (B3 deficiency), are the most common diseases worldwide, and are particularly frequent in developing countries.

2.6. Phytochemicals

Two main groups of phytochemicals, derived from different biosynthetic pathways, are present in wheat grain: phenolics and terpenoids [21].

The primary group of phytochemicals in wheat grain are phenolic acids, but numerous other phenolic compounds have been identified, including lignans, alkylresorcinols, and flavonoids [64]. Phenolic compounds are characterized by at least one aromatic ring carrying at least one hydroxyl group. Phenolic acids have vigorous antioxidant activity, and the total phenolic content is correlated with total antioxidant activity [69,70]. The importance of antioxidant properties for human health is widely discussed; however, evidence that phenolic compounds, including ferulic acid, the major phenolic acid in wheat, improve vascular function in humans is increasing [71–73].

Cereals are also significant sources of terpenoids in the form of sterols and tocols. Sterols in wheat and other plant materials can be classified according to structural and biosynthetic characteristics into desmethyl sterols, 4α-monomethyl sterols, and 4,4-dimethyl sterols [74–76]. Sterols have the capacity to low cholesterol in humans, the health benefits of which are broadly accepted in Europe. Tocols consist of a chromanol ring with a C16 phytol side chain, which can either be saturated (tocopherols, T) or contain three double bonds at carbons 3, 7, and 11 (tocotrienols, T-3). Although the name "vitamin E" is frequently applied to all tocols, they can be distinguished by their biological activity, with α-tocopherol being the most active form [21]. Vitamin E is the most essential lipid-soluble antioxidant in the human body, and together with other antioxidants such as vitamin C they provide an efficient protective network against oxidative stress. α-tocopherol is the most reactive vitamer and is unstable and the first to be broken down [77]. This may reduce its capacity as a long-term antioxidant in food systems, and a combination of tocols is regularly preferred to ensure antioxidant protection. Other studies have shown that tocotrienols might be similar to or even have more potential than tocopherols as antioxidants [78,79].

2.7. Wholegrain

Since the 1900s, when Dr. Thomas Allinson promoted Allinson's bread as a healthier lifestyle, the intake of wholegrain wheat has been promoted for its health benefits [80]. The interest in studying the wholegrain wheat composition has been increasing in order to identify compounds that promote health and better recognize the full potential of wheat in disease avoidance and health [81]. Wholegrain wheat products include various components with recognized or proposed health benefits, including dietary fiber, phenolic acids, carotenoids, flavonoids, sterols, lignans, selenium, magnesium, alkylresorcinols, tocopherols, and B-complex vitamins [7,69,81–85], which are mainly presented in the bran. Consequently, they are either absent or present in lower amounts in white flour, which is almost exclusively derived from endosperm starch cells [7]. While other whole grains may contain similar components, wheat eminence in the diet potentially makes this cereal a more significant contributor to the intake of these compounds [86,87].

Wholegrain-based foods are a valuable source of dietary fiber. About 1 g of dietary fiber is delivered by a simple side of 40 g of white wheat bread, and a similar serving of wholemeal bread would provide 3–4.5 g of dietary fiber [19,88]. This means that the choice of bread alone has a significant effect on dietary fiber intake [19]. However, cereals are mainly consumed as refined products with lower dietary fiber contents, such that today the daily dietary fiber intake is less than the recommended (25–35 g recommended per day) [89].

Wholegrain wheat may protect against the development of diseases related to chronic diet. Extensive cohort studies have described a noticeably reduced risk of cardiovascular disease [15,90,91], type 2 diabetes [14,17,92], and certain forms of cancer [93–95] with increased consumption of wholegrains, and wheat has been identified as a critical food in creating these results [96].

3. Wheat/Gluten-Related Disorders

Wheat/gluten-related diseases can be classified into three different disorders: autoimmune, allergic, and neither autoimmune nor allergic (Figure 2). Celiac disease is the most prominent autoimmune gluten-related disorder (CD). It is a condition of the small intestine caused by gluten and gluten-related proteins and influenced by environmental and genetic factors [97,98]. An IgE and non-IgE mediated immune response characterize wheat allergy (WA), resulting in an allergic reaction in some individuals upon contact, inhalation, or uptake of foods containing wheat but not necessarily other grains as barley or rye. However, IgE-cross reactivity to other cereals is possible in some people [99–101]. Patients with non-celiac wheat/gluten sensitivity (NCWGS) experience identical symptoms to CD, but they do not test positive for CD [102].

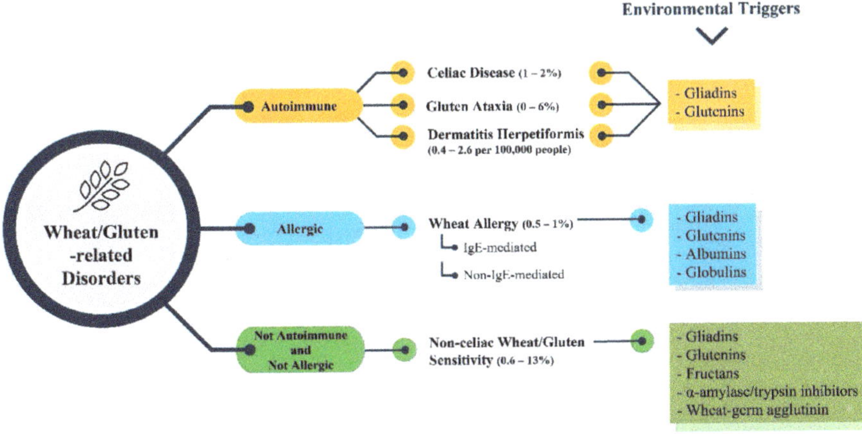

Figure 2. Wheat/gluten-related disorders, their prevalence and environmental triggers.

3.1. Celiac Disease

The binding of gluten peptides to T cells triggers CD in some individuals expressing human leukocyte antigen (HLA) DQ2 or DQ8 in cells specialized in presenting antigens. Specific CD4+ T cells then recognize the presented peptides releasing inflammatory cytokines, leading to changes in the architecture of intestinal mucosa with atrophy and flattening of villi that can lead to total villous degeneration and enteropathy. Moreover, gliadin peptides are responsible for the activation of innate immunity of the intestinal epithelial cells [103,104]. Hence, the gliadin peptides can directly stimulate the immune response of macrophages and dendritic cells through pattern recognition receptors (PRR), such as toll-like receptors (TLRs) 4 [105]. It has also been demonstrated that tissue transglutaminase, an enzyme involved in the deamidation of glutamine residues to glutamate, present in the intestinal epithelium, plays an important role by increasing the binding affinity of gluten peptides to HLA-DQ2 and DQ8 heterodimeric receptors [7].

The expression of the major histocompatibility complex (MHC) class II molecules is related to genetic risk factors. HLA-DQ2 and HLA-DQ8 are the most potent genetic risk factor in CD since they are critical in initiating detrimental immune responses [98]. Nevertheless, additional genetic variations are reported as risk factors in CD, which means HLA-DQ2 and HLA-DQ8 do not account for all the genetic susceptibility to CD [106]. For example, two cytokines implicated in CD pathogenesis are encoded by specific gene polymorphisms of IL-2/IL-21 [107]. In the past, non-HLA genetic risk factors were reported [108]. The consumption of gluten and gluten-related proteins is the main environmental risk factor for CD, which employs a strong immunodominant function in people with a genetic susceptibility to CD [98]. Studies of the timing of gluten introduction in infants' diets suggest that infants who began to receive gluten either before four months or after seven months of age have more risk of developing CD. This conclusion supports the notion that there is a time-space between four and seven months of age during which the introduction of gluten might induce the tolerance to CD [109–111]. However, more recent studies refute this statement, showing no evidence that avoidance of either early (at four months of age) or late (at/after six or even twelve months) gluten introduction puts children at risk of CD [112–115]. Environmental factors can also play a pathogenetic role in the disease. Studies conducted on twins showed that, in 25% of the cases, one of the two twins did not develop CD, supporting this environmental hypothesis [116].

It is well known that the gut microbiota, which is the microorganisms colonizing the human gut, contribute to the development and function of the immune system, being vital for the development of adequate protective immune responses against harmful agents and tolerance to harmless antigens [117]. Various studies already associated gut microbiota imbalances with immune homeostasis disruption and the risk of developing immune-mediated diseases, such as celiac disease (CD), among others [118]. In addition, modifications of the intestinal microbiota during pediatric age [119,120], neonatal infections [121], and recurrent infections involving rotavirus [122,123] were associated with an increased prevalence of CD. A recent longitudinal prospective cohort study developed by Leonard et al. have analysed the gut microbiota of infants at risk of CD to track shifts in the microbiota before CD development. Comparing 10 infants who developed CD and 10 infants who did not, the researchers identified complex patterns of increased abundances of proinflammatory species and decreased abundances of protective and anti-inflammatory species at various time points preceding the onset of the disease. They believe these microbiome shifts, coupled with metabolome findings, may represent potential biomarkers of CD development [124]. However, further studies are necessary since the number of patients examined was low.

3.1.1. Diagnosis

The most common symptoms in adults and children with CD are diarrhea, which may be persistent or intermittent, abdominal pain, fatigue, abdominal distension, vomiting and nausea, constipation, bloating and gas, and weight loss [125]. Gain of gluten intolerance can arise at any time in life due to additional triggers apart from gluten. Trigger factors such as α-interferon, gastrointestinal infections, medications, and surgery have all been indicated [126–128]. For CD diagnosis, a combination of clinical, serological tests and duodenal biopsies is required to check for damage in the intestine caused by the disease [98,129]. Patients with a clinical demonstration of CD should be submitted to serological tests [11,130]. The recommended serological test for the detection of CD is IgA-tTG. As IgA-deficiency affects 2–3% of CD patients and leads to false-negative results, total IgA levels also need to be measured [131,132]. Other antibodies can be used such as anti-deamidated gliadin (anti-DGP), anti-tissue transglutaminase (tTG), and anti-endomysial (EM) antibodies [115,133–135]. In the presence of IgA-deficiency, IgG antibody-based tests (IgG-tTG and/or IgG-DGP) should be used [132]. Depending on the antibody test a duodenal biopsy is performed to establish the definitive diagnosis of CD [136,137]. In agreement with the Marsh classification villous atrophy, hyperplasia, and increased intraepithelial lymphocytes on the duodenal biopsy together to positive serological results confirm celiac disease diagnosis [138]. HLA typing can be used when the results of the serological and duodenal biopsies are inconclusive and the diagnosis of CD is uncertain [139,140].The determination of HLA forms is effective along with histological findings [141,142], knowing that HLA-DQ2 and HLA-DQ8 molecules are correlated with ~95% and 5% CD patients, respectively [142,143].

3.1.2. Treatment

The only currently available therapy for CD is adherence to a gluten-free diet (GFD). A GFD consists of complete avoidance of gluten and gluten-related proteins, which rules out wheat, rye, barley, and any products containing them [131]. Proper adherence to a GFD usually improves the clinical symptoms, increases bone density, and improves body weight distribution or nutritional status [144]. After long-term adherence to GFD, the intestinal villi can be significantly reconstituted [145]. However, GFD may be associated with digestive problems such as constipation due to a low fiber intake. In many cases, a strict GFD is onerous to follow due to gluten residues in certain food products [146,147].

Alternative treatments for CD are being development and the most promising strategies include mechanisms to improve the intestine's permeability, detoxify gluten, or induce modifications in the immune response to gluten [3,98].

3.2. Other Autoimmune Wheat/Gluten-Related Diseases

3.2.1. Gluten Ataxia

Gluten ataxia (GA) is a form of cerebellar ataxia, affecting mainly Purkinje cells, and is caused by antibodies released when digesting gluten that mistakenly attacks part of the brain in individuals that are sensitive and genetically susceptible [148]. The clinical symptoms of GA are identical to those of other ataxias. They include gait ataxia (100%), lower limb ataxia (90%), gaze-evoked nystagmus (84%), upper limb ataxia (75%), ocular signs like dysarthria (66%), and other movement disorders including chorea, myoclonus, opsoclonus myoclonus, and palatal tremor [149].

Diagnosis

GA diagnosis is supported when anti-tTG, anti-gliadin, and anti-TG6 (anti-transglutaminase 6) antibodies are found in the serum. The best diagnostic approach for patients with suspected GA remains unclear. Still, one study reported that considering the whole spectrum of gluten, the IgG anti-gliadin antibody performs better than the gluten ataxia marker because of its elevated sensitivity [149].

Studies of GA patients have shown that the brain presents anti-tTG antibodies. If CD serology is positive, it is necessary to look for evidence of CD through an intestinal biopsy [150]. Magnetic resonance imaging (MRI) can be utilized to diagnose GA. MRI results from up to 60% of GA patients show evidence of moderate cerebellar atrophy [151].

Treatment

Following a rigorous GFD should be the treatment to GA patients. Moreover, studies show that immunotherapy with steroid and intravenous immunoglobulins (IVIG) can be an efficient treatment for such patients [152].

3.2.2. Dermatitis Herpetiformis

Dermatitis herpetiformis (DH), repeatedly associated with CD, is an autoimmune, chronic, and recurrent cutaneous-intestinal disorder detected in genetically susceptible individuals [11,153,154]. Anti-tTG antibodies that also recognize epidermal transglutaminase (ETG) can be produced after exposure to gluten. ETG is homologous to tTG in terms of structure and is the primary antigen in DH [153]. IgA antibody deposition in dermal papillae causes pruritic, vesiculobullous, and localized lesions in DH patients. DH affects the extensor surfaces such as knees, buttocks, elbows, and scapular areas [153,155,156].

Diagnosis

Patients with clinical symptoms are advised to undergo direct immunofluorescence (DIF) tests on perilesional skin. If the test result is negative, new material is collected, and it is determined whether the patient is on a GFD, which could lead to false-negative results [157]. Other confirmatory tests such as the dosage of anti-tTG can be used in patients with symptoms suggesting DH but with negative direct immunofluorescence [158].

Treatment

Like CD patients, DH patients have the same HLA haplotypes (DQ2 and DQ8) and following a GFD improves the symptoms [154]. Drug therapy with dapsone or sulfonamides is also a possible treatment [153].

3.3. Wheat Allergy

Allergens cause allergic reactions, and wheat is one of the five most frequent foods causing them in children. After milk and eggs, wheat is the most common allergen in Japan, Germany, and Finland [159]. In children and adults, wheat allergy (WA) prevalence is approximately 1% depending on age and region [160,161]. In contrast to CD, distinct wheat components such as water-insoluble proteins (gliadin and glutenin) and water/saline-soluble proteins (albumin and globulin) contribute to the development of WA [11,162,163].

3.3.1. IgE-Mediated Wheat Allergy

IgE-mediated WA is triggered by allergen ingestion (food allergy), inhalation (respiratory allergy), or skin contact (dermal allergy). The antigen is introduced by dendritic cells that trigger CD4+ T cells to differentiate into T helper type 2 (Th2) cells. These cells produce cytokines such as IL-4, IL-5, and IL-13 that stimulate B cells to produce IgE [142,164]. When a new exposure to wheat allergens occurs, the IgE antibodies that are bound to their high-affinity receptor (FcεRI) on basophils or mast cells, recognize specific epitopes in wheat allergens [165]. The recognition results in IgE-crosslinking that triggers the release of vasoactive mediators like histamine from mast cells or basophils, leading to allergic responses, including WA [166,167]. The most common symptoms of WA due to these mechanisms include gastrointestinal symptoms (nausea, abdominal pain, vomiting, diarrhea), dermal (itching, eczema, redness), respiratory (rhinitis, asthma), circulatory (flushing, angioedema), and cerebral (disturbed thinking, headache, dizziness) which typically manifest minutes to hours after exposure [164,168].

Wheat-dependent exercise-induced anaphylaxis (WDEIA) is a particular type of IgE-mediated WA. This condition gives rise to severe anaphylactic reactions to wheat when intense exercise is practiced soon after being consumed [169]. Symptoms of WDEIA include angioedema, chest pain, diarrhea, dysphagia, dyspnea, flushing, headache, hoarseness, nausea, pruritus, and syncope [142,170].

Baker's asthma is also an IgE-mediated WA that develops after allergen inhalation, especially cereal flour dust present in the work environment, and affects 0.03–0.24% of pastry factory workers, cereal handlers, confectioners, and bakery workers. It is considered one of the most frequent occupational, cereal-induced allergic asthmas [171–173]. Consuming cooked wheat or products containing it does not manifest symptoms in these patients, but they may react after eating products contaminated with raw wheat flour [174].

IgE-mediated wheat allergens are widely distributed in wheat's different protein fractions. Currently, 28 allergens have been identified in wheat, according to WHO/IUIS Allergen Nomenclature Sub-Committee (Table 2).

The heat-resistant α-amylase/trypsin inhibitor is an allergen that binds to specific IgE and is involved in anaphylaxis, in some cases of WDEIA [175], and baker's asthma [176]. Wheat seeds highly express Tri a 37, which is a plant defence protein. It is also resistant to digestion and heat and can act as a powerful allergen. Individuals who have IgE antibodies against Tri a 37 have a high risk of severe allergic symptoms upon wheat intake [177,178]. ω-5-gliadin, also known as Tri a 19, is involved in anaphylactic reactions to wheat and WDEIA in children [179,180].

Table 2. Wheat proteins implicated in IgE-mediated wheat allergy. Source: www.allergen.org (accessed on 19 July 2021).

Wheat Allergen	Biochemical Name	Molecular Weight (kDa)	Route of Allergen Exposure
Tri a 12	Profilin	14	Food
Tri a 14	Non-specific lipid transfer protein 1	9	Food
Tri a 15	Monomeric alpha-amylase inhibitor 0.28		Airway
Tri a 17	Beta-amylase 56	56	Food
Tri a 18	Agglutinin isolectin 1		Food
Tri a 19	Omega-5 gliadin, seed storage protein	65	Food
Tri a 20	Gamma gliadin	35 to 38	Food
Tri a 21	Alpha-beta-gliadin		Food
Tri a 25	Thioredoxin		Food
Tri a 26	High molecular weight glutenin	88	Food
Tri a 27	Thiol reductase homolog	27	Food
Tri a 28	Dimeric alpha-amylase inhibitor 0.19	13	Food
Tri a 29	Tetrameric alpha-amylase inhibitor CM1/CM2	13	Airway
Tri a 30	Tetrameric alpha-amylase inhibitor CM3	16	Airway
Tri a 31	Triosephosphate isomerase		Airway
Tri a 32	1-Cys-peroxiredoxin		Airway
Tri a 33	Serpin		Airway
Tri a 34	Glyceraldehyde-3-phosphate dehydrogenase		Airway
Tri a 35	Dehydrin		Airway
Tri a 36	Low molecular weight glutenin GluB3-23	40	Food
Tri a 37	Alpha purothionin	12	Food
Tri a 39	Serine protease inhibitor-like protein		Airway
Tri a 40	Chloroform/methanol-soluble (CM) 17 protein [alpha-amylase inhibitor]	15.96	Airway
Tri a 41	Mitochondrial ubiquitin ligase activator of NFKB 1		Food
Tri a 42	Hypothetical protein from cDNA		Food
Tri a 43	Hypothetical protein from cDNA		Food
Tri a 44	Endosperm transfer cell specific PR60 precursor		Food
Tri a 45	Elongation factor 1 (EIF1)		Food

Diagnosis

The diagnosis of WA in its various clinical presentations (whether allergy associated with wheat ingestion, baker's asthma, or WDEIA) depend on taking a detailed clinical history, physical examination, and selecting the proper tests.

The first examinations include a skin prick test and measurement of specific IgE to wheat allergens in wheat extracts and blood serum. In case of wheat allergy due to ingestion, the results of the tests and the clinical history may prompt an oral food challenge. The double-blind placebo-controlled wheat challenge continues to be the gold standard. Oral food challenges are generally considered secure, but experts must perform them prudently because anaphylactic reactions may happen [181]. For diagnosis of WDEIA, in addition to establishing an accurate clinical history, tests for specific IgE against wheat and specific wheat allergens, such as ω-5-gliadin, are performed. These patients may also need to complete a placebo-controlled wheat/exercise challenge, which involves the controlled ingestion of wheat followed 30 min later by 15–20 min of exercise on the treadmill [168,170]. The standard gold diagnosis of baker's asthma is the bronchial challenge test in which patients test positive. In addition, experts establish the clinical history followed by the confirmation of specific IgE to wheat in serum and/or by skin prick test [168].

Treatment

Complete elimination of wheat from the diet is the only available therapy to treat IgE-mediated WA. In allergy associated with wheat ingestion, patients should follow an adequate wheat elimination diet and be trained in the correct interpretation of product labels [168]. To prevent WDEIA, patients must avoid wheat consumption in any circumstance, but if not, they can only exercise 6 h after the consumption of wheat or wheat-containing products [170]. In the case of baker's asthma, a total restriction of exposure to wheat flours is recommended [168].

However, in many cases, strict avoidance of wheat is challenging because wheat is present in so many distinct food products, and involuntary exposure to small traces can occur. Currently, new approaches to treat IgE-mediated WA are actively being sought. Immunotherapy is a promising treatment based on the administration of increasing amounts of an allergenic source to regulate the immune system and achieve remission of allergic symptoms [182]. Three distinct types of immunotherapy are currently being tested: sublingual immunotherapy (SLIT), oral immunotherapy (OIT), and epicutaneous immunotherapy (EPIT). In SLIT and OIT the amount of food ingested is gradually increased to avoid the induction of systemic reactions, while EPIT involves delivering the allergen to the patient using a skin patch [183].

3.3.2. Non-IgE-Mediated Wheat Allergy

Non-IgE-mediated wheat allergy usually occurs 2 h after ingestion of wheat. It is strongly associated with eosinophilic esophagitis (EoE) or eosinophilic gastritis (EG), which occur when eosinophils infiltrate the gastrointestinal tract [168]. Typical manifestations of this type of WA are indigestion, diarrhea, vomiting, arthralgia, and headaches that can appear numerous hours or days after consumption of allergens [11].

Diagnosis

To diagnose EoE, when suspected, an esophageal biopsy is performed by an esophagogastroduodenoscopy (EGD), and it is necessary to find 15 eosinophils per high-power field (eos/hpf) at least. However, the identification of which food causes EoE is more difficult. The gold standard remains: an EGD performance eight weeks after an elimination diet to assess the significance of a food allergen in EoE pathogenesis [168,184,185]. The EG diagnosis is made when clinical symptoms suggest it. Then to confirm a positive diagnosis, a biopsy must show eosinophilic inflammation whit 30 eos/hpf in the stomach and 50 eos/hpf in the duodenum [168,186].

Treatment

The currently accepted treatment to EoE is similar to other atopic diseases and is based on corticosteroid use and allergen avoidance. To treat EoE, steroid treatment for an IgE-mediated food allergy is one convenient approach. Three accepted nutritional strategies can also be used to treat this disease: (1) an elemental diet in which only essential formulas are ingested; (2) avoidance of specific antigens according to allergy testing results and/or diet history; and (3) empiric food elimination of the most common food antigens [187,188]. The adaptation of an EG diet through empiric dietary elimination therapy, consisting of exclusion of common food triggers established for EoE, and very restrictive therapies, consisting of amino acid-based formula ingestion with a few foods, has been tried by pediatric and adult patients, and found to be effective in the majority of pediatric patients [189]. Nevertheless, diet alone is infrequently an effective therapy due to the severity of the symptoms and steroids are necessary to rapidly reduce them. For this reason, the majority of patients are primarily treated with systemic steroids (0.5–1 mg/kg/day for 5–14 days) followed by a gradual decrease over 2–4 weeks [168].

3.4. Non-Celiac Wheat/Gluten Sensitivity

Non-celiac wheat/gluten sensitivity (NCWGS) makes people experience symptoms similar to CD and WA. However, patients with NCWGS do not have specific IgE against wheat proteins or IgA anti-TG2 autoantibodies. The symptoms develop in a few hours or days after wheat/gluten consumption and include abdominal distension, abdominal pain, diarrhea, gas, among others. Patients also experience extraintestinal symptoms, including headache, fatigue, pain in muscles and joints, and eczema [190]. Recent studies have given rise to the idea that other wheat components, such as oligosaccharides like fructans [191], α-amylase/trypsin inhibitors [192], and wheat-germ agglutinin [193], may contribute to the development of NCWGS.

The pathogenic mechanisms of NCWGS are far from understood. Preliminary data indicate that activation of innate immunity triggers NCWGS without the involvement of adaptive immunity, which would be a crucial factor in CD development [194–196]. The increased expression of toll-like-receptors (TLRs), a protein class that plays a vital role in innate immunity, in the small intestine is the evidence supporting the hypothesis of the activation of innate immunity in NCWGS. TLR2, TLR1, and TLR4 have been identified in the intestinal mucosa and some cells of the lamina propria of patients with NCWGS [194]. There is diverging information on intestinal permeability in NCWGS. A study conducted in 2011 determined the gut permeability of NCWGS and CD patients using the urine lactulose/mannitol test. The small intestines of NCWGS patients were significantly less permeable than those of CD patients and controls. Moreover, duodenal biopsies of NCWGS patients found higher expression of claudin-4 mRNA, a marker of reduced permeability [194]. By comparison, another study reported a subgroup of HLA-DQ2/DQ8+ patients with diarrhea-predominant irritable bowel syndrome following a gluten challenge that had increased intestinal permeability [197]. Moreover, Hollon et al. (2015), in an ex vivo study, evaluated alterations in transepithelial electrical resistance (TEER) of tissue biopsies from NCWGS patients, active CD patients, CD patients in remission and controls subjected to pepsin-trypsin digested gliadin. This study has shown that exposure to gliadin increases intestinal permeability and decreases TEER in all patient groups compared to controls [198]. This discrepancy suggests that further studies are required to define the small intestine's permeability in NCWGS and improve our overall knowledge about it.

3.4.1. Diagnosis

Currently, the lack of diagnostic biomarkers for NCWGS means that diagnosis depends on a clinical symptoms evaluation and elimination of CD and WA. According to the Salerno Experts' criteria, first, patients have to adhere to a wheat/gluten exclusion or wheat/gluten-reduced diet in order to reduce the symptoms. Then, to confirm the diagnosis, a double-blind, placebo-controlled gluten challenge must be performed to determine if symptoms were indeed related to wheat/gluten ingestion [199,200]. About half of patients with NCWGS present the first generation antibody to gliadin (AGA) which is considered the only serological marker [199,201,202]. Nevertheless, testing for the presence of AGA is not a specific analysis to diagnose NCWGS. Still, for the moment, its positivity, particularly at a high titer, in suspected NCWGS patients can support the diagnosis [203].

Not long ago, Kabbani et al. reported a diagnostic algorithm based on the specific combination of the presence or absence of several histological, serological, and clinical markers to identify NCWGS and distinguish it from CD. The authors concluded that patients with negative celiac serologies (no IgA/IgG deaminated gliadin peptide or IgA tTG antibody) on a regular diet are improbable to have CD. Those with negative serology who also do not have a clinical indication of malabsorption and CD risk factors are likely to have NCWGS and may not necessitate additional examination. Those with ambiguous serology should be subjected to an HLA typing to establish the requirement for biopsy [204].

A recent discovery has given hope to future NCWGS diagnoses. This study developed by Barbaro et al. verified that NCWGS and CD patients had significantly increased levels of zonulin compared with asymptomatic controls and diarrhea-predominant irritable bowel syndrome (IBS-D) patients. They came to the conclusion that zonulin can be considered a diagnostic biomarker in NCWGS and combined with demographic and clinical data, differentiates NCWGS from IBS-D with high efficiency. Moreover, wheat withdrawal was associated with reducing zonulin levels only in NCWGS carrying HLA genotype [205]. However, further studies are necessary since the number of patients examined was low.

3.4.2. Treatment

The guidelines to treat NCWGS patients are not established yet. The specialists advise these patients to adjust their dietary preferences and begin a GFD [203]. In some cases, any progress after GFD is only partial. In these situations, a low FODMAP (fermentable oligosaccharides, disaccharides, monosaccharides, and polyols) diet together with gluten removal can enhance the clinical condition considerably [206].

Significant research efforts are being made to manage NCWGS. For example, multiple studies have concentrated on analyzing the toxicity of different varieties of wheat. Intriguingly, *Triticum monococcum* ssp. *monococcum*, an ancient diploid wheat, does not activate distinct immune cells involved in gluten-related disorders as much [207]. While clinical studies have shown that CD patients cannot consume these varieties, it has been indicated that they would be safe for patients with NCWGS [208]. Innovative hybridized cereals such as tritordeum have also been shown to be an alternative for NCWGS patients due to their low gliadin content [209]. CRISPR/Cas9 technology has been used to produce wheat with less α-gliadin, translating into an 85% reduction in immunoreactivity [210]. Nevertheless, it should be mentioned that most of this research is designed to tackle gluten proteins, particularly gliadins, and their post-ingestion downstream effects in CD. In NCWGS, the environmental culprit is yet to be well defined.

4. Gluten-Related Misconceptions

GFDs are commonly recognized as the treatment for CD and other gluten-related disorders (GRD), as mentioned above. However, nowadays, the number of people without any GRD who adopt a GFD is rising [211]. The prevalence of adherence to a GFD in the overall adult population can reach 7% in a few countries [212,213]. As of 7th December 2020, a Google search for "gluten-free diet" generated over 4.5 million results. The general population's principal reason for purchasing gluten-free foods is that they are supposed to be healthier than their gluten-containing equivalents [214]. Recommendations from a multitude of books, celebrities, and other media have unquestionably supported the increased consciousness of the potential health benefits of gluten avoidance, such as weight loss [215].

There are three significant misconceptions by the general population leading them to follow a GFD. (1) A gluten-free diet is a healthier option. (2) Eating gluten-free will help them lose weight. (3) The wheat we consume today contains more gluten than older varieties.

Considering the first misconception, claims of the potential benefits of following a GFD include increased energy, better sleep, clearer skin, faster weight loss, and improved medical conditions such as autism and rheumatoid arthritis [214]. Evidence of the health benefits of a GFD for GRD patients is incontrovertible. However, no published experimental evidence supports similar claims for the overall population [216]. On the contrary, an issue associated with unnecessary gluten avoidance is the reduced consumption of whole grains, foregoing the likely benefit of lowering cardiovascular risk. The GFD promotion between people without CD must not be encouraged [217].

Regarding the second misconception, some studies of CD patients report a change in weight as an effect of following a GFD. In a survey of 369 adult patients with CD who followed a GFD for an average of 2.8 years, 22 of the 81 (27%) who were at first overweight increased weight [218]. In another study of 371 adults with CD who adhere to a GFD for two years, 55 of the 67 (82%) initially overweight patients earned weight [219]. Other researchers have reported that between 149 children with CD adhering to a GFD for at least 12 months, the percentage of overweight people almost duplicated (11% to 21%) [220]. These studies suggest that body weight may increase for a considerable portion of overweight celiac patients while on a GFD. However, it has not been established if people without CD or gluten sensitivity would gain weight if they followed a GFD. In this respect, it is essential to note that gluten-free does not imply fat-free or calorie-free, and some gluten-free products contain more calories and sugar than corresponding gluten-containing foods [214]. Indeed, a study carried out in 2018 analyzed the most recent surveys on the nutritional quality of gluten-free products and concluded that the key inadequacies of currently available GF products are a low protein content and a high fat and salt content compared with their gluten-containing counterparts. However, they also verified more acceptable levels of fiber and sugar than in the past [221]. In this way, we can affirm that gluten-free products are not adequated to people wishing to lose weight.

The third misconception that wheat breeding has led to the production of wheat varieties containing higher levels of gluten originated from successful books like "Wheat Belly" by William Davis and "Grain Brain" by David Perlmutter [222]. However, the level of gluten in wheat has actually remained unchanged over the years. A 2013 study reported that gluten levels in numerous varieties, on average, have slightly changed since the 1920s, and although there was actually an increase in CD in the second half of the century, the breeding of wheat for higher gluten content does not suggest to be the reason for that [223]. In 2010, van den Broeck, when studying old and modern wheat varieties toxicity, suggested that breeding practices may have influenced the increased CD prevalence. However, some evidence has shown that modern wheat is not more toxic for celiac patients and that breeding does not seem to be related to a higher prevalence of CD [224]. On the other hand, as nitrogen (N) fertilization of cereal crops has increased, another hypothesis has emerged. Intensified fertilization with N may increase the allergenic proteins content of wheat, which may be related to the increase in CD pathology. The study that put forward this hypothesis concluded, after a literature meta-analysis, that wheat grown under higher N availability in the soil produces not only higher yield but also grains and flour with higher concentrations of gliadin in all genotypes [225]. However, further experimental studies need to be done, and if this hypothesis stands, we will have an important lead to follow to prevent and control the spread of CD.

5. Conclusions

Wheat is the widest cultivated crop on Earth and has been consumed for 10,000 years by humans from its most primitive form to the current species. Wheat is a nutritious cereal, rich in dietary fiber. The nutritional recommendations of many countries emphasize cereals as the basis of a balanced diet. This is particularly true in low and medium-income countries where grain-based food is the main source of energy, carbohydrates, fibers, proteins, B vitamins, and minerals essential for human survival. The exclusive properties of dough made from wheat flour derive from the gluten protein complex and allow it to be processed into bread, pasta and noodles, and other diverse forms of food feeding most of the world population.

Considering the predominance of wheat, the challenge of the increasing incidence of wheat/gluten-related disorders like CD and NCWGS must be addressed now. Patients with CD should strictly follow GFD since they must avoid foods containing gluten, patients with a WA should prevent contact with any form of wheat, and NCWGS patients should follow a wheat/gluten exclusion diet as well. In some cases, adherence to a low FODMAP diet and gluten removal can drastically improve the clinical outlook. There have been many research advances in improving CD and WA diagnosis, but the same does not happen at NCWGS. Thus, first, we have to comprehend the fundamental mechanism behind the NCWGS pathogenicity to establish more sensitive diagnostic markers and therapeutics then.

Only a tiny percentage of the worldwide population is affected by these wheat/gluten-related disorders. Opting or promoting a GFD to improve well-being unwarranted by any medical suggestion is an unhealthy alternative since the consumption of wheat is more beneficial than its non-consumption. Thus, to answer the question "How healthy is to eat wheat?" and the take-home message is that wheat is an excellent food for people without any associated medical conditions because it is a very nutritious cereal, rich in macro and micronutrients that only beneficiate our health. The problem is that people are removing wheat from the diet without any medical indication or health/nutritional condition with a

Foods **2021**, 10, 1765

proven relationship and consequently are not consuming the necessary nutrients. This is a mistake that results from a growing number of misconceptions related to this cereal that should be avoided and clarified as they end up harming these people's health. Here we have presented the medical conditions and the nutritional benefits of consuming wheat, so readers can access unbiased information that clearly shows the best and the worst of this cereal in terms of nutrition and health.

Author Contributions: Conceptualization, M.R.; investigation, M.R. and C.S.; data curation, C.S.; writing—original draft preparation, C.S.; writing—review and editing, C.S., M.R., T.d.S. and G.I.; project administration, G.I.; funding acquisition, G.I., P.P. and A.S.B. All authors have read and agreed to the published version of the manuscript.

Funding: This work was funded by the R&D Project GLUTEN2TARGET-Optimizing natural low toxicity of wheat for celiac patients through a nano/microparticles detoxifying approach, reference POCI-01-0145-FEDER-029068 and PTDC/BAA-AGR/29068/2017, financed by the European Regional Development Fund (ERDF) through COMPETE 2020-Operational Program for Competitiveness and Internationalization (POCI) and by Foundation for Science and Technology (FCT).

Acknowledgments: The authors acknowledge the supported by the Associate Laboratory for Green Chemistry-LAQV, which is financed by FCT under the Partnership Agreement UIDB/50006/2020.

Conflicts of Interest: The authors declare no conflict of interest.

References

1. Food and Agriculture Organization of the United Nations. FAOSTAT Statistics Database Crops. Available online: http://www.fao.org/faostat/en/#data/QC (accessed on 6 October 2020).
2. Food and Agriculture Organization of the United Nations. FAOSTAT Statistics Database New Food Balances. Available online: http://www.fao.org/faostat/en/#data/FBS (accessed on 6 October 2020).
3. Ribeiro, M.; Nunes, F.M.; Rodriguez-Quijano, M.; Carrillo, J.M.; Branlard, G.; Igrejas, G. Next-generation therapies for celiac disease: The gluten-targeted approaches. *Trends Food Sci. Technol.* **2018**, *75*, 56–71. [CrossRef]
4. Khan, K.; Shrewry, P.R. *Wheat: Chemistry and Technology*, 4th ed.; AACC International, Inc.: St. Paul, MN, USA, 2009; ISBN 9780128104545.
5. Šramková, Z.; Gregová, E.; Šturdík, E. Chemical composition and nutritional quality of wheat grain. *Acta Chim. Slovaca* **2009**, *2*, 115–138.
6. Peña, R. Wheat for Bread and Other Foods. In *Bread Wheat Improvement and Production*; Food and Agriculture Organization of the United Nations: Rome, Italy, 2002.
7. Shewry, P.R. Wheat. *J. Exp. Bot.* **2009**, *60*, 1537–1553. [CrossRef]
8. Evans, K.E.; Hadjivassiliou, M.; Sanders, D.S. Is it time to screen for adult coeliac disease? *Eur. J. Gastroenterol. Hepatol.* **2011**, *23*, 833–838. [CrossRef]
9. Mustalahti, K.; Catassi, C.; Reunanen, A.; Fabiani, E.; Heier, M.; McMillan, S.; Murray, L.; Metzger, M.H.; Gasparin, M.; Bravi, E.; et al. The prevalence of celiac disease in Europe: Results of a centralized, international mass screening project. *Ann. Med.* **2010**, *42*, 587–595. [CrossRef]
10. Biesiekierski, J.R.; Newnham, E.D.; Irving, P.M.; Barrett, J.S.; Haines, M.; Doecke, J.D.; Shepherd, S.J.; Muir, J.G.; Gibson, P.R. Gluten Causes gastrointestinal symptoms in subjects without celiac disease: A double-blind randomized placebo-controlled trial. *Am. J. Gastroenterol.* **2011**, *106*, 508–514. [CrossRef]
11. Taraghikhah, N.; Ashtari, S.; Asri, N.; Shahbazkhani, B.; Al-Dulaimi, D.; Rostami-Nejad, M.; Rezaei-Tavirani, M.; Razzaghi, M.R.; Zali, M.R. An updated overview of spectrum of gluten-related disorders: Clinical and diagnostic aspects. *BMC Gastroenterol.* **2020**, *20*, 258. [CrossRef]
12. Kumar, P.; Yadava, R.; Gollen, B.; Kumar, S.; Verma, R.; Yadav, S. Nutritional Contents and Medicinal Properties of Wheat: A Review. *Life Sci. Med. Res.* **2011**, *2011*, 22.
13. Wirfält, E.; McTaggart, A.; Pala, V.; Gullberg, B.; Frasca, G.; Panico, S.; Bueno-de-Mesquita, H.; Peeters, P.; Engeset, D.; Skeie, G.; et al. Food sources of carbohydrates in a European cohort of adults. *Public Health Nutr.* **2002**, *5*, 1197–1215. [CrossRef]
14. De Munter, J.S.L.; Hu, F.B.; Spiegelman, D.; Franz, M.; Van Dam, R.M. Whole grain, bran, and germ intake and risk of type 2 diabetes: A prospective cohort study and systematic review. *PLoS Med.* **2007**, *4*, e261. [CrossRef]
15. Mellen, P.B.; Walsh, T.F.; Herrington, D.M. Whole grain intake and cardiovascular disease: A meta-analysis. *Nutr. Metab. Cardiovasc. Dis.* **2008**, *18*, 283–290. [CrossRef]
16. Nettleton, J.A.; McKeown, N.M.; Kanoni, S.; Lemaitre, R.N.; Hivert, M.F.; Ngwa, J.; Van Rooij, F.J.A.; Sonestedt, E.; Wojczynski, M.K.; Ye, Z.; et al. Interactions of dietary whole-grain intake with fasting glucose- and insulin-related genetic loci in individuals of European descent: A meta-analysis of 14 cohort studies. *Diabetes Care* **2010**, *33*, 2684–2691. [CrossRef]

17. Hu, Y.; Ding, M.; Sampson, L.; Willett, W.C.; Manson, J.A.E.; Wang, M.; Rosner, B.; Hu, F.B.; Sun, Q. Intake of whole grain foods and risk of type 2 diabetes: Results from three prospective cohort studies. *BMJ* **2020**, *370*, m2206. [CrossRef]
18. Zhang, P.; Ma, G.; Wang, C.; Lu, H.; Li, S.; Xie, Y.; Ma, D.; Zhu, Y.; Guo, T. Effect of irrigation and nitrogen application on grain amino acid composition and protein quality in winter wheat. *PLoS ONE* **2017**, *12*, e0178494. [CrossRef] [PubMed]
19. Poutanen, K. Past and future of cereal grains as food for health. *Trends Food Sci. Technol.* **2012**, *25*, 58–62. [CrossRef]
20. Vogel, K.P.; Johnson, V.A.; Mattern, P.J. Protein and Lysine Contents of Endosperm and Bran of the Parents and Progenies of Crosses of Common Wheat 1. *Crop Sci.* **1978**, *18*, 751–754. [CrossRef]
21. Shewry, P.R.; Hey, S.J. The contribution of wheat to human diet and health. *Food Energy Secur.* **2015**, *4*, 178–202. [CrossRef] [PubMed]
22. Davis, K.; Cain, R.F.; Peters, L.; Tourneau, D.L.; Mcginnis, J. Evaluation of the nutrient composition of wheat. II. Proximate analysis, thiamin, riboflavin, niacin, and pyridoxine. *Cereal Chem.* **1981**, *58*, 116–120.
23. Osborne, T.B. *The Proteins of the Wheat Kernel*; Carnegie Inst.: Washington, DC, USA, 1907.
24. Belitz, H.; Grosch, W.; Schieberle, P. Cereals and Cereal Products. In *Food Chemistry*; Springer: Berlin/Heidelberg, Germany, 2009; pp. 670–745.
25. Yu, S.; Tian, L. Breeding Major Cereal Grains through the Lens of Nutrition Sensitivity. *Mol. Plant* **2018**, *11*, 23–30. [CrossRef] [PubMed]
26. Galili, G.; Amir, R. Fortifying plants with the essential amino acids lysine and methionine to improve nutritional quality. *Plant Biotechnol. J.* **2013**, *11*, 211–222. [CrossRef] [PubMed]
27. Food and Agriculture Organization of the United Nations. *Dietary Protein Quality Evaluation in Human Nutrition Report of an FAO Expert Consultation*; Food and Agriculture Organization of the United Nations: Auckland, New Zealand, 2013.
28. Mares, D.J.; Stone, B.A. Studies on wheat endosperm i. Chemical composition and ultrastructure of the cell walls. *Aust. J. Biol. Sci.* **1973**, *26*, 793–812. [CrossRef]
29. Uthayakumaran, S.; Wrigley, C. Wheat: Grain-Quality Characteristics and Management of Quality Requirements. In *Cereal Grains: Assessing and Managing Quality*, 2nd ed.; Elsevier Inc.: Amsterdam, The Netherlands, 2017; pp. 91–134, ISBN 9780081007198.
30. Chibbar, R.N.; Jaiswal, S.; Gangola, M.; Båga, M. Carbohydrate Metabolism. In *Encyclopedia of Food Grains*, 2nd ed.; Elsevier Inc.: Amsterdam, The Netherlands, 2015; Volume 2–4, pp. 161–173, ISBN 9780123947864.
31. Guragain, Y.N.; Probst, K.V.; Vadlani, P.V. Fuel Alcohol Production. In *Encyclopedia of Food Grains*, 2nd ed.; Elsevier Inc.: Amsterdam, The Netherlands, 2015; Volume 3–4, pp. 235–244, ISBN 9780123947864.
32. Stone, B.; Morell, M.K. Chapter 9: Carbohydrates. In *WHEAT: Chemistry and Technology*; AACC International, Inc.: St. Paul, MN, USA, 2009; pp. 299–362.
33. WHO; FAO. Global trends in production and consumption of carbohydrate foods. In *Carbohydrates in Human Nutrition*; Food and Agriculture Organization of the United Nations: Rome, Italy, 1998.
34. Jones, J.M.; Peña, R.J.; Korczak, R.; Braun, H.J. Carbohydrates, grains, and wheat in nutrition and health: An overview part I. Role of carbohydrates in health. *Cereal Foods World* **2015**, *60*, 224–233. [CrossRef]
35. Raichle, M.E.; Gusnard, D.A. Appraising the brain's energy budget. *Proc. Natl. Acad. Sci. USA* **2002**, *99*, 10237–10239. [CrossRef]
36. Kahleova, H.; Dort, S.; Holubkov, R.; Barnard, N.D. A plant-based high-carbohydrate, low-fat diet in overweight individuals in a 16-week randomized clinical trial: The role of carbohydrates. *Nutrients* **2018**, *10*, 1302. [CrossRef] [PubMed]
37. WHO; FAO. The role of carbohydrates in maintenance of health. In *Carbohydrates in Human Nutrition*; Food and Agriculture Organization of the United Nations: Rome, Italy, 1998.
38. Chung, O. Lipid-protein interactions in wheat flour, dough, gluten, and protein fractions. *Cereal Foods World* **1986**, *31*, 242–256.
39. Chung, O.K.; Ohm, J.-B.; Ram, M.S.; Howitt, C.A. Chapter 10: Wheat Lipids. In *WHEAT: Chemistry and Technology*; AACC International, Inc.: St. Paul, MN, USA, 2009; pp. 363–399.
40. WHO. *The World Health Report 2002—Reducing Risks, Promoting Healthy Life*; WHO: Geneva, Switzerland, 2002.
41. Hotz, C.; Brown, K. Assessment of the risk of zinc deficiency in populations. *Food Nutr. Bull.* **2004**, *25*, S130–S162.
42. Bailey, R.L.; West, K.P.; Black, R.E. The epidemiology of global micronutrient deficiencies. *Ann. Nutr. Metab.* **2015**, *66*, 22–33. [CrossRef] [PubMed]
43. Shankar, A.H. Mineral Deficiencies. In *Hunter's Tropical Medicine and Emerging Infectious Disease*, 9th ed.; Elsevier Inc.: Amsterdam, The Netherlands, 2012; pp. 1003–1010, ISBN 9781416043904.
44. Neal, A.L.; Geraki, K.; Borg, S.; Quinn, P.; Mosselmans, J.F.; Brinch-Pedersen, H.; Shewry, P.R. Iron and zinc complexation in wild-type and ferritin-expressing wheat grain: Implications for mineral transport into developing grain. *J. Biol. Inorg. Chem.* **2013**, *18*, 557–570. [CrossRef]
45. Vasconcelos, M.W.; Gruissem, W.; Bhullar, N.K. Iron biofortification in the 21st century: Setting realistic targets, overcoming obstacles, and new strategies for healthy nutrition. *Curr. Opin. Biotechnol.* **2017**, *44*, 8–15. [CrossRef]
46. Zhang, Y.Q.; Sun, Y.X.; Ye, Y.L.; Karim, M.R.; Xue, Y.F.; Yan, P.; Meng, Q.F.; Cui, Z.L.; Cakmak, I.; Zhang, F.S.; et al. Zinc biofortification of wheat through fertilizer applications in different locations of China. *Field Crops Res.* **2012**, *125*, 1–7. [CrossRef]
47. Balk, J.; Connorton, J.M.; Wan, Y.; Lovegrove, A.; Moore, K.L.; Uauy, C.; Sharp, P.A.; Shewry, P.R. Improving wheat as a source of iron and zinc for global nutrition. *Nutr. Bull.* **2019**, *44*, 53–59. [CrossRef]
48. Connorton, J.M.; Jones, E.R.; Rodríguez-Ramiro, I.; Fairweather-Tait, S.; Uauy, C.; Balk, J. Wheat vacuolar iron transporter TaVIT2 transports Fe and Mn and is effective for biofortification. *Plant Physiol.* **2017**, *174*, 2434–2444. [CrossRef]

49. Menguer, P.K.; Vincent, T.; Miller, A.J.; Brown, J.K.M.; Vincze, E.; Borg, S.; Holm, P.B.; Sanders, D.; Podar, D. Improving zinc accumulation in cereal endosperm using HvMTP1, a transition metal transporter. *Plant Biotechnol. J.* **2018**, *16*, 63–71. [CrossRef]

50. Stadtman, T.C. Selenocysteine. *Annu. Rev. Biochem.* **1996**, *65*, 83–100. [CrossRef]

51. Whanger, P.D. Selenocompounds in Plants and Animals and their Biological Significance. *J. Am. Coll. Nutr.* **2002**, *21*, 223–232. [CrossRef]

52. Schwarz, K.; Foltz, C.M. Selenium as an Integral Part of Factor 3 Against Dietary Necrotic Liver Degeneration. *J. Am. Chem. Soc.* **1957**, *79*, 3292–3293. [CrossRef]

53. FAO; WHO. *Human Vitamin and Mineral Requirements Report of a Joint FAO/WHO Expert Consultation Bangkok, Thailand*; Food and Agriculture Organization of the United Nations: Rome, Italy, 2001.

54. Combs, G.F. Selenium in global food systems. *Br. J. Nutr.* **2001**, *85*, 517–547. [CrossRef]

55. Tan, J.; Huang, Y. Selenium in geo-ecosystem and its relation to endemic diseases in China. *Water Air Soil Pollut.* **1991**, *57–58*, 59–68. [CrossRef]

56. Hawkesford, M.J.; Zhao, F.J. Strategies for increasing the selenium content of wheat. *J. Cereal Sci.* **2007**, *46*, 282–292. [CrossRef]

57. Gissel-Nielsen, G.; Gupta, U.C.; Lamand, M.; Westermarck, T. Selenium in Soils and Plants and Its Importance in Livestock and Human Nutrition. *Adv. Agron.* **1984**, *37*, 397–460.

58. Ros, G.H.; van Rotterdam, A.M.D.; Bussink, D.W.; Bindraban, P.S. Selenium fertilization strategies for bio-fortification of food: An agro-ecosystem approach. *Plant Soil* **2016**, *404*, 99–112. [CrossRef]

59. Eurola, M.H.; Ekholm, P.I.; Ylinen, M.E.; Varo, P.T.; Koivistoinen, P.E. Selenium in Finnish foods after beginning the use of selenate-supplemented fertilisers. *J. Sci. Food Agric.* **1991**, *56*, 57–70. [CrossRef]

60. Alfthan, G.; Aspila, P.; Ekholm, P.; Eurola, M.; Hartikainen, H.; Hero, H.; Hietaniemi, V.; Root, T.; Salminen, P.; Venäläinen, E.R.; et al. Nationwide supplementation of sodium selenate to commercial fertilizers: History and 25-year results from the finnish selenium monitoring programme. In *Combating Micronutrient Deficiencies: Food-Based Approaches*; Published Jointly by CABI and FAO: Rome, Italy, 2010; pp. 312–337, ISBN 9781845937140.

61. Smrkolj, P.; Stibilj, V.; Kreft, I.; Germ, M. Selenium species in buckwheat cultivated with foliar addition of Se(VI) and various levels of UV-B radiation. *Food Chem.* **2006**, *96*, 675–681. [CrossRef]

62. White, P.J. Selenium accumulation by plants. *Ann. Bot.* **2016**, *117*, 217–235. [CrossRef] [PubMed]

63. Batifoulier, F.; Verny, M.A.; Chanliaud, E.; Rémésy, C.; Demigné, C. Variability of B vitamin concentrations in wheat grain, milling fractions and bread products. *Eur. J. Agron.* **2006**, *25*, 163–169. [CrossRef]

64. Piironen, V.; Salmenkallio-Marttila, M. Chapter 7: Micronutrients and Phytochemicals in Wheat Grain. In *WHEAT: Chemistry and Technology*; AACC International, Inc.: St. Paul, MN, USA, 2009; pp. 179–222.

65. Van den Berg, H. Bioavailability of niacin—PubMed. *Eur. J. Clin. Nutr.* **1997**, *51*, S64–S65. [PubMed]

66. Shewry, P.R.; Van Schaik, F.; Ravel, C.; Charmet, G.; Rakszegi, M.; Bedo, Z.; Ward, J.L. Genotype and environment effects on the contents of vitamins B1, B2, B3, and B6 in wheat grain. *J. Agric. Food Chem.* **2011**, *59*, 10564–10571. [CrossRef]

67. Carter, E.G.A.; Carpenter, K.J. The bioavailability for humans of bound niacin from wheat bran. *Am. J. Clin. Nutr.* **1982**, *36*, 855–861. [CrossRef] [PubMed]

68. Winichagoon, P.; Kachondham, Y.; Attig, G.A.; Tontisirin, K. *Integrating Food and Nutrition into Development: Thailand's Experiences and Future Visions*; Institute of Nutrition, Mahidon University: Bangkok, Thailand, 1992.

69. Adom, K.K.; Sorrells, M.E.; Liu, R.H. Phytochemical Profiles and Antioxidant Activity of Wheat Varieties. *J. Agric. Food Chem.* **2003**, *51*, 7825–7834. [CrossRef] [PubMed]

70. Beta, T.; Nam, S.; Dexter, J.E.; Sapirstein, H.D. Phenolic Content and Antioxidant Activity of Pearled Wheat and Roller-Milled Fractions. *Cereal Chem. J.* **2005**, *82*, 390–393. [CrossRef]

71. Katz, D.L.; Nawaz, H.; Boukhalil, J.; Chan, W.; Ahmadi, R.; Giannamore, V.; Sarrel, P.M. Effects of oat and wheat cereals on endothelial responses. *Prev. Med.* **2001**, *33*, 476–484. [CrossRef]

72. Vauzour, D.; Houseman, E.J.; George, T.W.; Corona, G.; Garnotel, R.; Jackson, K.G.; Sellier, C.; Gillery, P.; Kennedy, O.B.; Lovegrove, J.A.; et al. Moderate Champagne consumption promotes an acute improvement in acute endothelial-independent vascular function in healthy human volunteers. *Br. J. Nutr.* **2010**, *103*, 1168–1178. [CrossRef]

73. Rodriguez-Mateos, A.; Rendeiro, C.; Bergillos-Meca, T.; Tabatabaee, S.; George, T.W.; Heiss, C.; Spencer, J.P.E. Intake and time dependence of blueberry flavonoid-induced improvements in vascular function: A randomized, controlled, double-blind, crossover intervention study with mechanistic insights into biological activity. *Am. J. Clin. Nutr.* **2013**, *98*, 1179–1191. [CrossRef] [PubMed]

74. Piironen, V.I. Determination of tocopherols and tocotrienols in foods and tissues. In *Modern Analytical Methodologies in Fat- and Water-Soluble Vitamins*; Wiley-Interscience: New York, NY, USA, 2000.

75. Moreau, R.A.; Whitaker, B.D.; Hicks, K.B. Phytosterols, phytostanols, and their conjugates in foods: Structural diversity, quantitative analysis, and health-promoting uses. *Prog. Lipid Res.* **2002**, *41*, 457–500. [CrossRef]

76. Salo, P.; Wester, I.; Hopia, A. Phytosterols. In *Lipids for Functional Foods and Nutraceuticals*; Elsevier: Amsterdam, The Netherlands, 2012; pp. 183–224.

77. Bramley, P.; Elmadfa, I.; Kafatos, A.; Kelly, F.; Manios, Y.; Roxborough, H.; Schuch, W.; Sheehy, P.; Wagner, K.-H. Vitamin E. *J. Sci. Food Agric.* **2000**, *80*, 913–938. [CrossRef]

78. Wagner, K.-H.; Wotruba, F.; Elmadfa, I. Antioxidative potential of tocotrienols and tocopherols in coconut fat at different oxidation temperatures. *Eur. J. Lipid Sci. Technol.* **2001**, *103*, 746–751. [CrossRef]

79. Yoshida, Y.; Niki, E.; Noguchi, N. Comparative study on the action of tocopherols and tocotrienols as antioxidant: Chemical and physical effects. *Chem. Phys. Lipids* **2003**, *123*, 63–75. [CrossRef]

80. Hoevenaars, F.; van der Kamp, J.-W.; van den Brink, W.; Wopereis, S. Next Generation Health Claims Based on Resilience: The Example of Whole-Grain Wheat. *Nutrients* **2020**, *12*, 2945. [CrossRef]

81. Shewry, P.R. The HEALTHGRAIN programme opens new opportunities for improving wheat for nutrition and health. *Nutr. Bull.* **2009**, *34*, 225–231. [CrossRef]

82. Ward, J.L.; Poutanen, K.; Gebruers, K.; Piironen, V.; Lampi, A.M.; Nyström, L.; Andersson, A.A.M.; Åman, P.; Boros, D.; Rakszegi, M.; et al. The HEALTHGRAIN cereal diversity screen: Concept, results, and prospects. *J. Agric. Food Chem.* **2008**, *56*, 9699–9709. [CrossRef] [PubMed]

83. Adom, K.K.; Sorrells, M.E.; Rui, H.L. Phytochemicals and antioxidant activity of milled fractions of different wheat varieties. *J. Agric. Food Chem.* **2005**, *53*, 2297–2306. [CrossRef]

84. Milder, I.E.J.; Arts, I.C.W.; Putte, B.v.d.; Venema, D.P.; Hollman, P.C.H. Lignan contents of Dutch plant foods: A database including lariciresinol, pinoresinol, secoisolariciresinol and matairesinol. *Br. J. Nutr.* **2005**, *93*, 393–402. [CrossRef]

85. García-Estepa, R.M.; Guerra-Hernández, E.; García-Villanova, B. Phytic acid content in milled cereal products and breads. *Food Res. Int.* **1999**, *32*, 217–221. [CrossRef]

86. Nyström, L.; Paasonen, A.; Lampi, A.M.; Piironen, V. Total plant sterols, steryl ferulates and steryl glycosides in milling fractions of wheat and rye. *J. Cereal Sci.* **2007**, *45*, 106–115. [CrossRef]

87. Adom, K.K.; Liu, R.H. Antioxidant activity of grains. *J. Agric. Food Chem.* **2002**, *50*, 6182–6187. [CrossRef] [PubMed]

88. Fineli. Available online: https://fineli.fi/fineli/en/index (accessed on 3 July 2021).

89. WHO. *Diet, Nutrition and the Prevention of Chronic Disease: Report of a Joint WHO/FAO Expert Consultation*; World Health Organization: Geneva, Switzerland, 2003.

90. Cho, S.S.; Qi, L.; Fahey, G.C.; Klurfeld, D.M. Consumption of cereal fiber, mixtures of whole grains and bran, and whole grains and risk reduction in type 2 diabetes, obesity, and cardiovascular disease. *Am. J. Clin. Nutr.* **2013**, *98*, 594–619. [CrossRef]

91. Jacobs, D.R.; Gallaher, D.D. Whole grain intake and cardiovascular disease: A review. *Curr. Atheroscler. Rep.* **2004**, *6*, 415–423. [CrossRef]

92. Åberg, S.; Mann, J.; Neumann, S.; Ross, A.B.; Reynolds, A.N. Whole-grain processing and glycemic control in type 2 diabetes: A randomized crossover trial. *Diabetes Care* **2020**, *43*, 1717–1723. [CrossRef]

93. Zong, G.; Gao, A.; Hu, F.B.; Sun, Q. Whole grain intake and mortality from all causes, cardiovascular disease, and cancer. *Circulation* **2016**, *133*, 2370–2380. [CrossRef] [PubMed]

94. Chen, G.-C.; Tong, X.; Xu, J.-Y.; Han, S.-F.; Wan, Z.-X.; Qin, J.-B.; Qin, L.-Q. Whole-grain intake and total, cardiovascular, and cancer mortality: A systematic review and meta-analysis of prospective studies. *Am. J. Clin. Nutr.* **2016**, *104*, 164–172. [CrossRef] [PubMed]

95. Chan, J.M.; Wang, F.; Holly, E.A. Whole Grains and Risk of Pancreatic Cancer in a Large Population-based Case-Control Study in the San Francisco Bay Area, California. *Am. J. Epidemiol.* **2007**, *166*, 1174–1185. [CrossRef] [PubMed]

96. Venn, B.J.; Mann, J.I. Cereal grains, legumes and diabetes. *Eur. J. Clin. Nutr.* **2004**, *58*, 1443–1461. [CrossRef]

97. Gujral, N.; Freeman, H.J.; Thomson, A.B.R. Celiac disease: Prevalence, diagnosis, pathogenesis and treatment. *World J. Gastroenterol.* **2012**, *18*, 6036–6059. [CrossRef] [PubMed]

98. Cabanillas, B. Gluten-related disorders: Celiac disease, wheat allergy, and nonceliac gluten sensitivity. *Crit. Rev. Food Sci. Nutr.* **2020**, *60*, 2606–2621. [CrossRef]

99. Pourpak, Z.; Mesdaghi, M.; Mansouri, M.; Kazemnejad, A.; Toosi, S.B.; Farhoudi, A. Which cereal is a suitable substitute for wheat in children with wheat allergy? *Pediatr. Allergy Immunol.* **2005**, *16*, 262–266. [CrossRef] [PubMed]

100. Jin, Y.; Acharya, H.G.; Acharya, D.; Jorgensen, R.; Gao, H.; Secord, J.; Ng, P.K.W.; Gangur, V. Advances in molecular mechanisms of wheat allergenicity in animal models: A comprehensive review. *Molecules* **2019**, *24*. [CrossRef]

101. Czaja-Bulsa, G.; Bulsa, M. What do we know now about IgE-mediated wheat allergy in children? *Nutrients* **2017**, *9*, 35. [CrossRef] [PubMed]

102. Pinto-Sanchez, M.I.; Verdu, E.F. Non-celiac gluten or wheat sensitivity: It's complicated! *Neurogastroenterol. Motil.* **2018**, *30*, e13392. [CrossRef]

103. Maiuri, L.; Ciacci, C.; Ricciardelli, I.; Vacca, L.; Raia, V.; Auricchio, S.; Picard, J.; Osman, M.; Quaratino, S.; Londei, M. Association between innate response to gliadin and activation of pathogenic T cells in coeliac disease. *Lancet* **2003**, *362*, 30–37. [CrossRef]

104. Londei, M.; Ciacci, C.; Ricciardelli, I.; Vacca, L.; Quaratino, S.; Maiuri, L. Gliadin as a stimulator of innate responses in celiac disease. *Mol. Immunol.* **2005**, *42*, 913–918. [CrossRef]

105. Lammers, K.M.; Chieppa, M.; Liu, L.; Liu, S.; Omatsu, T.; Janka-Junttila, M.; Casolaro, V.; Reinecker, H.C.; Parent, C.A.; Fasano, A. Gliadin induces neutrophil migration via engagement of the formyl peptide receptor, FPR1. *PLoS ONE* **2015**, *10*, e0138338. [CrossRef]

106. Gutierrez-Achury, J.; Zhernakova, A.; Pulit, S.L.; Trynka, G.; Hunt, K.A.; Romanos, J.; Raychaudhuri, S.; Van Heel, D.A.; Wijmenga, C.; De Bakker, P.I.W. Fine mapping in the MHC region accounts for 18% additional genetic risk for celiac disease. *Nat. Genet.* **2015**, *47*, 577–578. [CrossRef]

107. Guo, C.C.; Huang, W.H.; Zhang, N.; Dong, F.; Jing, L.P.; Liu, Y.; Ye, X.G.; Xiao, D.; Ou, M.L.; Zhang, B.H.; et al. Association between IL2/IL21 and SH2B3 polymorphisms and risk of celiac disease: A meta-analysis. *Genet. Mol. Res.* **2015**, *14*, 13221–13235. [CrossRef]

108. Romanos, J.; Rosén, A.; Kumar, V.; Trynka, G.; Franke, L.; Szperl, A.; Gutierrez-Achury, J.; Van Diemen, C.C.; Kanninga, R.; Jankipersadsing, S.A.; et al. Improving coeliac disease risk prediction by testing non-HLA variants additional to HLA variants. *Gut* **2014**, *63*, 415–422. [CrossRef] [PubMed]

109. Norris, J.M.; Barriga, K.; Hoffenberg, E.J.; Taki, I.; Miao, D.; Haas, J.E.; Emery, L.M.; Sokol, R.J.; Erlich, H.A.; Eisenbarth, G.S.; et al. Risk of celiac disease autoimmunity and timing of gluten introduction in the diet of infants at increased risk of disease. *J. Am. Med. Assoc.* **2005**, *293*, 2343–2351. [CrossRef]

110. Norris, J.M.; Barriga, K.; Klingensmith, G.; Hoffman, M.; Eisenbarth, G.S.; Erlich, H.A.; Rewers, M. Timing of Initial Cereal Exposure in Infancy and Risk of Islet Autoimmunity. *J. Am. Med. Assoc.* **2003**, *290*, 1713–1720. [CrossRef] [PubMed]

111. Agostoni, C.; Decsi, T.; Fewtrell, M.; Goulet, O.; Kolacek, S.; Koletzko, B.; Michaelsen, K.F.; Moreno, L.; Puntis, J.; Rigo, J.; et al. Complementary feeding: A commentary by the ESPGHAN Committee on Nutrition. *J. Pediatr. Gastroenterol. Nutr.* **2008**, *46*, 99–110. [CrossRef] [PubMed]

112. Lionetti, E.; Castellaneta, S.; Francavilla, R.; Pulvirenti, A.; Tonutti, E.; Amarri, S.; Barbato, M.; Barbera, C.; Barera, G.; Bellantoni, A.; et al. Introduction of Gluten, HLA Status, and the Risk of Celiac Disease in Children. *N. Engl. J. Med.* **2014**, *371*, 1295–1303. [CrossRef] [PubMed]

113. Vriezinga, S.L.; Auricchio, R.; Bravi, E.; Castillejo, G.; Chmielewska, A.; Crespo Escobar, P.; Kolaček, S.; Koletzko, S.; Korponay-Szabo, I.R.; Mummert, E.; et al. Randomized Feeding Intervention in Infants at High Risk for Celiac Disease. *N. Engl. J. Med.* **2014**, *371*, 1304–1315. [CrossRef]

114. Szajewska, H.; Shamir, R.; Chmielewska, A.; Pieścik-Lech, M.; Auricchio, R.; Ivarsson, A.; Kolacek, S.; Koletzko, S.; Korponay-Szabo, I.; Mearin, M.L.; et al. Systematic review with meta-analysis: Early infant feeding and coeliac disease-update 2015. *Aliment. Pharmacol. Ther.* **2015**, *41*, 1038–1054. [CrossRef]

115. Al-Toma, A.; Volta, U.; Auricchio, R.; Castillejo, G.; Sanders, D.S.; Cellier, C.; Mulder, C.J.; Lundin, K.E.A. European Society for the Study of Coeliac Disease (ESsCD) guideline for coeliac disease and other gluten-related disorders. *United Eur. Gastroenterol. J.* **2019**, *7*, 583–613. [CrossRef]

116. Greco, L.; Romino, R.; Coto, I.; Di Cosmo, N.; Percopo, S.; Maglio, M.; Paparo, F.; Gasperi, V.; Limongelli, M.G.; Cotichini, R.; et al. The first large population based twin study of coeliac disease. *Gut* **2002**, *50*, 624–628. [CrossRef] [PubMed]

117. Sommer, F.; Bäckhed, F. The gut microbiota-masters of host development and physiology. *Nat. Rev. Microbiol.* **2013**, *11*, 227–238. [CrossRef] [PubMed]

118. Belkaid, Y.; Hand, T.W. Role of the microbiota in immunity and inflammation. *Cell* **2014**, *157*, 121–141. [CrossRef]

119. De Palma, G.; Nadal, I.; Medina, M.; Donat, E.; Ribes-Koninckx, C.; Calabuig, M.; Sanz, Y. Intestinal dysbiosis and reduced immunoglobulin-coated bacteria associated with coeliac disease in children. *BMC Microbiol.* **2010**, *10*, 63. [CrossRef] [PubMed]

120. Collado, M.C.; Donat, E.; Ribes-Koninckx, C.; Calabuig, M.; Sanz, Y. Specific duodenal and faecal bacterial groups associated with paediatric coeliac disease. *J. Clin. Pathol.* **2009**, *62*, 264–269. [CrossRef]

121. Myléus, A.; Hernell, O.; Gothefors, L.; Hammarström, M.L.; Persson, L.Å.; Stenlund, H.; Ivarsson, A. Early infections are associated with increased risk for celiac disease: An incident case-referent study. *BMC Pediatr.* **2012**, *12*, 194. [CrossRef]

122. Pavone, P.; Nicolini, E.; Taibi, R.; Ruggieri, M. Rotavirus and celiac disease. *Am. J. Gastroenterol.* **2007**, *102*, 1831. [CrossRef] [PubMed]

123. Stene, L.C.; Honeyman, M.C.; Hoffenberg, E.J.; Haas, J.E.; Sokol, R.J.; Emery, L.; Taki, I.; Norris, J.M.; Erlich, H.A.; Eisenbarth, G.S.; et al. Rotavirus infection frequency and risk of celiac disease autoimmunity in early childhood: A longitudinal study. *Am. J. Gastroenterol.* **2006**, *101*, 2333–2340. [CrossRef]

124. Leonard, M.M.; Valitutti, F.; Karathia, H.; Pujolassos, M.; Kenyon, V.; Fanelli, B.; Troisi, J.; Subramanian, P.; Camhi, S.; Colucci, A.; et al. Microbiome signatures of progression toward celiac disease onset in at-risk children in a longitudinal prospective cohort study. *Proc. Natl. Acad. Sci. USA* **2021**, *118*, e2020322118. [CrossRef]

125. Guandalini, S.; Assiri, A. Celiac disease: A review. *JAMA Pediatr.* **2014**, *168*, 272–278. [CrossRef]

126. Welander, A.; Tjernberg, A.R.; Montgomery, S.M.; Ludvigsson, J.; Ludvigsson, J.F. Infectious disease and risk of later celiac disease in childhood. *Pediatrics* **2010**, *125*, e530–e536. [CrossRef] [PubMed]

127. Riddle, M.S.; Murray, J.A.; Porter, C.K. The incidence and risk of celiac disease in a healthy US adult population. *Am. J. Gastroenterol.* **2012**, *107*, 1248–1255. [CrossRef] [PubMed]

128. Kahrs, C.R.; Chuda, K.; Tapia, G.; Stene, L.C.; Mårild, K.; Rasmussen, T.; Rønningen, K.S.; Lundin, K.E.A.; Kramna, L.; Cinek, O.; et al. Enterovirus as trigger of coeliac disease: Nested case-control study within prospective birth cohort. *BMJ* **2019**, *364*, l231. [CrossRef] [PubMed]

129. Kaswala, D.; Veeraraghavan, G.; Kelly, C.; Leffler, D. Celiac Disease: Diagnostic Standards and Dilemmas. *Diseases* **2015**, *3*, 86–101. [CrossRef]

130. Leffler, D. Celiac disease diagnosis and management: A 46-year-old woman with anemia. *JAMA* **2011**, *306*, 1582–1592. [CrossRef] [PubMed]

131. Rubio-Tapia, A.; Hill, I.D.; Kelly, C.P.; Calderwood, A.H.; Murray, J.A. ACG Clinical Guidelines: Diagnosis and Management of Celiac Disease. *Am. J. Gastroenterol.* **2013**, *108*, 656–676. [CrossRef]

132. Rostom, A.; Dubé, C.; Cranney, A.; Saloojee, N.; Sy, R.; Garritty, C.; Sampson, M.; Zhang, L.; Yazdi, F.; Mamaladze, V.; et al. The diagnostic accuracy of serologic tests for celiac disease: A systematic review. *Gastroenterology* **2005**, *128*. [CrossRef]

133. Tye-Din, J.A.; Galipeau, H.J.; Agardh, D. Celiac disease: A review of current concepts in pathogenesis, prevention, and novel therapies. *Front. Pediatr.* **2018**, *6*, 350. [CrossRef]

134. Elli, L.; Branchi, F.; Tomba, C.; Villalta, D.; Norsa, L.; Ferretti, F.; Roncoroni, L.; Bardella, M.T. Diagnosis of gluten related disorders: Celiac disease, wheat allergy and non-celiac gluten sensitivity. *World J. Gastroenterol.* **2015**, *21*, 7110–7119. [CrossRef] [PubMed]

135. Volta, U.; Granito, A.; Parisi, C.; Fabbri, A.; Fiorini, E.; Piscaglia, M.; Tovoli, F.; Grasso, V.; Muratori, P.; Pappas, G.; et al. Deamidated gliadin peptide antibodies as a routine test for celiac disease: A prospective analysis. *J. Clin. Gastroenterol.* **2010**, *44*, 186–190. [CrossRef]

136. Lewis, N.R.; Scott, B.B. Meta-analysis: Deamidated gliadin peptide antibody and tissue transglutaminase antibody compared as screening tests for coeliac disease. *Aliment. Pharmacol. Ther.* **2010**, *31*, 73–81. [CrossRef]

137. Van Der Windt, D.A.W.M.; Jellema, P.; Mulder, C.J.; Kneepkens, C.M.F.; Van Der Horst, H.E. Diagnostic testing for celiac disease among patients with abdominal symptoms: A systematic review. *JAMA* **2010**, *303*, 1738–1746. [CrossRef]

138. Marsh, M.N. Gluten, Major and the Histocompatibility Small Intestine Complex, A Molecular and Immunobiologic Approach to Spectrum of Gluten Sensitivity ('Celiac Sprue'). *Gastroenterology* **1992**, *102*, 330–354. [CrossRef]

139. Catassi, C.; Fasano, A. Celiac disease diagnosis: Simple rules are better than complicated algorithms. *Am. J. Med.* **2010**, *123*, 691–693. [CrossRef] [PubMed]

140. Ludvigsson, J.F.; Bai, J.C.; Biagi, F.; Card, T.R.; Ciacci, C.; Ciclitira, P.J.; Green, P.H.R.; Hadjivassiliou, M.; Holdoway, A.; Van Heel, D.A.; et al. Diagnosis and management of adult coeliac disease: Guidelines from the British society of gastroenterology. *Gut* **2014**, *63*, 1210–1228. [CrossRef] [PubMed]

141. Kaukinen, K.; Partanen, J.; Maki, M.; Collin, P. HLA-DQ typing in the diagnosis of celiac disease. *Am. J. Gastroenterol.* **2002**, *97*, 695–699. [CrossRef]

142. Sharma, N.; Bhatia, S.; Chunduri, V.; Kaur, S.; Sharma, S.; Kapoor, P.; Kumari, A.; Garg, M. Pathogenesis of Celiac Disease and Other Gluten Related Disorders in Wheat and Strategies for Mitigating Them. *Front. Nutr.* **2020**, *7*. [CrossRef] [PubMed]

143. Leffler, D.A.; Green, P.H.R.; Fasano, A. Extraintestinal manifestations of coeliac disease. *Nat. Rev. Gastroenterol. Hepatol.* **2015**, *12*, 561–571. [CrossRef]

144. Kelly, C.P.; Bai, J.C.; Liu, E.; Leffler, D.A. Advances in diagnosis and management of celiac disease. *Gastroenterology* **2015**, *148*, 1175–1186. [CrossRef]

145. Rubio-Tapia, A.; Rahim, M.W.; See, J.A.; Lahr, B.D.; Wu, T.T.; Murray, J.A. Mucosal recovery and mortality in adults with celiac disease after treatment with a gluten-free diet. *Am. J. Gastroenterol.* **2010**, *105*, 1412–1420. [CrossRef]

146. Allen, K.J.; Turner, P.J.; Pawankar, R.; Taylor, S.; Sicherer, S.; Lack, G.; Rosario, N.; Ebisawa, M.; Wong, G.; Mills, E.N.C.; et al. Precautionary labelling of foods for allergen content: Are we ready for a global framework? *World Allergy Organ. J.* **2014**, *7*, 10. [CrossRef]

147. Sharma, G.M.; Pereira, M.; Williams, K.M. Gluten detection in foods available in the United States—A market survey. *Food Chem.* **2015**, *169*, 120–126. [CrossRef] [PubMed]

148. Hadjivassiliou, M.; Sanders, D.D.; Aeschlimann, D.P. Gluten-related disorders: Gluten ataxia. *Dig. Dis.* **2015**, *33*, 264–268. [CrossRef] [PubMed]

149. Hadjivassiliou, M.; Grünewald, R.; Sharrack, B.; Sanders, D.; Lobo, A.; Williamson, C.; Woodroofe, N.; Wood, N.; Davies-Jones, A. Gluten ataxia in perspective: Epidemiology, genetic susceptibility and clinical characteristics. *Brain* **2003**, *126*, 685–691. [CrossRef]

150. Hadjivassiliou, M.; Mäki, M.; Sanders, D.S.; Williamson, C.A.; Grünewald, R.A.; Woodroofe, N.M.; Korponay-Szabó, I.R. Autoantibody targeting of brain and intestinal transglutaminase in gluten ataxia. *Neurology* **2006**, *66*, 373–377. [CrossRef] [PubMed]

151. Ghezzi, A.; Filippi, M.; Falini, A.; Zaffaroni, M. Cerebral involvement in celiac disease: A serial MRI study in a patient with brainstem and cerebellar symptoms. *Neurology* **1997**, *49*, 1447–1450. [CrossRef] [PubMed]

152. Nanri, K.; Mitoma, H.; Ihara, M.; Tanaka, N.; Taguchi, T.; Takeguchi, M.; Ishiko, T.; Mizusawa, H. Gluten Ataxia in Japan. *Cerebellum* **2014**, *13*, 623–627. [CrossRef]

153. Clarindo, M.V.; Possebon, A.T.; Soligo, E.M.; Uyeda, H.; Ruaro, R.T.; Empinotti, J.C. Dermatitis herpetiformis: Pathophysiology, clinical presentation, diagnosis and treatment. *An. Bras. Dermatol.* **2014**, *89*, 865–877. [CrossRef]

154. Antiga, E.; Maglie, R.; Quintarelli, L.; Verdelli, A.; Bonciani, D.; Bonciolini, V.; Caproni, M. Dermatitis herpetiformis: Novel perspectives. *Front. Immunol.* **2019**, *10*, 1290. [CrossRef]

155. Hull, C.M.; Liddle, M.; Hansen, N.; Meyer, L.J.; Schmidt, L.; Taylor, T.; Jaskowski, T.D.; Hill, H.R.; Zone, J.J. Elevation of IgA anti-epidermal transglutaminase antibodies in dermatitis herpetiformis. *Br. J. Dermatol.* **2008**, *159*, 120–124. [CrossRef] [PubMed]

156. Mendes, F.B.R.; Hissa-Elian, A.; De Abreu, M.A.M.M.; Gonçalves, V.S. Review: Dermatitis herpetiformis. *An. Bras. Dermatol.* **2013**, *88*, 594–599. [CrossRef]

157. Caproni, M.; Antiga, E.; Melani, L.; Fabbri, P. Guidelines for the diagnosis and treatment of dermatitis herpetiformis. *J. Eur. Acad. Dermatol. Venereol.* **2009**, *23*, 633–638. [CrossRef]

158. Antiga, E.; Caproni, M. The diagnosis and treatment of dermatitis herpetiformis. *Clin. Cosmet. Investig. Dermatol.* **2015**, *8*, 257–265. [CrossRef] [PubMed]

159. Longo, G.; Berti, I.; Burks, A.W.; Krauss, B.; Barbi, E. IgE-mediated food allergy in children. *Lancet* **2013**, *382*, 1656–1664. [CrossRef]

160. Zuidmeer, L.; Goldhahn, K.; Rona, R.J.; Gislason, D.; Madsen, C.; Summers, C.; Sodergren, E.; Dahlstrom, J.; Lindner, T.; Sigurdardottir, S.T.; et al. The prevalence of plant food allergies: A systematic review. *J. Allergy Clin. Immunol.* **2008**, *121*, 1210–1218. [CrossRef] [PubMed]

161. Nwaru, B.I.; Hickstein, L.; Panesar, S.S.; Roberts, G.; Muraro, A.; Sheikh, A. Prevalence of common food allergies in Europe: A systematic review and meta-analysis. *Allergy Eur. J. Allergy Clin. Immunol.* **2014**, *69*, 992–1007. [CrossRef] [PubMed]

162. Pasha, I.; Saeed, F.; Sultan, M.T.; Batool, R.; Aziz, M.; Ahmed, W. Wheat Allergy and Intolerence; Recent Updates and Perspectives. *Crit. Rev. Food Sci. Nutr.* **2016**, *56*, 13–24. [CrossRef]

163. Marchioni Beery, R.M.; Birk, J.W. Wheat-related disorders reviewed: Making a grain of sense. *Expert Rev. Gastroenterol. Hepatol.* **2015**, *9*, 851–864. [CrossRef]

164. Ortiz, C.; Valenzuela, R.; Lucero Alvarez, Y. Celiac disease, non celiac gluten sensitivity and wheat allergy: Comparison of 3 different diseases triggered by the same food. *Rev. Chil. Pediatr.* **2017**, *88*, 417–423. [CrossRef]

165. Tordesillas, L.; Berin, M.C.; Sampson, H.A. Immunology of Food Allergy. *Immunity* **2017**, *47*, 32–50. [CrossRef]

166. Tanabe, S. Analysis of food allergen structures and development of foods for allergic patients. *Biosci. Biotechnol. Biochem.* **2008**, *72*, 649–659. [CrossRef]

167. Galli, S.J.; Tsai, M. IgE and mast cells in allergic disease. *Nat. Med.* **2012**, *18*, 693–704. [CrossRef]

168. Cianferoni, A. Wheat allergy: Diagnosis and management. *J. Asthma Allergy* **2016**, *9*, 13–25. [CrossRef] [PubMed]

169. Christensen, M.J.; Eller, E.; Mortz, C.G.; Brockow, K.; Bindslev-Jensen, C. Wheat-Dependent Cofactor-Augmented Anaphylaxis: A Prospective Study of Exercise, Aspirin, and Alcohol Efficacy as Cofactors. *J. Allergy Clin. Immunol. Pract.* **2019**, *7*, 114–121. [CrossRef] [PubMed]

170. Scherf, K.A.; Brockow, K.; Biedermann, T.; Koehler, P.; Wieser, H. Wheat-dependent exercise-induced anaphylaxis. *Clin. Exp. Allergy* **2016**, *46*, 10–20. [CrossRef]

171. Quirce, S.; Diaz-Perales, A. Diagnosis and management of grain-induced asthma. *Allergy Asthma Immunol. Res.* **2013**, *5*, 348–356. [CrossRef] [PubMed]

172. Kim, J.H.; Kim, J.E.; Choi, G.S.; Hwang, E.K.; An, S.; Ye, Y.M.; Park, H.S. A case of occupational rhinitis caused by rice powder in the grain industry. *Allergy Asthma Immunol. Res.* **2010**, *2*, 141–143. [CrossRef]

173. Brant, A. Baker's asthma. *Curr. Opin. Allergy Clin. Immunol.* **2007**, *7*, 152–155. [CrossRef]

174. Armentia, A.; Díaz-Perales, A.; Castrodeza, J.; Dueñas-Laita, A.; Palacin, A.; Fernández, S. Why can patients with baker's asthma tolerate wheat flour ingestion? Is wheat pollen allergy relevant? *Allergol. Immunopathol.* **2009**, *37*, 203–204. [CrossRef]

175. Pastorello, E.A.; Farioli, L.; Conti, A.; Pravettoni, V.; Bonomi, S.; Iametti, S.; Fortunato, D.; Scibilia, J.; Bindslev-Jensen, C.; Ballmer-Weber, B.; et al. Wheat IgE-mediated food allergy in european patients: α-amylase inhibitors, lipid transfer proteins and low-molecular-weight glutenins—Allergenic molecules recognized by double-blind, placebo-controlled food challenge. *Int. Arch. Allergy Immunol.* **2007**, *144*, 10–22. [CrossRef]

176. Salcedo, G.; Quirce, S.; Diaz-Perales, A. Wheat Allergens Associated With Baker's Asthma. *J. Investig. Allergol. Clin. Immunol.* **2011**, *21*, 81–92.

177. Pahr, S.; Selb, R.; Weber, M.; Focke-Tejkl, M.; Hofer, G.; Dordić, A.; Keller, W.; Papadopoulos, N.G.; Giavi, S.; Mäkelä, M.; et al. Biochemical, biophysical and IgE-epitope characterization of the wheat food allergen, Tri a 37. *PLoS ONE* **2014**, *9*, e111483. [CrossRef]

178. Pahr, S.; Constantin, C.; Papadopoulos, N.G.; Giavi, S.; Mäkelä, M.; Pelkonen, A.; Ebner, C.; Mari, A.; Scheiblhofer, S.; Thalhamer, J.; et al. α-Purothionin, a new wheat allergen associated with severe allergy. *J. Allergy Clin. Immunol.* **2013**, *132*, 1000–1003. [CrossRef]

179. Le, T.A.; Al Kindi, M.; Tan, J.A.; Smith, A.; Heddle, R.J.; Kette, F.E.; Hissaria, P.; Smith, W.B. The clinical spectrum of omega-5-gliadin allergy. *Intern. Med. J.* **2016**, *46*, 710–716. [CrossRef] [PubMed]

180. Rongfei, Z.; Wenjing, L.; Nan, H.; Guanghui, L. Wheat—Dependent exercise-induced anaphylaxis occurred with a delayed onset of 10 to 24 h after wheat ingestion: A case report. *Allergy Asthma Immunol. Res.* **2014**, *6*, 370–372. [CrossRef] [PubMed]

181. Sicherer, S.H.; Sampson, H.A. Food allergy: A review and update on epidemiology, pathogenesis, diagnosis, prevention, and management. *J. Allergy Clin. Immunol.* **2018**, *141*, 41–58. [CrossRef]

182. Burks, A.W.; Jones, S.M.; Wood, R.A.; Fleischer, D.M.; Sicherer, S.H.; Lindblad, R.W.; Stablein, D.; Henning, A.K.; Vickery, B.P.; Liu, A.H.; et al. Oral Immunotherapy for Treatment of Egg Allergy in Children. *N. Engl. J. Med.* **2012**, *367*, 233–243. [CrossRef] [PubMed]

183. Jones, S.M.; Burks, A.W.; Dupont, C. State of the art on food allergen immunotherapy: Oral, sublingual, and epicutaneous. *J. Allergy Clin. Immunol.* **2014**, *133*, 318–323. [CrossRef] [PubMed]

184. Merves, J.; Muir, A.; Modayur Chandramouleeswaran, P.; Cianferoni, A.; Wang, M.L.; Spergel, J.M. Eosinophilic esophagitis. *Ann. Allergy Asthma Immunol.* **2014**, *112*, 397–403. [CrossRef]

185. Liacouras, C.A.; Spergel, J.; Gober, L.M. Eosinophilic esophagitis: Clinical presentation in children. *Gastroenterol. Clin. N. Am.* **2014**, *43*, 219–229. [CrossRef]

186. Prussin, C. Eosinophilic gastroenteritis and related eosinophilic disorders. *Gastroenterol. Clin. N. Am.* **2014**, *43*, 317–327. [CrossRef]

187. Liacouras, C.A.; Furuta, G.T.; Hirano, I.; Atkins, D.; Attwood, S.E.; Bonis, P.A.; Burks, A.W.; Chehade, M.; Collins, M.H.; Dellon, E.S.; et al. Eosinophilic esophagitis: Updated consensus recommendations for children and adults. *J. Allergy Clin. Immunol.* **2011**, *128*, 3–20.e6. [CrossRef]

188. Greenhawt, M.; Aceves, S.S.; Spergel, J.M.; Rothenberg, M.E. The management of eosinophilic esophagitis. *J. Allergy Clin. Immunol. Pract.* **2013**, *1*, 332–340. [CrossRef] [PubMed]
189. Ko, H.M.; Morotti, R.A.; Yershov, O.; Chehade, M. Eosinophilic gastritis in children: Clinicopathological correlation, disease course, and response to therapy. *Am. J. Gastroenterol.* **2014**, *109*, 1277–1285. [CrossRef] [PubMed]
190. Leonard, M.M.; Sapone, A.; Catassi, C.; Fasano, A. Celiac disease and nonceliac gluten sensitivity: A review. *JAMA* **2017**, *318*, 647–656. [CrossRef]
191. Skodje, G.I.; Sarna, V.K.; Minelle, I.H.; Rolfsen, K.L.; Muir, J.G.; Gibson, P.R.; Veierød, M.B.; Henriksen, C.; Lundin, K.E.A. Fructan, Rather Than Gluten, Induces Symptoms in Patients With Self-Reported Non-Celiac Gluten Sensitivity. *Gastroenterology* **2018**, *154*, 529–539.e2. [CrossRef]
192. Junker, Y.; Zeissig, S.; Kim, S.J.; Barisani, D.; Wieser, H.; Leffler, D.A.; Zevallos, V.; Libermann, T.A.; Dillon, S.; Freitag, T.L.; et al. Wheat amylase trypsin inhibitors drive intestinal inflammation via activation of toll-like receptor 4. *J. Exp. Med.* **2012**, *209*, 2395–2408. [CrossRef]
193. Dalla Pellegrina, C.; Perbellini, O.; Scupoli, M.T.; Tomelleri, C.; Zanetti, C.; Zoccatelli, G.; Fusi, M.; Peruffo, A.; Rizzi, C.; Chignola, R. Effects of wheat germ agglutinin on human gastrointestinal epithelium: Insights from an experimental model of immune/epithelial cell interaction. *Toxicol. Appl. Pharmacol.* **2009**, *237*, 146–153. [CrossRef] [PubMed]
194. Sapone, A.; Lammers, K.M.; Casolaro, V.; Cammarota, M.; Giuliano, M.T.; De Rosa, M.; Stefanile, R.; Mazzarella, G.; Tolone, C.; Russo, M.I.; et al. Divergence of gut permeability and mucosal immune gene expression in two gluten-associated conditions: Celiac disease and gluten sensitivity. *BMC Med.* **2011**, *9*, 23. [CrossRef]
195. Sapone, A.; Lammers, K.M.; Mazzarella, G.; Mikhailenko, I.; Cartenì, M.; Casolaro, V.; Fasano, A. Differential mucosal IL-17 expression in two gliadin-induced disorders: Gluten sensitivity and the autoimmune enteropathy celiac disease. *Int. Arch. Allergy Immunol.* **2010**, *152*, 75–80. [CrossRef] [PubMed]
196. Volta, U.; Villanacci, V. Celiac disease: Diagnostic criteria in progress. *Cell. Mol. Immunol.* **2011**, *8*, 96–102. [CrossRef]
197. Vazquez-Roque, M.I.; Camilleri, M.; Smyrk, T.; Murray, J.A.; Marietta, E.; O'Neill, J.; Carlson, P.; Lamsam, J.; Janzow, D.; Eckert, D.; et al. A controlled trial of gluten-free diet in patients with irritable bowel syndrome-diarrhea: Effects on bowel frequency and intestinal function. *Gastroenterology* **2013**, *144*, 903–911.e3. [CrossRef] [PubMed]
198. Hollon, J.; Puppa, E.L.; Greenwald, B.; Goldberg, E.; Guerrerio, A.; Fasano, A. Effect of gliadin on permeability of intestinal biopsy explants from celiac disease patients and patients with Non-Celiac gluten sensitivity. *Nutrients* **2015**, *7*, 1565–1576. [CrossRef] [PubMed]
199. Carroccio, A.; Mansueto, P.; Iacono, G.; Soresi, M.; D'Alcamo, A.; Cavataio, F.; Brusca, I.; Florena, A.M.; Ambrosiano, G.; Seidita, A.; et al. Non-celiac wheat sensitivity diagnosed by double-blind placebo-controlled challenge: Exploring a new clinical entity. *Am. J. Gastroenterol.* **2012**, *107*, 1898–1906. [CrossRef]
200. Catassi, C.; Elli, L.; Bonaz, B.; Bouma, G.; Carroccio, A.; Castillejo, G.; Cellier, C.; Cristofori, F.; de Magistris, L.; Dolinsek, J.; et al. Diagnosis of non-celiac gluten sensitivity (NCGS): The salerno experts' criteria. *Nutrients* **2015**, *7*, 4966–4977. [CrossRef]
201. Sapone, A.; Bai, J.C.; Ciacci, C.; Dolinsek, J.; Green, P.H.R.; Hadjivassiliou, M.; Kaukinen, K.; Rostami, K.; Sanders, D.S.; Schumann, M.; et al. Spectrum of gluten-related disorders: Consensus on new nomenclature and classification. *BMC Med.* **2012**, *10*, 13. [CrossRef]
202. Volta, U.; Tovoli, F.; Cicola, R.; Parisi, C.; Fabbri, A.; Piscaglia, M.; Fiorini, E.; Caio, G. Serological tests in gluten sensitivity (nonceliac gluten intolerance). *J. Clin. Gastroenterol.* **2012**, *46*, 680–685. [CrossRef] [PubMed]
203. Volta, U.; Caio, G.; Tovoli, F.; De Giorgio, R. Non-celiac gluten sensitivity: An emerging syndrome with many unsettled issues. *Ital. J. Med.* **2014**, *8*, 225–231. [CrossRef]
204. Kabbani, T.A.; Vanga, R.R.; Leffler, D.A.; Villafuerte-Galvez, J.; Pallav, K.; Hansen, J.; Mukherjee, R.; Dennis, M.; Kelly, C.P. Celiac disease or non-celiac gluten sensitivity? an approach to clinical differential diagnosis. *Am. J. Gastroenterol.* **2014**, *109*, 741–746. [CrossRef]
205. Barbaro, M.R.; Cremon, C.; Morselli-Labate, A.M.; Di Sabatino, A.; Giuffrida, P.; Corazza, G.R.; Di Stefano, M.; Caio, G.; Latella, G.; Ciacci, C.; et al. Serum zonulin and its diagnostic performance in non-coeliac gluten sensitivity. *Gut* **2020**, *69*, 1966–1974. [CrossRef]
206. Biesiekierski, J.R.; Peters, S.L.; Newnham, E.D.; Rosella, O.; Muir, J.G.; Gibson, P.R. No effects of gluten in patients with self-reported non-celiac gluten sensitivity after dietary reduction of fermentable, poorly absorbed, short-chain carbohydrates. *Gastroenterology* **2013**, *145*, 320–328.e3. [CrossRef]
207. Molberg, Ø.; Uhlen, A.K.; Jensen, T.; Flæte, N.S.; Fleckenstein, B.; Arentz-Hansen, H.; Raki, M.; Lundin, K.E.A.; Sollid, L.M. Mapping of gluten T-cell epitopes in the bread wheat ancestors: Implications for celiac disease. *Gastroenterology* **2005**, *128*, 393–401. [CrossRef] [PubMed]
208. Zanini, B.; Villanacci, V.; De Leo, L.; Lanzini, A. Triticum monococcum in patients with celiac disease: A phase II open study on safety of prolonged daily administration. *Eur. J. Nutr.* **2015**, *54*, 1027–1029. [CrossRef] [PubMed]
209. Vaquero, L.; Comino, I.; Vivas, S.; Rodríguez-Martín, L.; Giménez, M.J.; Pastor, J.; Sousa, C.; Barro, F. Tritordeum: A novel cereal for food processing with good acceptability and significant reduction in gluten immunogenic peptides in comparison with wheat. *J. Sci. Food Agric.* **2018**, *98*, 2201–2209. [CrossRef]
210. Sánchez-León, S.; Gil-Humanes, J.; Ozuna, C.V.; Giménez, M.J.; Sousa, C.; Voytas, D.F.; Barro, F. Low-gluten, nontransgenic wheat engineered with CRISPR/Cas9. *Plant Biotechnol. J.* **2018**, *16*, 902–910. [CrossRef] [PubMed]

211. Kim, H.S.; Patel, K.G.; Orosz, E.; Kothari, N.; Demyen, M.F.; Pyrsopoulos, N.; Ahlawat, S.K. Time trends in the prevalence of celiac disease and gluten-free diet in the US population: Results from the national health and nutrition examination surveys 2009-2014. *JAMA Intern. Med.* **2016**, *176*, 1716–1717. [CrossRef]

212. Ontiveros, N.; Rodríguez-Bellegarrigue, C.I.; Galicia-Rodríguez, G.; Vergara-Jiménez, M.d.J.; Zepeda-Gómez, E.M.; Arámburo-Galvez, J.G.; Gracia-Valenzuela, M.H.; Cabrera-Chávez, F. Prevalence of self-reported gluten-related disorders and adherence to a gluten-free diet in salvadoran adult population. *Int. J. Environ. Res. Public Health* **2018**, *15*, 786. [CrossRef]

213. Cabrera-Chávez, F.; Dezar, G.V.A.; Islas-Zamorano, A.P.; Espinoza-Alderete, J.G.; Vergara-Jiménez, M.J.; Magaña-Ordorica, D.; Ontiveros, N. Prevalence of self-reported gluten sensitivity and adherence to a gluten-free diet in argentinian adult population. *Nutrients* **2017**, *9*, 81. [CrossRef]

214. Marcason, W. Is there evidence to support the claim that a gluten-free diet should be used for weight loss? *J. Am. Diet. Assoc.* **2011**, *111*, 1786. [CrossRef]

215. Hasselbeck, E. *The G-Free Diet: A Gluten-Free Survival Guide*; Center Street: New York, NY, USA, 2009; ISBN 1599951894.

216. Gaesser, G.A.; Angadi, S.S. Gluten-free diet: Imprudent dietary advice for the general population? *J. Acad. Nutr. Diet.* **2012**, *112*, 1330–1333. [CrossRef]

217. Lebwohl, B.; Cao, Y.; Zong, G.; Hu, F.B.; Green, P.H.R.; Neugut, A.I.; Rimm, E.B.; Sampson, L.; Dougherty, L.W.; Giovannucci, E.; et al. Long term gluten consumption in adults without celiac disease and risk of coronary heart disease: Prospective cohort study. *BMJ* **2017**, *357*, j1892. [CrossRef]

218. Cheng, J.; Brar, P.S.; Lee, A.R.; Green, P.H.R. Body mass index in celiac disease: Beneficial effect of a gluten-free diet. *J. Clin. Gastroenterol.* **2010**, *44*, 267–271. [CrossRef]

219. Dickey, W.; Kearney, N. Overweight in celiac disease: Prevalence, clinical characteristics, and effect of a gluten-free diet. *Am. J. Gastroenterol.* **2006**, *101*, 2356–2359. [CrossRef]

220. Valletta, E.; Fornaro, M.; Cipolli, M.; Conte, S.; Bissolo, F.; Danchielli, C. Celiac disease and obesity: Need for nutritional follow-up after diagnosis. *Eur. J. Clin. Nutr.* **2010**, *64*, 1371–1372. [CrossRef] [PubMed]

221. Melini, V.; Melini, F. Gluten-free diet: Gaps and needs for a healthier diet. *Nutrients* **2019**, *11*, 170. [CrossRef] [PubMed]

222. Pearlman, M.; Casey, L. Who Should Be Gluten-Free? A Review for the General Practitioner. *Med. Clin. N. Am.* **2019**, *103*, 89–99. [CrossRef]

223. Kasarda, D.D. Can an increase in celiac disease be attributed to an increase in the gluten content of wheat as a consequence of wheat breeding? *J. Agric. Food Chem.* **2013**, *61*, 1155–1159. [CrossRef] [PubMed]

224. Ribeiro, M.; Nunes, F.M. We might have got it wrong: Modern wheat is not more toxic for celiac patients. *Food Chem.* **2019**, *278*, 820–822. [CrossRef] [PubMed]

225. Penuelas, J.; Gargallo-Garriga, A.; Janssens, I.A.; Ciais, P.; Obersteiner, M.; Klem, K.; Urban, O.; Zhu, Y.-G.; Sardans, J. Could Global Intensification of Nitrogen Fertilisation Increase Immunogenic Proteins and Favour the Spread of Coeliac Pathology? *Foods* **2020**, *9*, 1602. [CrossRef] [PubMed]

 *foods*

Review

The 10,000-Year Success Story of Wheat!

Telma de Sousa [1,2,3], **Miguel Ribeiro** [1,2,3], **Carolina Sabença** [1,2,3] and **Gilberto Igrejas** [1,2,3,*]

1   Department of Genetics and Biotechnology, University of Trás-os-Montes and Alto Douro,
    5000-801 Vila Real, Portugal; telmaslsousa@hotmail.com (T.d.S.); jmribeiro@utad.pt (M.R.);
    anacarolina@utad.pt (C.S.)
2   Functional Genomics and Proteomics Unity, University of Trás-os-Montes and Alto Douro,
    5000-801 Vila Real, Portugal
3   LAQV-REQUIMTE, Faculty of Science and Technology, University Nova of Lisbon,
    2825-149 Lisbon, Caparica, Portugal
*   Correspondence: gigrejas@utad.pt; Tel.: +351-259350-930

Abstract: Wheat is one of the most important cereal crops in the world as it is used in the production of a diverse range of traditional and modern processed foods. The ancient varieties einkorn, emmer, and spelt not only played an important role as a source of food but became the ancestors of the modern varieties currently grown worldwide. Hexaploid wheat (*Triticum aestivum* L.) and tetraploid wheat (*Triticum durum* Desf.) now account for around 95% and 5% of the world production, respectively. The success of this cereal is inextricably associated with the capacity of its grain proteins, the gluten, to form a viscoelastic dough that allows the transformation of wheat flour into a wide variety of staple forms of food in the human diet. This review aims to give a holistic view of the temporal and proteogenomic evolution of wheat from its domestication to the massively produced high-yield crop of our day.

Keywords: wheat; *Triticum aestivum* L.; *Triticum durum* Desf.; gluten; breadmaking; celiac disease

Citation: de Sousa, T.; Ribeiro, M.; Sabença, C.; Igrejas, G. The 10,000-Year Success Story of Wheat!. *Foods* **2021**, *10*, 2124. https://doi.org/10.3390/foods10092124

Academic Editors: Barbara Laddomada and Weiqun Wang

Received: 21 August 2021
Accepted: 6 September 2021
Published: 8 September 2021

Publisher's Note: MDPI stays neutral with regard to jurisdictional claims in published maps and institutional affiliations.

1. Introduction

Wheat is a high-yielding crop that is easy to store and is very adaptable to different climates. From the most primitive form of wheat to the species currently grown, these and other desirable characteristics have been selected and developed by human societies since ancient times [1,2]. Its domestication is thought to have occurred in the Fertile Crescent about 10,000 years ago and has spread to all parts of the world through the first farmers, adapting the domesticated populations to different environments [3]. Wheat is able to grow in temperate, Mediterranean, and subtropical regions of the two hemispheres, mainly due to its enormous genetic diversity. For example, there are more than 25,000 varieties of *Triticum aestivum* L. adapted to different temperate environments [4]. Wheat can be classified according to when it is sown. Winter wheat is sown in autumn because the seedlings need a period during the vegetative phase when temperatures are between 0 and 5 °C. About 80% of the world's wheat is winter wheat. Spring wheat is planted in spring and harvested in late summer or fall in South Asian countries or North Africa [5].

The most important varieties of modern wheat are tetraploid durum wheat (*Triticum durum*) and hexaploid bread wheat (*Triticum aestivum* L.), which have different attributes in terms of genomic composition, grain composition, and end use [6]. The progressive domestication of these wheat species led to the restructuring of the rachis and glumes, converting the brittle ears of wild species into non-brittle and bare grain ears in cultivated species [7]. These morphological effects of domestication facilitated the cropping of wheat, leading to an exponential increase in its economic importance [8].

As well as being a major source of carbohydrate in the form of starch, wheat seeds also provide a great source of protein in the human diet. Wheat endosperm proteins, the prolamins, can be divided into gliadins and glutenins according to their polymerization properties. They are the main components of gluten in wheat flour. The elasticity and extensibility of dough are the basic characteristics of breadmaking and are largely determined by the glutenin and gliadin content of the flour, respectively [9]. Indeed one of wheat's main advantages is the unique ability of wheat doughs to be processed into different types of bread and other bakery products (including cakes and biscuits), pasta, and other processed foods [10]. The markets and the processing industry depend on different varieties of wheat with different quality attributes that meet the needs of specific products. Wheat is generally traded according to specific characteristics, namely, the grain protein content and hardness [11]. These market requirements have led to a more comprehensive and thoughtful approach to wheat breeding so as not to neglect essential protein content and protein quality [12,13].

The present work aims to describe the history of wheat with a particular focus on its singular characteristics making it almost ubiquitous in the human diet, the ability to form the gluten viscoelastic network. Worldwide production, market requirements, consumption habits, and research trends are also summarized in this work.

2. Origin and Evolution

The history of wheat from about 10,000 B.C. is an important part of the history of agriculture [2]. Wheat is thought to have first been cultivated in the Fertile Crescent, an area in the Middle East spreading from Jordan, Palestine, and Lebanon to Syria, Turkey, Iraq, and Iran [14]. The earliest cultivated species are hulled (glumed) wheats and comprise all three polyploidy levels known in *Triticum* spp., diploid, tetraploid, and hexaploid [14]. These species are einkorn (*Triticum monococcum* L.), emmer (*Triticum Dicoccum*), and spelt (*Triticum spelta*), and their corresponding wild ancestors are still found in these territories [15,16]. However, the domestication of these three species caused considerable genetic erosion that has since been reinforced through modern breeding practices with the unwanted consequence of increased susceptibility or vulnerability to environmental stresses, pests, and diseases [8].

Wheat belongs to the *Poaceae* family [17,18], which includes the *Triticeae* tribe, the division with the most economically important cereals. The *Triticeae* tribe includes 14 genera, grouped into the subtribes *Triticinae* and *Hordeinae*, and the production of amphiploids and interspecific hybrids suggests there is genetic or cytoplasmic compatibility between the different genera [19]. The most representative wheat species of this tribe are hexaploid *Triticum aestivum* L. (A^uA^uBBDD; 2n = 6x = 42), tetraploid *Triticum turgidum* L. (A^uA^uBB; 2n = 4x = 28), and diploid einkorn *Triticum monococcum* L. (A^mA^m; 2n = 2x = 14) [8]. Currently, the most economically important species are tetraploid durum wheat (*Triticum turgidum* subsp. *durum* (A^uA^uBB; 2n = 4x = 28) and hexaploidy bread wheat [8,11]. The origin of the latter two wheat species was the result of two polyploidization events (Figure 1).

The first of these occurred with the association of the genomes of two diploid species: one related to the wild species *Triticum urartu* (A^uA^u; 2n = 2x = 14) provided the A genome and the other yet unknown species from the *Sitopsis* provided the B genome. This produced the allotetraploid wild emmer wheat (*Triticum Dicoccum*; A^uA^uBB; 2n = 4x = 28) [20,21]. With the genetic resources of both ancestral diploids, this cultivated allotetraploid wheat is generally more vigorous, giving higher yields and adapting to a wider range of environmental conditions compared to its parents [22]. The second event occurred between *T. turgidum* and *Aegilops tauschii* (DD), a wild diploid species, which produced the allohexaploid early spelt (*Triticum aestivum subsp. spelta*; A^uA^uBBDD; 2n = 6x = 42) [8,20]. Initially, *T. aestivum* L. and *T. turgidum* produced sterile hybrids; however, a duplication of chromosomes in gametes or progeny gave rise to fertile species [23].

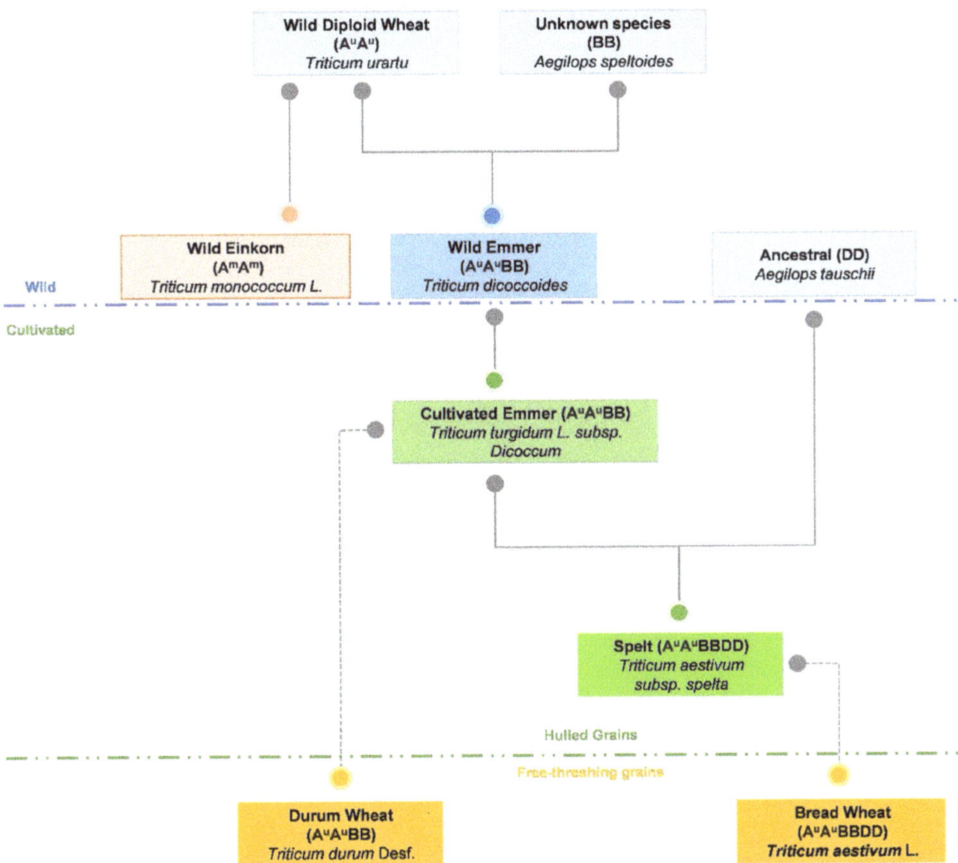

Figure 1. Phylogeny of the domesticated species of durum wheat (*Triticum durum* Desf.) and common wheat (*Triticum aestivum* L.).

Initial cytogenetic studies of the origin of wheat proposed the diploid species *T. monococcum* L. as the donor of the A genome in the tetraploid species of both evolutionary lines [24]. However, recent studies have shown that variations in the repetitive nucleotide sequences in the A genome, present in both wheat tetraploid species, were more related to the *T. urartu* genome than to the *T. monococcum* L. genome [25,26]. The origin of the B genome remains unresolved. In polyploid wheat, the B genome constitutes the majority of the genomic component and as the genetic variation at the DNA level is high, the true donor of the B genome since polyploid formation has been very difficult to establish [26]. There are many strands of evidence (morphological, geographical, cytological, genetic, biochemical, and molecular) that suggest the section *Sitopsis* of *Aegilops*, namely *Aegilops speltoides* (genome S), as the B genome donor of both soft and hard wheat [26,27]. The ambiguous nature of the B genome remains a subject of study with several possible theories: (1) the parent is extinct; (2) the diploid parent still exists but has yet to be discovered; (3) in the formation of the allotetraploid, the diploid B genome donor has undergone several changes; and (4) in the condition of allopolyploidy, the B genome evolves rapidly due to several chromosomal structural changes, namely, due to the introgression of chromosomal segments of other allopolyploid or diploid species [27]. In contrast, the D genome is derived from the diploid progenitor *Aegilops tauschii* [16]. This genome carried genes and alleles that favored adaptation to the more continental climate of central Asia, thus allowing hexaploid wheat to be cultivated more widely than emmer wheat [28]. Khorasan wheat (*Triticum turgidum* ssp. *turanicum*) is also an important ancient species for the appearance of durum wheat, as it has morphological characteristics similar to those of durum wheat. There is no concrete data on its origin; however, through DNA analysis, it is likely that the species originated from a spontaneous cross between durum wheat and wild wheat [29].

In fact, wheat is a useful model organism in which to study the evolutionary theory of allopolyploid speciation, and consequently, the study of its adaptation to domestication [26]. The domestication of *T. aestivum* L. and *T. durum* Desf. began a selection process that increased the adaptation of these wheat species as well as their cultivation on a large scale [30]. This is attributable, in part, to the allopolyploid genome structures [20]. Allopolyploids contain two or more diverged homoeologous genomes, a fusion of different genomes [31]. Alloploidization accelerated the genome evolution of wheat through changes like gene loss, gene silencing, gene activation, and duplications [31,32]. Two evolutionary processes in particular were crucial in domestication [2]. A natural mutation at the *Q* locus of chromosome 5A modified the effects of recessive mutations at the *Tg* (*tenacious glume*) locus and thus made the wild emmer wheat and early spelt easier to thresh, later evolving into the free-threshing ears of durum wheat and common wheat, respectively, constituting a bridge between primitive wheat and modern wheat [8,14,16]. The fragility of the glumes of the two modern wheats is a technological advantage that facilitates the grinding of grain into flour, whereas the grains of more primitive species are covered by glumes (shells) that even after harvest make grinding difficult [14]. The other evolutionary process was the loss of ear fragmentation at maturity, resulting from mutations in loci, such as the *brittle rachis* (*Br*) gene, which allows the dispersion of seeds in natural populations [22,33]. The chromosomal location of *Br* in emmer wheat is on chromosomes 3A and 3B, but the equivalent in einkorn wheat is still unknown. In this regard, the selection of the non-breaking characteristic led to modern wheat varieties superseding einkorn, emmer, and spelt wheats [33].

3. Production

Despite modern wheat varieties (durum wheat and common wheat) having a higher yield potential, they have disadvantages compared to old varieties, particularly in terms of reduced tolerance of abiotic and biotic stresses, such as diseases, pests, drought, heat, cold, salinity, pollution, and shortage of soil nutrients [14]. The varieties we know today are the outcome of breeding programs that mainly selected and developed grains with specific characteristics, such as higher yield, superior breadmaking quality, efficient use of nutrients, and resistance to biotic and abiotic stresses. Selection is done by phenotyping and/or genotyping, using techniques such as sequencing or molecular markers [34–36].

In the 1960s, the world's ability to deal with the increasing population and food demand was desperate (especially in developing countries), but this phenomenon stimulated a dramatic increase in cereal production in many of these countries, and large-scale hunger and the associated social and economic turmoil were averted [37]. Several genetic traits were selected to improve the yield, stability, and large-scale adaptability of rice, corn, and wheat [38]. The introduction of high-yielding wheat varieties was accompanied by massive application of chemical fertilizers and pesticides. The impact of these advances was so huge that it was called the Green Revolution [39].

The traditional and older varieties of wheat are tall and leafy, with very brittle stems. The Green Revolution brought greater productivity to wheat, mainly because of the introduction of cereal dwarfing genes [40]. The insertion of dwarfing genes from the Japanese variety 'Norin 10' made it possible to simultaneously increase the yield potential, harvest rate, responsiveness to fertilizers, and resistance to biotic and abiotic diseases [38,41]. In fact, more than 70% of commercial wheat varieties currently grown have the 'Norin 10' dwarfing genes in their genomes [39]. 'Norin 10' contains two dwarfing genes, *Rht1* and *Rht2*, which are semi-dominant alleles of homoeologous genes on chromosomes B and D, respectively [39]. The effect of each gene on plant height is similar and their combined effect is additive [42,43]. This comes from the fact that these gene products act as transcription factors in gibberellin signaling. Gibberellin is an essential endogenous regulator of plant growth [44] and the alleles that regulate dwarf genes not only reduce plant height, but also reduce responsiveness to gibberellin levels [45]. The dwarf genes also decrease the length of the coleoptile, which reduces the seedling survival rate and hence population density, a disadvantage of their use [46,47].

Nowadays, about 95% of the wheat grown in the world is common wheat, used mainly as whole flour and refined flour to produce a wide variety of flat and fermented breads and for the manufacture of a wide variety of other bakery products. The remaining 5% is mostly durum wheat used to produce semolina (coarse flour), the main raw material for the manufacture of a wide variety of different baking product [6,11]. Figure 2 shows the production of wheat worldwide. The top five wheat producers are China, India, Russia, USA, and France [1].

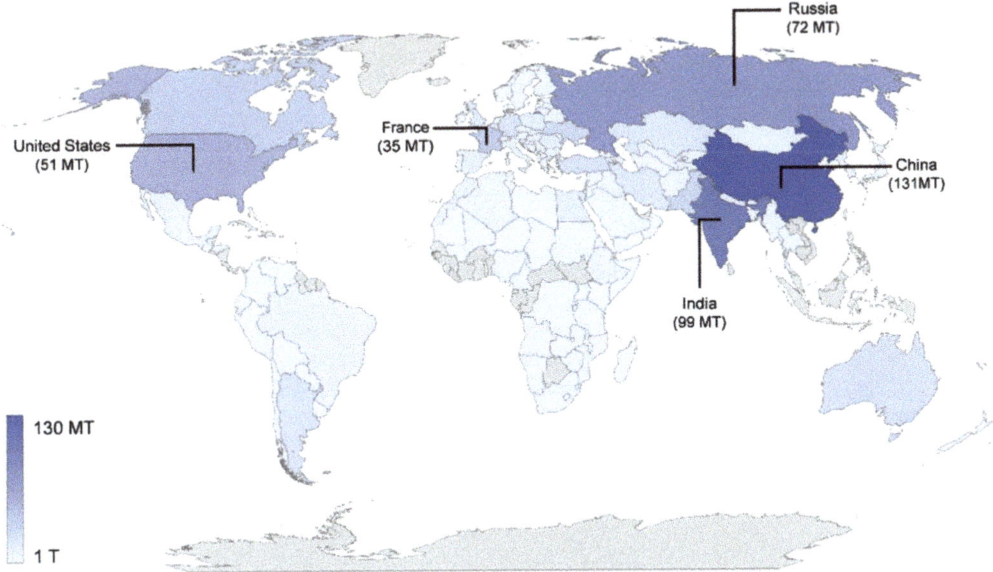

Figure 2. Worldwide wheat production. Data indicate the total wheat production in megaton (MT) in the year 2018 per country FAOSTAT [1].

The world's wheat harvested area has remained relatively stable since 1960, but in the same period, production has risen significantly. Combining these data clearly shows that wheat production became much more efficient over the years, in part due to the introduction of hybrid crops that allowed an increase in yield (Figure 3) [1,48].

Sustainable food production requires environmentally friendly agricultural practices. Limiting the number of species or varieties of a species results in a huge loss of genetic diversity with associated negative impacts on the vulnerability of ecosystems and species extinction, and limiting the ability of farmers to respond to future agricultural needs [14]. The disadvantage of today's high-yield high-protein wheat varieties is that they require large amounts of agrochemical input, such as fertilizers, pesticides, and herbicides. Additionally, climate change and global warming can have adverse effects on both the yield and quality of wheat. An alternative to overcome these obstacles is to use older varieties of wheat and landraces, like einkorn, emmer, and spelt, that have never been subjected to breeding practices [12,13].

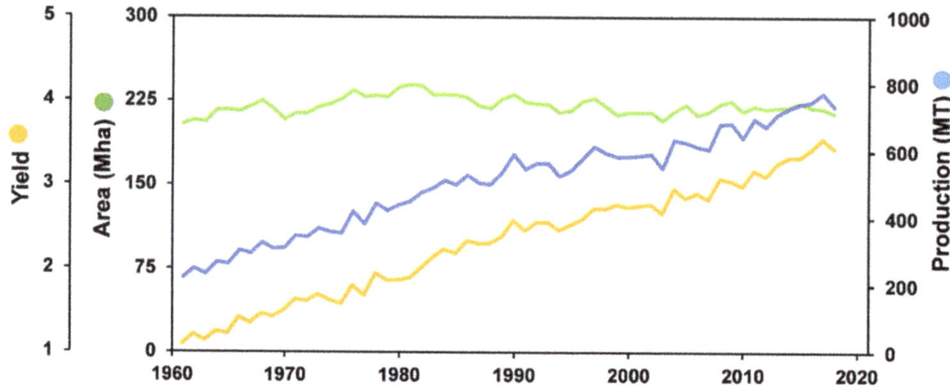

Figure 3. The world wheat production, harvested area, and yield between 1960 and 2018 FAOSTAT [1]. Through the indexmundi database, we were able to verify that China, India, and Russia had a positive production growth rate between the years 2018–2020 while, on the contrary, the USA had a decreased production growth rate [49].

4. Wheat Consumption Habits

The consumption of bread, especially bread made with whole wheat flour and multi-grain flour, is tending to increase in developed countries, especially due to the increased awareness of the need to reduce the consumption of simple carbohydrates, fats, and cholesterol and increase the consumption of complex carbohydrates, dietary fiber, and vegetable protein [50]. Each type of bread has its own characteristics, processing conditions, and specific requirements for the final product [11].

Durum wheat is widely used in regional foods (pasta, flat bread, couscous, and hamburgers) in Europe, North Africa, and Western Asia [10,16]. In Italy, durum wheat is traditionally used in the production of pasta, now consumed worldwide [51]. Durum wheat flour is used alone or mixed with other flours and is widely used in Mediterranean countries to make bread [10]. Couscous is one of the major food staples in North African countries, such as Egypt, Libya, Tunisia, Algeria, and Morocco [52]. Durum wheat has also been used in North America to make a ready-to-eat puffed breakfast cereal and in Germany, the noodles in kugel, a sweet pudding eaten as a dessert, are made of durum wheat [53].

Common wheat is generally milled into flour (refined and whole meal) and is used in breads (fermented, crushed, and steamed), pasta, cookies, and cakes [11]. Indeed, fermented breads are popular worldwide. The medium-hard to hard wheats that yield strong doughs are more suitable for producing rolls for hamburgers and hot dogs, for the semi-mechanized or manual production of typical French bread, as well as flat bread types like Arab baladi bread, Indian chapati, and Mexican tortilla [54,55]. The light dough made from soft wheat is suitable for Asian bread cooked with steam [56]. Waxy wheat has been developed with low levels of amylose or no amylose starch, which makes it possible to characterize the impact of starch on the quality and functionality of end-use food. Waxy wheat is often used as a source of blended flour to improve the shelf-life stability, processing quality, or palatability of baked and sheeted wheat products [57]. It has been shown that adding 40% waxy wheat flour to Chinese noodles improved the quality of dry white Chinese noodles and exhibited the shortest cooking time [58]. Another example of the use of waxy wheat is in the refrigeration and freezing of food products. This type of wheat has specific effects on gelatinization and retrogradation. To produce products for freezing purposes, waxy wheat is the best choice since it has a slow retrogradation rate [59].

The old wheat varieties are still being explored today as healthy alternatives to bread wheat [4]. Mainly for human consumption as different types of fermented bread, unleavened bread such as chapatti, and pancakes, old wheat varieties are also used as animal feed [4]. Ancient wheats have a nutritional composition that is different from modern varieties, such as resistant starch, carotenoids, phytochemicals, and antioxidants, that offer numerous health benefits [60]. These days, there is often a mixture of flours from these old wheats with modern varieties to make up for the lack of some components. For ex-ample, it has been observed that, compared to modern wheat, there is a reasonably high amount of protein in the emmer grain and a higher amount of lysine in the einkorn wheat [61]. Thus, baking bread with flours from these ancestral wheats can be more beneficial to health, as the high content of lysine and other health-promoting compounds in these wheats can complement those of modern wheat flours to achieve a better dietary balance [61,62]. Another reason why old wheat is important is that consumption trends are constantly changing, and there is an increasing demand for sustainable, regional, and artisanal products that have been manufactured in a way that conserves resources (promote biodiversity and reduce the ecological footprint) [63]. Old wheat products, such as breads made from Khorasan wheat, have good sensory properties and bread volumes almost as high as modern varieties. The old wheat species (einkorn, emmer, and spelt) are also used as whole grains for salads and used differently for processing, for example spelt used mainly for bakery products and the einkorn and emmer used mainly for pasta products [64,65]. These choices of using specific varieties to the manufacturing of specific products try to respond to the growing demand for a wider variety of products [65,66].

Bread made with emmer wheat flour is used worldwide, but especially in Switzerland. In Italy, it is used for *pane di farro* and to produce some pasta, but only in small quantities. In some rural areas of Italy and Iran, emmer is used as a source of carbohydrates in meals in the same way as rice [60].

Spelt flour and bread made with spelt have become increasingly popular in many countries due to the nutritional properties of this grain. Specifically, spelt bread has a higher protein content, a higher lipid content, a more desirable fatty acid profile, and higher percentages of several nutrients compared to common wheat [67]. From an agronomic point of view, compared to other wheats, spelt is more resistant to diseases and various pathogenic fungi, so it requires less fertilization, pesticides, and herbicides and, as the seeds are covered by the husk, no chemicals are needed before sowing [68,69]. With the increasing interest in organic farming, these characteristics have led to a higher global interest in the production of this species that may have several impacts on human nutrition, health, and sustainability [67]. On the other hand, general consumption habits are shifting to a gluten-free diet, as people are removing wheat from their diet without any medical advice and, consequently, are not consuming the necessary nutrients of wheat [70]. This demand for gluten-free products can arise from the belief that gluten-free and sugar-free foods help people overcome problems, such as bloating, indigestion, and others. Thus, bakery products that claim to be fat-free, sugar-free, gluten-free, whole wheat, and salt-free are popular with health-conscious consumers. According to the Agriculture and Horticulture Development Council, an average increase of 10% each year is expected for the next three years in the free-food market, including bakery products [71]. In the case of people diagnosed with disorders associated with gluten, such as celiac disease, some work has been carried out exploring the potentialities of sourdough bread. Basically, wheat is leavened using a long-time fermentation by sourdough, a cocktail of acidifying and proteolytic lactic acid bacteria, capable of hydrolyzing Pro-rich peptides, including the 33-mer peptide. Thus, a wheat product fermented with sourdough reduces its immunogenicity for celiac patients [72–74].

Considering the worldwide consumption of wheat products and the great impact of the growing world population, several techniques have been used to address these demands. One of the alternatives is the production of frozen dough products as they have a considerably longer shelf life. However, during freezing, there are structural changes induced in the gluten network [75]. Some studies indicate that these changes occur mainly in the α-helices, β-turns, and antiparallel-β-sheet structures, while the β-sheet content is not affected by the duration of frozen storage [76,77]. Other studies have led to the conclusion that changes in protein structure are essentially due to protein aggregation and, consequently, to gluten deterioration [78]. Thus, freezing, the frozen storage temperature, and temperature fluctuations during storage lead to a loss in dough and bread quality, resulting from lower yeast viability and bread volume and increased firmness of the bread crumb and mass weight loss [79]. To overcome this obstacle in freezing, different compounds are often added to improve the quality of the frozen dough. For example, the addition of NaI leads to an increase in the β-turns at the expense of intra- and intermolecular β-sheets. This promotes the increase in water–protein interactions, reducing protein–protein interactions, and, consequently, improving the quality of frozen dough [76,80].

5. Nutritional Value and Health Impact

Wheat is a good source of complex carbohydrates, with the starchy endosperm being the major storage tissue and accounting for about 80% of the wheat grain. Unlike simple carbohydrates, such as sugars, complex carbohydrates are preferred as they provide a more sustainable energy source as they gradually release energy according to the body's needs [50]. Additionally, diets high in these complex carbohydrates lead to fewer health problems [81] and wheat starch provides a good source of energy in nutrition (1550 kJ/100 g) [50,82]. Compared to ancient wheats, the carbohydrate contents in durum (71%) and common wheat (75%) are slightly higher than, for example, spelt (68%) and einkorn (67%), which provide lower carbohydrate contents [65]. Fiber is also a very important dietary factor. The majority of dietary fiber in wheat is found in the white layers, and after wheat milling, the resulting white flour contains 2.0–2.5% fiber [83]. Thus, 40 g of white wheat bread can provide about 1 g of dietary fiber, whereas a similar serving of whole wheat bread can provide 3–4.5 g of dietary fiber. A high-fiber diet is positively associated with health benefits [84,85]. As well as the carbohydrate content, the ancient varieties, spelt (12%), einkorn (9.8%), and emmer (9.8%), have a lower fiber content than common wheat (13.4%) [13,65,86].

Lipids are a relatively minor component (2.5–3.3%) of wheat caryopsis, and 35–45% of wheat lipids are in the endosperm. Wheat has significant levels of lipids of various types, such as acylglycerols, fatty acids, and phospholipids, between others [87]. Lipids play an important role in the human diet as they are a high energy source and form the structure of cell membranes [88]. Although grain lipidomics is rather complex, there is experimental evidence that ancient wheats (einkorn, emmer, and spelt) have characteristic lipidomic profiles that are different from each other [89]. In general, ancient wheats, except for eikorn, have a higher lipid content than common wheat [90,91], but einkorn has higher monounsaturated fatty acids, lower polyunsaturated fatty acids, and lower saturated fatty acids than durum wheat, which has been proven to be good for human health [61].

Wheat, depending on the wheat genetic background and fertilizer inputs, can provide a protein content. Nevertheless, durum wheats provide more protein than most other cereals and the distribution of essential amino acids is at least as good or better than other cereals [50]. The protein content of wheat can, in general, vary between 9% and 18% of the grain weight [16,92]. For example, common wheat intended for the production of cakes and/or biscuits will ideally have a protein content of 7–11%, while wheat intended for the production of high-volume loaf bread typically requires a 12% or higher protein content (Figure 4) [83]. When compared to ancient wheats, modern wheat has a possibly lower protein content, as the grain of modern wheat is larger and heavier, which produces a larger starchy endosperm, which, in turn, lowers its protein content. In fact, in einkorn, the protein content is 15.5–22.8% and the percentage of amylose is 23.8%, whereas in common wheat, the protein content and amylose percentages are 12.9–19.9% and 28.4%, respectively [14].

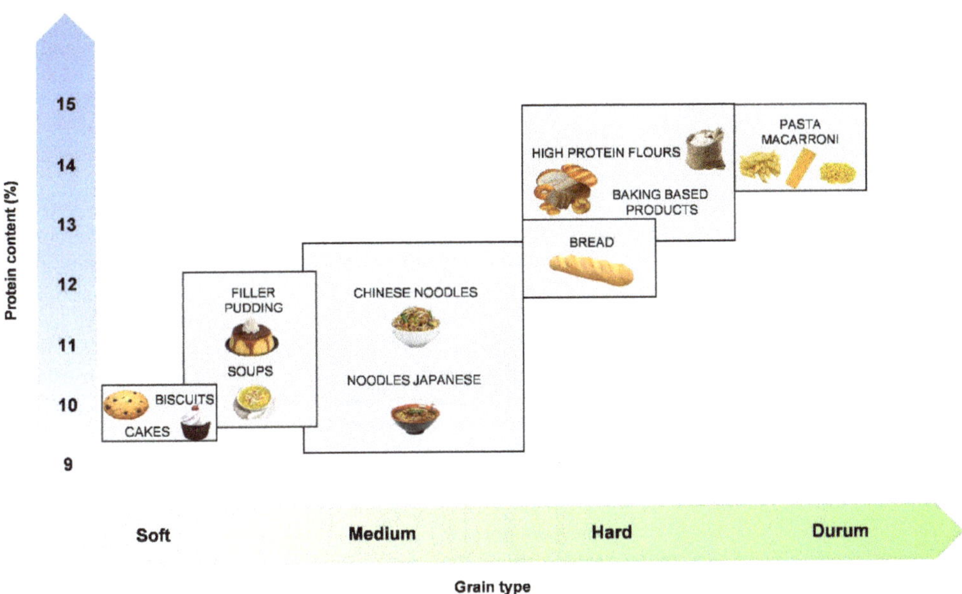

Figure 4. Examples of the most appropriated varieties according to the hardness of the grain and protein content for specific final products.

Furthermore, the amino acid composition of wheat lacks essential amino acids, such as lysine, threonine, and methionine [48], exhibiting a higher content of glutamine (>30%) and proline (>10%) residues. For example, glutamine and glutamic acid are involved in several important biological processes, such as gene expression regulation, intracellular protein turnover, nutrient metabolism, and oxidative defense [93]. On the other hand, proline residues confer some resistance to gluten proteins to proteolysis by digestive enzymes in the gastrointestinal tract [94]. As gluten is rich in proline, it produces a compact and hard structure that usually presents difficulties in digestion and elimination [95]. Some of these digestion-resistant peptides are believed to be related to adverse immune processes in susceptible people [96]. In celiac disease, some of these peptides are responsible for the onset of the disease, triggering an abnormal immune system response in genetically predisposed individuals [97,98]. There is evidence that the D genome of bread wheat has more immunogenic epitopes than the A and B genomes. Thus, it is expected that diploid and tetraploid wheats (*T. monococcum* and *Triticum durum*, for example) tend to contain less immunogenic epitopes for celiac patients than bread wheat [99]. In general, gliadins are considered the main cause of problems for celiac patients and the most immunogenic sequences occur in the N-terminal repetitive domain of α-/β-gliadins, which mainly consist of glutamine, proline, and aromatic amino acids [72,100]. Currently, several studies have explored the immunotoxicity for celiac patients of ancient and modern common wheat varieties and also other wheat species [101]. A study focused on the content of celiac-related epitopes of ancient and modern *T. aestivum* and *T. durum* wheat varieties, and also spelt showed the following relationship: spelt > wheat landraces > modern wheat varieties = tetraploid varieties [102]. It was hypothesized that the breeding programs could have led to the increase of the glutenin content in wheat, therefore decreasing the gliadin to glutenin ratio [102]. In the same way, Prandi et al., analyzing the gluten peptides released after in vitro digestion of different old and modern *Triticum* varieties, found that older varieties compared to modern varieties had higher amounts of immunogenic peptides, thus concluding that both ancient and modern varieties are not safe for consumption in the case of celiac patients [103]. Furthermore, other authors concluded that the total quantitative amount of immunogenic peptides is not directly related to the ploidy level or their origin (ancient/ modern). For example, common wheat (hexaploid) showed the same amount of immunogenic peptides as emmer and durum wheat (tetraploids) [103,104]. Nevertheless, all the results point to a natural inter-varietal genetic difference affecting the amount of celiac-related epitopes, which makes it possible to identify accessions presenting a lower amount of T cell stimulatory epitope sequences and therefore are more suitable for wheat breeding programs [101,105].

In addition to the main components of wheat, proteins, carbohydrates, and lipids, wheat grain is also an important source of phytochemicals, vitamins, antioxidants, and macro and micronutrients that are important components of human health [83].

6. Functional and Technological Properties

6.1. Grain Hardness

The hardness of the wheat grain refers to its resistance to milling. The harder the wheat, the greater its resistance and the more difficult it is to grind into flour [83]. Between common wheat and durum wheat varieties, there are significant differences in the composition of the grain and in the quality of wheat processing, i.e., a variety may be suitable for the production of certain foods but unsuitable for the processing of others [11,50]. For example, as summarized in Figure 4, for a product like leavened bread, medium to hard grains with 11% to 13% of protein are normally used, while for cookies, cakes, and Chinese noodles, soft to medium grains with 8% to 12% protein are more suitable [11].

The discovery of the starch granule protein, friabilin, which influences the texture and quality of the wheat grain, provided a biochemical basis for evaluating grain texture. The friabilin protein complex regulates the degree of adhesion of starch granules to the protein matrix, a factor of great importance in determining grain hardness [106]. The variation in grain texture (hardness or softness) is a characteristic inherited and controlled by a single locus referred to as *Hardness* (*Ha*), which comprises three genes: *Pina* codes for puroindoline a, *Pinb* codes for puroindoline b, and *Gsp*-1 codes for a less abundant protein called grain softness protein [28].

In wheat grain, the accumulation of friabilin is dependent on the softness/hardness genes *Ha/ha* [106]. Soft wheats have a wild-type form (*Ha*) whereas hard wheats have a recessive (*ha*) or mutated form [107]. Additionally, variation in *Pin* genes significantly affects the grinding characteristics and quality of the final product, for example, the wild-type *Pina* and *Pinb* genotype results in a soft-textured phenotype, while mutation in one or both genes results in a more rigid phenotype [108]. Durum wheat represents a generally tougher class of wheat, but the relationship to *Ha/ha* is still unclear [106].

The texture of the endosperm is also a crucial factor for wheat producers as this characteristic influences certain physical properties, such as the particle size and density of the flour, risk of damaging starch, water absorption, and milling yield [109]. The cell contents of common wheat endosperm fracture, leaving the starch granules intact, and resulting in a wide particle size distribution. The main physical difference between hard and soft wheat endosperm is that the starch granules of soft wheat are surrounded by a protein matrix. On the other hand, in hard grains, crushing and grinding the grain is more difficult and produces flour with a coarse texture because more of the starch is damaged. In turn, endosperms of harder wheats have a greater capacity to absorb water and, consequently, are more easily hydrolyzed by alpha-amylase [107].

It is difficult to reduce the particle size of durum wheat flour since the grain tends to fragment along the lines of the cell boundaries, and on average, particles tend to be relatively uniform and larger than particles from soft wheat flours [110].

6.2. Wheat Proteins

6.2.1. Non-Gluten Proteins

Wheat proteins can be divided into two groups: gluten proteins and non-gluten proteins. The group of non-gluten proteins is divided into albumins (soluble in water) and globulins (soluble in neutral saline solutions) and play mainly structural and metabolic functions and a minor role in wheat quality [111]. For example, albumins and globulins include enzymes and enzyme inhibitors that regulate development at different stages in wheat growth [112]. These proteins are also involved in IgE-mediated food allergies [113,114]. Albumins are soluble in water and globulins are soluble in neutral saline solutions. Non-gluten proteins are mostly monomeric; however, albumins and globulins tend to form polymers through the formation of inter-chain disulphide bonds [115]. Tomic et al. developed a study in which they focused on finding the relationship between particular albumin fractions and the enzymatic activity of wheat flour and the rheological properties of the dough [116]. The study included albumins like α-amylase, α-amylase/protease inhibitors (13 and 16 kDa), as well as enzymes with different physiological functions (62 kDa). This study showed that the 15–30 kDa albumin fraction was shown to influence the rheological properties of the dough, especially water absorption and resistance to extension. With regard to the 5–15 kDa, 30–50 kDa, and 30–60 kDa albumin fractions and the proteolytic and α-amylolytic activity, no relationship was found regarding the enzymatic state of the flour [116].

Recently, other wheat proteins, such as ALP (avenin-like protein), have been associated with wheat processing properties and bread-making quality [117]. ALP-coding genes were mapped at the long arm of chromosome 4A and at the short arms of chromosomes 7A and 7D in bread wheat. The ALP proteins, in addition to improving the dough mixing properties, also possess antifungal functions, showing an important potential for wheat breeding [118]. On the other hand, a novel gene involved in breadmaking properties was reported. The wbm (wheat bread making) gene was demonstrated to have a significant effect on gluten quality, gluten strength, gluten extensibility, and bakery quality [119].

6.2.2. Gluten

Gluten, one of the first protein components to be chemically analyzed, was first described by Giacomo Beccari in 1728 [16]. Beccari reported preparing a water-insoluble fraction of wheat flour that he called "glutinis" [120]. The gluten was later defined as being composed of prolamins and glutelins by Thomas Burr Osborne between the years 1886 and 1928 based on his study of plant proteins. It was Osborne who classified proteins into groups based on their solubility [121]. For a long time, based on Osborne's definition, gliadins were thought to be the prolamins and distinctly different from glutenins thought to be glutelins. However, biochemical and molecular studies, corroborated by genetic studies, have proven that all gluten proteins are structurally and evolutionarily related and can be collectively defined as prolamins [122]. According to Osborne, the gliadin and glutenin together represent up to 85% of the total protein content within wheat endosperm (gliadins ~40%, high- and low-molecular-weight glutenin sub-units ~10% and ~30%, respectively) [123].

Gliadins are mostly monomeric proteins with a molecular weight that ranges from 28 to 55 kDa. They are separated into four groups on the basis of mobility at low pH in gel electrophoresis: α-, β-, γ-, and ω-gliadins [124]. α-, β-, and γ-gliadins are rich in cysteine and methionine residues through which intramolecular disulphide bonds form, imposing a specific conformation on the polypeptide [125]. By contrast, ω-gliadins usually do not have any cysteine residues. An exception is a group of mutant ω-gliadin forms that have a single cysteine residue, which allows them to establish intermolecular disulphide bonds and to be incorporated into the glutenin polymer [126]. Based on analysis of the complete or partial amino acid sequence, amino acid composition, and molecular weight, gliadins can be redivided into four different types: ω5-, ω1,2-, α/β-, and γ-gliadins. The α-, β-, and some γ-gliadins are also encoded by the short arms of chromosome 6 (*Gli-2*) whereas the majority of γ- and ω-gliadins are encoded by the *Gli-1* genes on the short arms of chromosome 1 [127]. Specifically, they are located on chromosomes 1A, 1B, and 1D at the three homeologous *Gli-A1*, *Gli-B1*, and *Gli-D1* loci and on chromosomes 6A, 6B, and 6D at the *Gli-A2*, *Gli-B2*, and *Gli-D2* loci [128].

Glutenins are polymeric proteins with a molecular weight that can exceed 34,000 kDa [129]. Based on their mobility when separated through polyacrylamide gels, they were classified into high-molecular-weight glutenins (HMW-GSs; 70–90 kDa) and low-molecular-weight glutenins (LMW-GSs; 30–45 kDa) [129]. The genes encoding HMW glutenins are located on the long arm of chromosomes 1A, 1B, and 1D at the *Glu-A1*, *Glu-B1*, and *Glu-D1* loci, respectively [130]. The *Glu-A1* locus can present more than 21 allelic variants in *T. durum* Desf. In *T. aestivum* L., the *Glu-B1* locus has more than 69 alleles and the *Glu-D1* locus more than 29 [131]. In each common wheat variety, three to five different HMW-GSs can be found. *Glu-A1* encodes one subunit, *Glu-B1* encodes one or two subunits, and *Glu-D1* encodes two subunits. These subunits can be further classified into x-type and y-type. In particular, y-type subunits do not have a particularly positive effect on dough consistency. On the other hand, x-type subunits are responsible for the formation of high-molecular-weight protein aggregates resulting from their ability to form linear polymers through cysteine residues [128,132,133].

According to molecular weight and composition, LMW-GSs can be divided into groups B, C, and D [134]; and LMW-GS B type glutenin divided into three further categories according to the first amino acid in the peptide chain, LMW-m (methionine), LMW-s (serine), and LMW-i (isoleucine) [134,135]. The genes that encode them are on the short arms of chromosomes 1A, 1B, and 1D at the *Glu-A3*, *Glu-B3*, and *Glu-D3 loci*, respectively [130].

6.3. Effect of Gluten Proteins on Wheat Functionality

Gluten accounts for about 85% of the protein component of wheat endosperm [136]. The gluten matrix and the functions it performs determine the end-use quality of wheat [137]. The viscosity and extensibility of the dough is mainly related to the properties of hydrated gliadins, while the strength and elasticity of the dough comes essentially from the cohesive properties of hydrated glutenins [138]. Furthermore, the gluten protein fractions contain many individual components with much allelic variation between varieties. For example, a typical wheat variety can have up to 60 different gluten proteins [139].

In 1981, Payne and colleagues demonstrated that HMW-GS variation influences the breadmaking quality of wheat flour. They correlated the presence of two HMW-GS with breadmaking quality. Subunit 1 expressed from the *Glu-A1* locus and subunits 5 + 10 expressed from the *Glu-D1* locus were related to good quality, when compared to null and 2 + 12 sub-units, respectively [140]. Years later, in 1987, when analyzing 84 varieties of British-grown wheat, they developed a scoring system (based on the SDS-sedimentation test) to assess the individual contribution of each subunit expressed from the *Glu-1* locus to the flour quality. The score for a variety could be obtained by adding the points attributed to each individual subunit [141]. The relative contribution of each HMW-GS locus to the characteristics of the flour is *Glu-D1* > *Glu-B1* > *Glu-A1*. The protein content is independent of this contribution ratio, but the contribution of each locus depends on the expressed subunits and wheat variety in question [135,142].

The effects of gliadins and LMW-GSs on the characteristics of flour are not yet fully understood as the genetic link is not fully established and it is difficult to distinguish these proteins in polyacrylamide gels [143]. For example, the genes encoding the monomeric prolamins are strongly linked to the genes of the LMW glutenins at the *Glu-3* locus and the genes that encode gliadins and LMW glutenins (type C and D) are interspersed within the *Gli-1* loci [143,144]. However, there is evidence that allelic variation between the LMW glutenin subunits has an influence on the breadmaking quality, with the alleles of the *Glu-D3 locus* having a lesser influence on the quality characteristics than the alleles of *Glu-A3* and *Glu-B3*, even though the *Glu-D3 locus* is larger than the others [145,146].

The effect of specific LMW-GSs on gluten strength has been determined in durum wheat. Glutenin subunit LMW-2 and its variants confers stronger gluten characteristics than LMW-1 subunits. There are still many unknowns about the relationship between LMW-GS composition in bread wheat and gluten strength. This is partly due to the larger number of LMW-GSs in bread wheat than in durum wheat (the D genome is not present in durum wheat) [11].

In addition to the individual effects of gliadins and glutenins, the gluten network cannot be formed without the presence of these two proteins, and the ratio between glutenins and gliadins is an essential quality determinant for its end use [147]. Studies revealed inter-varietal variation in the Gli/Glu ratio. Nevertheless, whereas common wheat is typically 1.5–3.1, that of ancient wheats was much higher (spelt: 2.8–4.0; emmer: 3.6–6.7; einkorn: 4.2–12.0) [148]. This ratio is fundamental for the breadmaking quality, such as the high volume of bread. When the content of gliadins is higher than that of glutenins, the mixing time is shortened and the stability of the dough reduced, while a greater amount of glutenins, namely HMW-GS and LMW-GS, is positively correlated with bread-making quality [13]. Rodríguez-Quijano et al., in one of their studies about common wheat and spelt, demonstrated that in rheological terms, the Gli/Glu ratio was positively associated with the viscosity and extensibility of the dough and negatively associated with the strength properties so important for breadmaking [13]. Another study conducted by Dhaka et al. demonstrated that the Gli/Glu ratio showed a significant negative relationship with specific bread volume (r = −0.73), dough development time (r = −0.73), and stability of mass (r = −0.79) and a positive relationship with LMW-GS quantity (r = 0.72) [9]. Barak et al. came to identical conclusions, where higher Gli/Glu ratios are negatively associated with general bread quality parameters.

6.4. Models of Gluten Structure and Function in Dough

The strength of the dough, its extensibility, and resistance to kneading are rheological properties that are key parameters in baking [149]. Using small-strain dynamic rheology, a lot of research has been done to define the theoretical basis for gluten structure, relating it to the rheological behavior of dough or gluten [83].

6.4.1. Pom-Pom Model

Proposed by MacLeish and Larson, the pom-pom model addresses the rheological behavior of HMW branched polymer melts and gluten [150]. It is thought that these polymers are a relatively flexible HMW backbone with several branches, the pom-poms, that protrude from each end of the backbone. Therefore, the interaction between the branches and the surrounding polymers creates entanglements and the stretching of the backbone between those entanglements leads to strain hardening. The predictions of this model have also shown that strain hardening is affected not only by the number of branches, but also the distance between the entanglements [149].

6.4.2. Loop-Train Model

The loop-train model was proposed by Belton in order to explain the elastic properties of gluten, emphasizing the role of HMW-GS [151]. According to this model, interchain hydrogen bonds (trains) in gluten hold some regions of the protein chain while some unbonded regions form loops. When gluten is stretched, it causes the loops to extend so that protein chains slide over one another. The reestablishment of the loop-train equilibrium of the unstretched protein creates the elastic restoring force. One of the major critiques of this model is the fact that it has yet to be applied to three-dimensional systems, and that HMW-GS would influence gluten behavior by affecting the properties and size of the gluten network instead of acting as distinct units [83].

6.4.3. Particle-Gel Model

Hamer and co-workers [152,153] proposed a model of glutenin in dough that is based on large aggregates of glutenin. These authors believe that the gluten macropolymer (GMP), the highest molecular weight and consequently least soluble glutenin fraction, affects dough behavior. The idea is that the GMP forms a gel (or particle network) that influences the dough's viscoelastic properties [83].

6.4.4. Linear Glutenin Hypothesis

This hypothesis is a model based on polymer science and is built around the concept of individual chain unfolding within assemblies of glutenin chains when under stress [154,155]. It is proposed the linear branches are composed of dimers of chain-extending LMW-GSs and dimers of chain-extending and chain-terminating LMW-GSs (predicted to be the final subunits in the structure). These dimers come from the head-to-tail arrangement of chain-extending LMW-GSs [156]. As a consequence, elastic properties come from the tendency of those chains to refold to the lowest free energy states. This hypothesis is consistent with the dominant linear molecular structure of glutenin proven by Ewart [157].

Although some models have achieved widespread acceptance, the structure of gluten and the relationship with its rheological properties still need to be further clarified [83,149].

7. Conclusions

Wheat is one of the most important grain crops in the world. The domestication of the ancient varieties of einkorn, emmer, and spelt was the basis for the appearance of durum and common wheat. This domestication increased the yield of wheat through genetic changes that affected characteristics, such as brittle rachis, tenacious glumes, and height. Modern varieties have proved to be highly productive with superior quality concerning end-use.

Protein content is the main factor determining the nutritional and technical quality of wheat, and the quantity and composition of wheat protein are good indicators of the quality of the final product. The gluten viscoelastic network is a fundamental property to produce fermented bread and other foods.

Given the importance of wheat, the continued study of this cereal on all fronts is essential. Plant species like wheat that are essential to human nutrition must evolve and adapt to the changing world climate. Disease, heat, and drought resistance are being addressed in current research. On this note, the demand for diversified, nutritious, and healthy wheat foods has led to a growing interest in nutrition research in ancient wheat, such as einkorn and spelt. These ancient wheats are a valuable source for improvement in opposition to modern wheat, which was already the subject of numerous alterations. Finally, although gluten-free habits have gained some adherence in recent years, often conditioned by some misconceptions associated with gluten, it should be noted that after 10,000 years, wheat is still one of the most representative foods in the human diet.

Author Contributions: The authors' contributions were as follows: conceptualization, M.R. and G.I.; investigation, M.R. and T.d.S.; data curation, T.d.S.; writing—original draft preparation, T.d.S.; writing—review and editing, T.d.S., M.R., C.S. and G.I.; project administration, G.I.; funding acquisition, G.I. All authors have read and agreed to the published version of the manuscript.

Funding: This work was funded by the R&D Project GLUTEN2TARGET-Optimizing natural lowtoxicity of wheat for celiac patients through a nano/microparticles detoxifying approach, referencePOCI-01-0145-FEDER-029068 and PTDC/BAA-AGR/29068/2017, financed by the European Regional Development Fund (ERDF) through COMPETE 2020-Operational Program for Competitiveness and Internationalization (POCI) and by Foundation for Science and Technology (FCT).

Institutional Review Board Statement: Not applicable.

Informed Consent Statement: Not applicable.

Data Availability Statement: Not applicable.

Acknowledgments: The authors acknowledge the supported by the Associate Laboratory for Green Chemistry-LAQV, which is financed by FCT under the Partnership Agreement UIDB/50006/2020.

Conflicts of Interest: The authors declare no conflict of interest.

References

1. FAOSTAT. Food and Agriculture Organization. Available online: http://www.fao.org/faostat/en/#home (accessed on 14 October 2020).
2. Venske, E.; Dos Santos, R.S.; Busanello, C.; Gustafson, P.; Costa de Oliveira, A. Bread wheat: A role model for plant domestication and breeding. *Hereditas* **2019**, *156*, 16. [CrossRef] [PubMed]
3. William, A.; Alain, B.; Maarten, V.G. *The World Wheat Book: A History of Wheat Breeding*; Lavoisier: Paris, France, 2011; Volume 2.
4. Shewry, P.R. Do ancient types of wheat have health benefits compared with modern bread wheat? *J. Cereal Sci.* **2018**, *79*, 469–476. [CrossRef]
5. Curtis, B. *Wheat in the World. Bread Wheat: Improvement and Production*; Food & Agriculture Organization of the UN: Rome, Italy, 2002.
6. Dubcovsky, J.; Dvorak, J. Genome plasticity a key factor in the success of polyploid wheat under domestication. *Science* **2007**, *316*, 1862–1866. [CrossRef]
7. Goncharov, N. Comparative-genetic analysis–a base for wheat taxonomy revision. *Czech J. Genet. Plant Breed.* **2005**, *41*, 52–55. [CrossRef]
8. Peng, J.H.; Sun, D.; Nevo, E. Domestication evolution, genetics and genomics in wheat. *Mol. Breed.* **2011**, *28*, 281. [CrossRef]
9. Dhaka, V.; Khatkar, B.J. Effects of gliadin/glutenin and HMW-GS/LMW-GS ratio on dough rheological properties and bread-making potential of wheat varieties. *J. Food Qual.* **2015**, *38*, 71–82. [CrossRef]
10. Peña, R.; Trethowan, R.; Pfeiffer, W.; Ginkel, M.V. Quality (end-use) improvement in wheat: Compositional, genetic, and environmental factors. *J. Crops Prod.* **2002**, *5*, 1–37. [CrossRef]
11. Peña, R. Wheat for bread and other foods. In *Bread Wheat Improvement and Production*; Food and Agriculture Organization of the United Nations: Rome, Italy, 2002; pp. 483–542.
12. Callejo, M.J.; Vargas-Kostiuk, M.-E.; Ribeiro, M.; Rodríguez-Quijano, M. *Triticum aestivum* ssp. *vulgare* and ssp. *spelta* cultivars: 2. Bread-making optimisation. *Eur. Food Res. Technol.* **2019**, *245*, 1399–1408. [CrossRef]
13. Rodríguez-Quijano, M.; Vargas-Kostiuk, M.-E.; Ribeiro, M.; Callejo, M.J. *Triticum aestivum* ssp. *vulgare* and ssp. *spelta* cultivars. 1. Functional evaluation. *Eur. Food Res. Technol.* **2019**, *245*, 1561–1570. [CrossRef]
14. Arzani, A.; Ashraf, M. Cultivated ancient wheats (*Triticum* spp.): A potential source of health-beneficial food products. *Compr. Rev. Food Sci. Food Saf.* **2017**, *16*, 477–488. [CrossRef]
15. Harlan, J.R.; Zohary, D. Distribution of wild wheats and barley. *Science* **1966**, *153*, 1074–1080. [CrossRef] [PubMed]
16. Shewry, P.R. Wheat. *J. Exp. Bot.* **2009**, *60*, 1537–1553. [CrossRef] [PubMed]
17. Braun, H.-J.; Atlin, G.; Payne, T. Multi-location testing as a tool to identify plant response to global climate change. *Clim. Chang. Crop Prod.* **2010**, *1*, 115–138.
18. Hawkesford, M.J.; Araus, J.L.; Park, R.; Calderini, D.; Miralles, D.; Shen, T.; Zhang, J.; Parry, M.A. Prospects of doubling global wheat yields. *Food Energy Secur.* **2013**, *2*, 34–48. [CrossRef]
19. Sakamoto, S. Patterns of phylogenetic differentiation in the tribe *Triticeae*. *Seiken Ziho* **1973**, *24*, 11–31.
20. Consortium, I.W. A chromosome-based draft sequence of the hexaploid bread wheat (*Triticum aestivum*) genome. *Science* **2014**, *345*, 1251788.
21. Petersen, G.; Seberg, O.; Yde, M.; Berthelsen, K. Phylogenetic relationships of *Triticum* and *Aegilops* and evidence for the origin of the A, B, and D genomes of common wheat (*Triticum aestivum*). *Mol. Phylogenet. Evol.* **2006**, *39*, 70–82. [CrossRef]
22. Feuillet, C.; Langridge, P.; Waugh, R. Cereal breeding takes a walk on the wild side. *Trends Genet.* **2008**, *24*, 24–32. [CrossRef]
23. Zohary, D.; Hopf, M.; Weiss, E. *Domestication of Plants in the Old World: The Origin and Spread of Domesticated Plants in Southwest Asia, Europe, and the Mediterranean Basin*; Oxford University Press on Demand: Oxford, UK, 2012.
24. Feldman, M.; Sears, E.R. The wild gene resources of wheat. *Sci. Am.* **1981**, *244*, 102–113. [CrossRef]
25. Ling, H.-Q.; Zhao, S.; Liu, D.; Wang, J.; Sun, H.; Zhang, C.; Fan, H.; Li, D.; Dong, L.; Tao, Y. Draft genome of the wheat A-genome progenitor *Triticum urartu*. *Nature* **2013**, *496*, 87–90. [CrossRef]
26. Gustafson, P.; Raskina, O.; Ma, X.; Nevo, E. Wheat evolution, domestication, and improvement. In *Wheat: Science and Trade*; Wiley: Danvers, MA, USA, 2009; pp. 5–30.
27. Feldman, M.; Levy, A. Allopolyploidy—A shaping force in the evolution of wheat genomes. *Cytogenet. Genome Res.* **2005**, *109*, 250–258. [CrossRef]
28. Chantret, N.; Salse, J.; Sabot, F.; Rahman, S.; Bellec, A.; Laubin, B.; Dubois, I.; Dossat, C.; Sourdille, P.; Joudrier, P.; et al. Molecular basis of evolutionary events that shaped the hardness locus in diploid and polyploid wheat species (*Triticum* and *Aegilops*). *Plant Cell* **2005**, *17*, 1033–1045. [CrossRef]
29. Ikanović, J.; Popović, V.; Janković, S.; Živanović, L.; Rakić, S.; Dončić, D.J.G. Khorasan wheat population researching (*triticum turgidum*, ssp *turanicum* (mckey) in the minimum tillage conditions. *Genetika* **2014**, *46*, 105–115. [CrossRef]
30. Brown, A.H. Variation under domestication in plants: 1859 and today. *Philos. Trans. R. Soc. Lond. B Biol. Sci.* **2010**, *365*, 2523–2530. [CrossRef]
31. Levy, A.A.; Feldman, M. The impact of polyploidy on grass genome evolution. *Plant Physiol.* **2002**, *130*, 1587–1593. [CrossRef]
32. Sabot, F.; Guyot, R.; Wicker, T.; Chantret, N.; Laubin, B.; Chalhoub, B.; Leroy, P.; Sourdille, P.; Bernard, M. Updating of transposable element annotations from large wheat genomic sequences reveals diverse activities and gene associations. *Mol. Genet. Genom.* **2005**, *274*, 119–130. [CrossRef] [PubMed]

33. Nalam, V.J.; Vales, M.I.; Watson, C.J.; Kianian, S.F.; Riera-Lizarazu, O. Map-based analysis of genes affecting the brittle rachis character in tetraploid wheat (*Triticum turgidum* L.). *Theor. Appl. Genet.* **2006**, *112*, 373–381. [CrossRef] [PubMed]

34. Siddique, K.; Belford, R.; Perry, M.; Tennant, D. Growth, development and light interception of old and modern wheat cultivars in a Mediterranean-type environment. *Aust. J. Agric. Res.* **1989**, *40*, 473–487.

35. Poland, J.; Endelman, J.; Dawson, J.; Rutkoski, J.; Wu, S.; Manes, Y.; Dreisigacker, S.; Crossa, J.; Sánchez-Villeda, H.; Sorrells, M. Genomic selection in wheat breeding using genotyping-by-sequencing. *Plant Genome* **2012**, *5*, 103–113. [CrossRef]

36. Shewry, P.R.; Hey, S. Do "ancient" wheat species differ from modern bread wheat in their contents of bioactive components? *J. Cereal Sci.* **2015**, *65*, 236–243. [CrossRef]

37. Evans, J.R.; Lawson, T. From green to gold: Agricultural revolution for food security. *J. Exp. Bot.* **2020**, *71*, 2211–2215. [CrossRef] [PubMed]

38. Khush, G.S. Green revolution: The way forward. *Nat. Rev. Genet.* **2001**, *2*, 815–822. [CrossRef] [PubMed]

39. Hedden, P. The genes of the Green Revolution. *Trends Genet.* **2003**, *19*, 5–9. [CrossRef]

40. Zhang, C.; Gao, L.; Sun, J.; Jia, J.; Ren, Z. Haplotype variation of Green Revolution gene Rht-D1 during wheat domestication and improvement. *J. Integr. Plant Biol.* **2014**, *56*, 774–780. [CrossRef] [PubMed]

41. Rajaram, S.; van Ginkel, M. *Increasing the Yield Potential in Wheat: Breaking the Barriers*; Cimmyt: Mex, Mexico, 1996.

42. Börner, A.; Plaschke, J.; Korzun, V.; Worland, A. The relationships between the dwarfing genes of wheat and rye. *Euphytica* **1996**, *89*, 69–75. [CrossRef]

43. Na, T.; Jiang, Y.; He, B.-R.; Hu, Y.-G. The effects of dwarfing genes (Rht-B1b, Rht-D1b, and Rht8) with different sensitivity to GA3 on the coleoptile length and plant height of wheat. *Agric. Sci. China* **2009**, *8*, 1028–1038.

44. Hooley, R. Gibberellins: Perception, transduction and responses. *Plant Mol. Biol.* **1994**, *26*, 1529–1555. [CrossRef]

45. Peng, J.; Richards, D.E.; Hartley, N.M.; Murphy, G.P.; Devos, K.M.; Flintham, J.E.; Beales, J.; Fish, L.J.; Worland, A.J.; Pelica, F. 'Green revolution' genes encode mutant gibberellin response modulators. *Nature* **1999**, *400*, 256–261. [CrossRef]

46. Allan, R. Agronomic comparisons between *Rht1* and *Rht2* semidwarf genes in winter wheat. *Crop Sci.* **1989**, *29*, 1103–1108. [CrossRef]

47. Rebetzke, G.; Richards, R.; Fettell, N.; Long, M.; Condon, A.G.; Forrester, R.; Botwright, T. Genotypic increases in coleoptile length improves stand establishment, vigour and grain yield of deep-sown wheat. *Field Crops Res.* **2007**, *100*, 10–23. [CrossRef]

48. Khan, K. *Wheat: Chemistry and Technology*; Elsevier: Amsterdam, The Netherlands, 2016.

49. Indexmundi. Available online: https://www.indexmundi.com/ (accessed on 23 July 2021).

50. Cauvain, S.P. *Breadmaking: Improving Quality*; Elsevier: Amsterdam, The Netherlands, 2012.

51. Vansteelandt, J.; Delcour, J. Physical behavior of durum wheat starch (*Triticum durum*) during industrial pasta processing. *J. Agric. Food Chem.* **1998**, *46*, 2499–2503. [CrossRef]

52. Kaup, S.; Walker, C. *Couscous in North Africa*; Cereal Foods World: St. Paul, MN, USA, 1986.

53. Elias, E. Durum wheat products. In *Durum Wheat Improvement in the Mediterranean Region: New Challenges, Serie A: Séminaires Méditerranéennes*; CIHEAM: Paris, France, 1995; Volume 40, pp. 23–31.

54. Pauly, A.; Pareyt, B.; Fierens, E.; Delcour, J.A. Wheat (*Triticum aestivum* L. and *T. turgidum* L. ssp. *durum*) kernel hardness: II. Implications for end-product quality and role of puroindolines therein. *Compr. Rev. Food Sci. Food Saf.* **2013**, *12*, 427–438. [CrossRef]

55. Qarooni, J. Wheat characteristics for flat breads: Hard or soft, white or red? *Cereal Foods World* **1996**, *41*, 391–395.

56. Nagao, S. Quality characteristics of soft wheats and their utilization in Japan. II. Evaluation of wheats from the United States, Australia, France and Japan. *Cereal Chem.* **1977**, *54*, 198–204.

57. Lee, M.R.; Swanson, B.G.; Baik, B.K.J. Influence of amylose content on properties of wheat starch and breadmaking quality of starch and gluten blends. *Cereal Chem.* **2001**, *78*, 701–706. [CrossRef]

58. Peng, Q.; Wu, R.-L.; Kong, Z.-Y.; Zhang, B.-Q.J. Effect of waxy wheat flour blends on the quality of fresh and stale bread. *Agric. Sci. China* **2009**, *8*, 401–409.

59. Graybosch, R.A.J. Technology. Waxy wheats: Origin, properties, and prospects. *Trends Food Sci. Technol.* **1998**, *9*, 135–142. [CrossRef]

60. Arzani, A. Emmer (*Triticum turgidum* spp. *dicoccum*) flour and breads. In *Flour and Breads and Their Fortification in Health and Disease Prevention*; Elsevier: Amsterdam, The Netherlands, 2011; pp. 69–78.

61. Hidalgo, A.; Brandolini, A.J. Agriculture. Nutritional properties of einkorn wheat (*Triticum monococcum* L.). *J. Sci. Food Agric.* **2014**, *94*, 601–612. [CrossRef]

62. Konvalina, P.; Moudrý, J., Jr.; Moudrý, J.J. Quality parametres of emmer wheat landraces. *J. Cent. Eur. Agric.* **2008**, *9*, 539–545.

63. Geisslitz, S.; Scherf, K.A. Rediscovering ancient wheats. *J. Cereal Sci.* **2020**, *65*, 236–243.

64. Benincasa, P.; Galieni, A.; Manetta, A.C.; Pace, R.; Guiducci, M.; Pisante, M.; Stagnari, F.J. Phenolic compounds in grains, sprouts and wheatgrass of hulled and non-hulled wheat species. *J. Sci. Food Agric.* **2015**, *95*, 1795–1803. [CrossRef] [PubMed]

65. Boukid, F.; Folloni, S.; Sforza, S.; Vittadini, E.; Prandi, B. Current Trends in Ancient Grains-Based Foodstuffs: Insights into Nutritional Aspects and Technological Applications. *Compr. Rev. Food Sci. Food Saf.* **2018**, *17*, 123–136. [CrossRef] [PubMed]

66. Pasqualone, A.; Piergiovanni, A.; Caponio, F.; Paradiso, V.M.; Summo, C.; Simeone, R. Evaluation of the technological characteristics and bread-making quality of alternative wheat cereals in comparison with common and durum wheat. *Food Sci. Technol. Int.* **2011**, *17*, 135–142. [CrossRef] [PubMed]

67. Frakolaki, G.; Giannou, V.; Topakas, E.; Tzia, C. Chemical characterization and breadmaking potential of spelt versus wheat flour. *J. Cereal Sci.* **2018**, *79*, 50–56. [CrossRef]
68. Gomez-Becerra, H.F.; Erdem, H.; Yazici, A.; Tutus, Y.; Torun, B.; Ozturk, L.; Cakmak, I. Grain concentrations of protein and mineral nutrients in a large collection of spelt wheat grown under different environments. *J. Cereal Sci.* **2010**, *52*, 342–349. [CrossRef]
69. Koenig, A.; Konitzer, K.; Wieser, H.; Koehler, P. Classification of spelt cultivars based on differences in storage protein compositions from wheat. *Food Chem.* **2015**, *168*, 176–182. [CrossRef]
70. Sabença, C.; Ribeiro, M.; Sousa, T.d.; Poeta, P.; Bagulho, A.S.; Igrejas, G.J.F. Wheat/Gluten-Related Disorders and Gluten-Free Diet Misconceptions: A Review. *Foods* **2021**, *10*, 1765. [CrossRef]
71. *Bakery Products Market—Growth, Trends, COVID-19 Impact, and Forecasts (2021–2026)*; Mordor Intelligence: Telangana, India, 2021.
72. Di Cagno, R.; Barbato, M.; Di Camillo, C.; Rizzello, C.G.; De Angelis, M.; Giuliani, G.; De Vincenzi, M.; Gobbetti, M.; Cucchiara, S.J. Gluten-free sourdough wheat baked goods appear safe for young celiac patients: A pilot study. *J. Pediatr. Gastroenterol. Nutr.* **2010**, *51*, 777–783. [CrossRef]
73. Di Cagno, R.; De Angelis, M.; Auricchio, S.; Greco, L.; Clarke, C.; De Vincenzi, M.; Giovannini, C.; D'Archivio, M.; Landolfo, F.; Parrilli, G.J.A.; et al. Sourdough bread made from wheat and nontoxic flours and started with selected lactobacilli is tolerated in celiac sprue patients. *Appl. Environ. Microbiol.* **2004**, *70*, 1088–1096. [CrossRef]
74. Laatikainen, R.; Koskenpato, J.; Hongisto, S.-M.; Loponen, J.; Poussa, T.; Huang, X.; Sontag-Strohm, T.; Salmenkari, H.; Korpela, R.J.N. Pilot study: Comparison of sourdough wheat bread and yeast-fermented wheat bread in individuals with wheat sensitivity and irritable bowel syndrome. *Nutrition* **2017**, *9*, 1215. [CrossRef]
75. Meziani, S.; Jasniewski, J.; Gaiani, C.; Ioannou, I.; Muller, J.-M.; Ghoul, M.; Desobry, S.J. Effects of freezing treatments on viscoelastic and structural behavior of frozen sweet dough. *J. Food Eng.* **2011**, *107*, 358–365. [CrossRef]
76. Wang, P.; Xu, L.; Nikoo, M.; Ocen, D.; Wu, F.; Yang, N.; Jin, Z.; Xu, X.J. Effect of frozen storage on the conformational, thermal and microscopic properties of gluten: Comparative studies on gluten-, glutenin-and gliadin-rich fractions. *Food Hydrocoll.* **2014**, *35*, 238–246. [CrossRef]
77. Wang, P.; Chen, H.; Mohanad, B.; Xu, L.; Ning, Y.; Xu, J.; Wu, F.; Yang, N.; Jin, Z.; Xu, X. Effect of frozen storage on physico-chemistry of wheat gluten proteins: Studies on gluten-, glutenin-and gliadin-rich fractions. *Food Hydrocoll.* **2014**, *39*, 187–194. [CrossRef]
78. Zhao, L.; Liu, X.; Hu, Z.; Li, L.; Li, B. Molecular structure evaluation of wheat gluten during frozen storage. *Food Biophys.* **2017**, *12*, 60–68. [CrossRef]
79. Phimolsiripol, Y.; Siripatrawan, U.; Tulyathan, V.; Cleland, D.J. Effects of freezing and temperature fluctuations during frozen storage on frozen dough and bread quality. *J. Food Eng.* **2008**, *84*, 48–56. [CrossRef]
80. Loveday, S.M.; Huang, V.T.; Reid, D.S.; Winger, R.J. Water dynamics in fresh and frozen yeasted dough. *Crit. Rev. Food Sci. Nutr.* **2012**, *52*, 390–409. [CrossRef]
81. Kahleova, H.; Dort, S.; Holubkov, R.; Barnard, N.D. A plant-based high-carbohydrate, low-fat diet in overweight individuals in a 16-week randomized clinical trial: The role of carbohydrates. *Nutrition* **2018**, *10*, 1302. [CrossRef]
82. Guragain, Y.N.; Ganesh, K.; Bansal, S.; Sathish, R.S.; Rao, N.; Vadlani, P.V. Low-lignin mutant biomass resources: Effect of compositional changes on ethanol yield. *Ind. Crops Prod.* **2014**, *61*, 1–8. [CrossRef]
83. Khan, K.; Shewry, P.R. *Wheat: Chemistry and Technology*; AACC International: Eagan, MN, USA, 2009.
84. Bushuk, W.; Rasper, V.F. *Wheat: Production, Properties and Quality*; Springer Science & Business Media: Berlin, Germany, 1994.
85. Poutanen, K.J. Past and future of cereal grains as food for health. *Trends Food Sci. Technol.* **2012**, *25*, 58–62. [CrossRef]
86. Escarnot, E.; Dornez, E.; Verspreet, J.; Agneessens, R.; Courtin, C.M. Quantification and visualization of dietary fibre components in spelt and wheat kernels. *J. Cereal Sci.* **2015**, *62*, 124–133. [CrossRef]
87. Finnie, S.; Jeannotte, R.; Faubion, J.M. Quantitative characterization of polar lipids from wheat whole meal, flour, and starch. *Cereal Chem. J.* **2009**, *86*, 637–645. [CrossRef]
88. Uthayakumaran, S.; Wrigley, C. Wheat: Grain-quality characteristics and management of quality requirements. In *Cereal Grains*; Elsevier: Amsterdam, The Netherlands, 2017; pp. 91–134.
89. Righetti, L.; Rubert, J.; Galaverna, G.; Folloni, S.; Ranieri, R.; Stranska-Zachariasova, M.; Hajslova, J.; Dall'Asta, C. Characterization and Discrimination of Ancient Grains: A Metabolomics Approach. *Int. J. Mol. Sci.* **2016**, *17*, 1217. [CrossRef]
90. Hidalgo, A.; Brandolini, A.; Pompei, C. Carotenoids evolution during pasta, bread and water biscuit preparation from wheat flours. *Food Chem.* **2010**, *121*, 746–751. [CrossRef]
91. Hidalgo, A.; Brandolini, A.; Ratti, S.J. Influence of genetic and environmental factors on selected nutritional traits of Triticum monococcum. *J. Agric. Food Chem.* **2009**, *57*, 6342–6348. [CrossRef]
92. Shewry, P.R.; Hey, S.J. The contribution of wheat to human diet and health. *Food Energy Secur.* **2015**, *4*, 178–202. [CrossRef] [PubMed]
93. Wu, G. Functional amino acids in growth, reproduction, and health. *Adv. Nutr.* **2010**, *1*, 31–37. [CrossRef]
94. Ribeiro, M.; Sousa, T.d.; Sabença, C.; Poeta, P.; Bagulho, A.S.; Igrejas, G. Advances in quantification and analysis of the celiac-related immunogenic potential of gluten. *Compr. Rev. Food Sci. Food Saf.* **2021**. [CrossRef] [PubMed]
95. Hansen, A.; Korner, R.; Ren, O.K. The intes-final T cell response to a gliadin in adult celiac disease is fo-cused on a single deamidated glutamine targeted by tissue transglutaminase. *JEM* **2000**, *191*, 603–612. [CrossRef] [PubMed]

96. Shan, L.; Molberg, Ø.; Parrot, I.; Hausch, F.; Filiz, F.; Gray, G.M.; Sollid, L.M.; Khosla, C. Structural basis for gluten intolerance in celiac sprue. *Science* **2002**, *297*, 2275–2279. [CrossRef] [PubMed]

97. Ronga, D.; Laviano, L.; Catellani, M.; Milc, J.; Prandi, B.; Boukid, F.; Sforza, S.; Dossena, A.; Graziano, S.; Gullì, M.; et al. Influence of environmental and genetic factors on content of toxic and immunogenic wheat gluten peptides. *Eur. J. Agron.* **2020**, *118*, 126091. [CrossRef]

98. Ribeiro, M.; Nunes, F.M.; Guedes, S.; Domingues, P.; Silva, A.M.; Carrillo, J.M.; Rodriguez-Quijano, M.; Branlard, G.; Igrejas, G. Efficient chemo-enzymatic gluten detoxification: Reducing toxic epitopes for celiac patients improving functional properties. *Sci. Rep.* **2015**, *5*, 1–17. [CrossRef]

99. Van Herpen, T.W.; Goryunova, S.V.; Van Der Schoot, J.; Mitreva, M.; Salentijn, E.; Vorst, O.; Schenk, M.F.; Van Veelen, P.A.; Koning, F.; Van Soest, L.J.; et al. Alpha-gliadin genes from the A, B, and D genomes of wheat contain different sets of celiac disease epitopes. *BMC Genom.* **2006**, *7*, 1–13. [CrossRef]

100. Malalgoda, M.; Simsek, S. Celiac disease and cereal proteins. *Food Hydrocoll.* **2017**, *68*, 108–113. [CrossRef]

101. Ribeiro, M.; Nunes, F.M. We might have got it wrong: Modern wheat is not more toxic for celiac patients. *Food Chem.* **2019**, *278*, 820–822. [CrossRef] [PubMed]

102. Ribeiro, M.; Rodriguez-Quijano, M.; Nunes, F.M.; Carrillo, J.M.; Branlard, G.; Igrejas, G. New insights into wheat toxicity: Breeding did not seem to contribute to a prevalence of potential celiac disease's immunostimulatory epitopes. *Food Chem.* **2016**, *213*, 8–18. [CrossRef]

103. Prandi, B.; Tedeschi, T.; Folloni, S.; Galaverna, G.; Sforza, S. Peptides from gluten digestion: A comparison between old and modern wheat varieties. *Food Res. Int.* **2017**, *91*, 92–102. [CrossRef]

104. Malalgoda, M.; Meinhardt, S.W.; Simsek, S. Detection and quantitation of immunogenic epitopes related to celiac disease in historical and modern hard red spring wheat cultivars. *Food Chem.* **2018**, *264*, 101–107. [CrossRef] [PubMed]

105. Spaenij–Dekking, L.; Kooy–Winkelaar, Y.; van Veelen, P.; Drijfhout, J.W.; Jonker, H.; van Soest, L.; Smulders, M.J.; Bosch, D.; Gilissen, L.J.; Koning, F. Natural variation in toxicity of wheat: Potential for selection of nontoxic varieties for celiac disease patients. *Gastroenterology* **2005**, *129*, 797–806. [CrossRef]

106. Pasha, I.; Anjum, F.; Morris, C. Grain hardness: A major determinant of wheat quality. *Food Sci. Technol. Int.* **2010**, *16*, 511–522. [CrossRef]

107. Gazza, L.; Nocente, F.; Ng, P.; Pogna, N. Genetic and biochemical analysis of common wheat cultivars lacking puroindoline a. *Theor. Appl. Genet.* **2005**, *110*, 470–478. [CrossRef]

108. Ribeiro, M.; Rodríguez-Quijano, M.; Giraldo, P.; Pinto, L.; Vázquez, J.F.; Carrillo, J.M.; Igrejas, G. Effect of allelic variation at glutenin and puroindoline loci on bread-making quality: Favorable combinations occur in less toxic varieties of wheat for celiac patients. *Eur. Food Res. Technol.* **2017**, *243*, 743–752. [CrossRef]

109. Greenblatt, G.; Bettge, A.; Morris, C. *Relationship between Endosperm Texture and the Occurrence of Friabilin and Bound Polar Lipids on Wheat Starch*; American Association of Cereal Chemists: St. Paul, MN, USA, 1995.

110. Pomeranz, Y. Composition and functionality of wheat-flour components. *Wheat Chem. Technol.* **1971**, *3*, 385–674.

111. Guerrieri, N.; Cavaletto, M. Cereals proteins. In *Proteins in Food Processing*; Elsevier: Amsterdam, The Netherlands, 2018; pp. 223–244.

112. Gao, L.; Wang, A.; Li, X.; Dong, K.; Wang, K.; Appels, R.; Ma, W.; Yan, Y. Wheat quality related differential expressions of albumins and globulins revealed by two-dimensional difference gel electrophoresis (2-D DIGE). *J. Proteom.* **2009**, *73*, 279–296. [CrossRef]

113. Boukid, F.; Prandi, B.; Faccini, A.; Sforza, S.J. A complete mass spectrometry (MS)-based peptidomic description of gluten peptides generated during in vitro gastrointestinal digestion of durum wheat: Implication for celiac disease. *J. Am. Soc. Mass Spectrom.* **2019**, *30*, 1481–1490. [CrossRef]

114. Platzer, B.; Baker, K.; Vera, M.P.; Singer, K.; Panduro, M.; Lexmond, W.S.; Turner, D.; Vargas, S.O.; Kinet, J.P.; Maurer, D.; et al. Dendritic cell-bound IgE functions to restrain allergic inflammation at mucosal sites. *Mucosal Immunol.* **2015**, *8*, 516–532. [CrossRef] [PubMed]

115. Veraverbeke, W.S.; Delcour, J.A. Wheat protein composition and properties of wheat glutenin in relation to breadmaking functionality. *Crit. Rev. Food Sci. Nutr.* **2002**, *42*, 179–208. [CrossRef] [PubMed]

116. Tomić, J.; Torbica, A.; Popović, L.; Strelec, I.; Vaštag, Ž.; Pojić, M.; Rakita, S.J. Albumins characterization in relation to rheological properties and enzymatic activity of wheat flour dough. *JAST* **2015**, *17*, 805–816.

117. Chen, X.; Cao, X.; Zhang, Y.; Islam, S.; Zhang, J.; Yang, R.; Liu, J.; Li, G.; Appels, R.; Keeble-Gagnere, G.; et al. Genetic characterization of cysteine-rich type-b avenin-like protein coding genes in common wheat. *Sci. Rep.* **2016**, *6*, 30692. [CrossRef] [PubMed]

118. Zhang, Y.; Hu, X.; Islam, S.; She, M.; Peng, Y.; Yu, Z.; Wylie, S.; Juhasz, A.; Dowla, M.; Yang, R.; et al. New insights into the evolution of wheat avenin-like proteins in wild emmer wheat (*Triticum dicoccoides*). *Proc. Natl. Acad. Sci. USA* **2018**, *115*, 13312–13317. [CrossRef] [PubMed]

119. Guzmán, C.; Xiao, Y.; Crossa, J.; González-Santoyo, H.; Huerta, J.; Singh, R.; Dreisigacker, S. Sources of the highly expressed wheat bread making (*wbm*) gene in CIMMYT spring wheat germplasm and its effect on processing and bread-making quality. *Euphytica* **2016**, *209*, 689–692. [CrossRef]

120. Beccari, J.B. *De Bononiensi Scientiarum et Artium Atque Academia Commentarii*; Part I; Nabu Press: Charleston, SC, USA, 1745; pp. 122–127.

121. Osborne, T.B. *The Vegetable Proteins*; Longmans, Green and Company: London, UK, 1924.
122. Shewry, P.R.; Tatham, A.S. The characteristics, structures and evolutionary relationships of prolamins. In *Seed Proteins*; Springer: Berlin/Heidelberg, Germany, 1999; pp. 11–33.
123. Osborne, T.B. *The Proteins of the Wheat Kernel*; Carnegie Institution: Washington, DC, USA, 1907.
124. Ribeiro, M.; Nunes-Miranda, J.D.; Branlard, G.; Carrillo, J.M.; Rodriguez-Quijano, M.; Igrejas, G. One hundred years of grain omics: Identifying the glutens that feed the world. *J. Proteome Res.* **2013**, *12*, 4702–4716. [CrossRef]
125. Ribeiro, M.; Nunes, F.M.; Rodriguez-Quijano, M.; Carrillo, J.M.; Branlard, G.; Igrejas, G. Next-generation therapies for celiac disease: The gluten-targeted approaches. *Trends Food Sci. Technol.* **2018**, *75*, 56–71. [CrossRef]
126. Tatham, A.S.; Shewry, P.R. The S-poor prolamins of wheat, barley and rye: Revisited. *J. Cereal Sci.* **2012**, *55*, 79–99. [CrossRef]
127. Payne, P.; Holt, L.M.; Jackson, E.A. Wheat storage proteins: Their genetics and their potential for manipulation by plant breeding. *Philos. Trans. R. Soc. Lond. B Biol. Sci.* **1984**, *304*, 359–371.
128. Payne, P.I. Genetics of wheat storage proteins and the effect of allelic variation on bread-making quality. *Annu. Rev. Plant Physiol.* **1987**, *38*, 141–153. [CrossRef]
129. Shewry, P.; Tatham, A. Disulphide bonds in wheat gluten proteins. *J. Cereal Sci.* **1997**, *25*, 207–227. [CrossRef]
130. Branlard, G.; Dardevet, M.; Saccomano, R.; Lagoutte, F.; Gourdon, J. Genetic diversity of wheat storage proteins and bread wheat quality. *Euphytica* **2001**, *119*, 59–67. [CrossRef]
131. Agarwal, M.; Shrivastava, N.; Padh, H. Advances in molecular marker techniques and their applications in plant sciences. *Plant Cell Rep.* **2008**, *27*, 617–631. [CrossRef]
132. Payne, P.I.; Lawrence, G.J. Catalogue of alleles for the complex gene loci, Glu-A1, Glu-B1, and Glu-D1 which code for high-molecular-weight subunits of glutenin in hexaploid wheat. *Cereal Res. Commun.* **1983**, *11*, 29–35.
133. Payne, P.I.; Holt, L.M.; Lawrence, G.J.; Law, C.N. The genetics of gliadin and glutenin, the major storage proteins of the wheat endosperm. *Plant Foods Hum. Nutr.* **1982**, *31*, 229–241. [CrossRef]
134. Jackson, E.A.; Holt, L.M.; Payne, P.I. Characterisation of high molecular weight gliadin and low-molecular-weight glutenin subunits of wheat endosperm by two-dimensional electrophoresis and the chromosomal localisation of their controlling genes. *Theor. Appl. Genet.* **1983**, *66*, 29–37. [CrossRef]
135. Rasheed, A.; Xia, X.; Yan, Y.; Appels, R.; Mahmood, T.; He, Z. Wheat seed storage proteins: Advances in molecular genetics, diversity and breeding applications. *J. Cereal Sci.* **2014**, *60*, 11–24. [CrossRef]
136. Ribeiro, M.; Sousa, T.; Poeta, P.; Bagulho, A.S.; Igrejas, G. Review of Structural Features and Binding Capacity of Polyphenols to Gluten Proteins and Peptides In Vitro: Relevance to Celiac Disease. *Antioxidants* **2020**, *9*, 463. [CrossRef] [PubMed]
137. Biesiekierski, J.R. What is gluten? *J. Gastroenterol. Hepatol.* **2017**, *32* (Suppl. 1), 78–81. [CrossRef]
138. Wieser, H. Chemistry of gluten proteins. *Food Microbiol.* **2007**, *24*, 115–119. [CrossRef]
139. Dupont, F.M.; Vensel, W.H.; Tanaka, C.K.; Hurkman, W.J.; Altenbach, S.B. Deciphering the complexities of the wheat flour proteome using quantitative two-dimensional electrophoresis, three proteases and tandem mass spectrometry. *Proteome Sci.* **2011**, *9*, 1–29. [CrossRef]
140. Payne, P.I.; Corfield, K.G.; Holt, L.M.; Blackman, J.A. Correlations between the inheritance of certain high-molecular weight subunits of glutenin and bread-making quality in progenies of six crosses of bread wheat. *J. Sci. Food Agric.* **1981**, *32*, 51–60. [CrossRef]
141. Payne, P.I.; Nightingale, M.A.; Krattiger, A.F.; Holt, L.M. The relationship between HMW glutenin subunit composition and the bread-making quality of British-grown wheat varieties. *J. Sci. Food Agric.* **1987**, *40*, 51–65. [CrossRef]
142. Igrejas, G.; Guedes-Pinto, H.; Carnide, V.; Clement, J.; Branlard, G. Genetical, biochemical and technological parameters associated with biscuit quality. II. Prediction using storage proteins and quality characteristics in a soft wheat population. *J. Cereal Sci.* **2002**, *36*, 187–197. [CrossRef]
143. Metakovsky, E.; Branlard, G.; Chernakov, V.; Upelniek, V.; Redaelli, R.; Pogna, N. Recombination mapping of some chromosome 1A-, 1B-, 1D-and 6B-controlled gliadins and low-molecular-weight glutenin subunits in common wheat. *Theor. Appl. Genet.* **1997**, *94*, 788–795. [CrossRef]
144. Masci, S.; Rovelli, L.; Kasarda, D.; Vensel, W.; Lafiandra, D. Characterisation and chromosomal localisation of C-type low-molecular-weight glutenin subunits in the bread wheat cultivar Chinese Spring. *Theor. Appl. Genet.* **2002**, *104*, 422–428. [CrossRef] [PubMed]
145. Wang, Y.; Zhen, S.; Luo, N.; Han, C.; Lu, X.; Li, X.; Xia, X.; He, Z.; Yan, Y. Low molecular weight glutenin subunit gene Glu-B3h confers superior dough strength and breadmaking quality in wheat (*Triticum aestivum* L.). *Sci. Rep.* **2016**, *6*, 27182. [CrossRef] [PubMed]
146. Huang, X.-Q.; Cloutier, S. Molecular characterization and genomic organization of low molecular weight glutenin subunit genes at the Glu-3 loci in hexaploid wheat (*Triticum aestivum* L.). *Theor. Appl. Genet.* **2008**, *116*, 953–966. [CrossRef] [PubMed]
147. Wrigley, C.; Békés, F.; Bushuk, W. Gluten: A balance of gliadin and glutenin. In *Gliadin and Glutenin: The Unique Balance of Wheat Quality*; Amer Assn of Cereal Chemists: St. Paul, MN, USA, 2006; pp. 3–32.
148. Geisslitz, S.; Longin, C.F.H.; Scherf, K.A.; Koehler, P. Comparative Study on Gluten Protein Composition of Ancient (Einkorn, Emmer and Spelt) and Modern Wheat Species (Durum and Common Wheat). *Foods* **2019**, *8*, 409. [CrossRef]
149. Dobraszczyk, B.; Morgenstern, M. Rheology and the breadmaking process. *J. Cereal Sci.* **2003**, *38*, 229–245. [CrossRef]

150. McLeish, T.; Larson, R. Molecular constitutive equations for a class of branched polymers: The pom-pom polymer. *J. Rheol.* **1998**, *42*, 81–110. [CrossRef]
151. Belton, P. Mini review: On the elasticity of wheat gluten. *J. Cereal Sci.* **1999**, *29*, 103–107. [CrossRef]
152. Don, C.; Lichtendonk, W.; Plijter, J.; Hamer, R. Glutenin macropolymer: A gel formed by glutenin particles. *J. Cereal Sci.* **2003**, *37*, 1–7. [CrossRef]
153. Hamer, R.; Vliet, T.v. Understanding the structure and properties of gluten: An overview. In Proceedings of the 7th International Workshop Gluten 2000, Bristol, UK, 2–6 April 2000; pp. 125–131.
154. Greenwood, C.; Ewart, J. Hypothesis for the structure of glutenin in relation to rheological properties of gluten and dough [Wheat]. *Cereal Chem.* **1975**, *52*, 146–153.
155. Ewart, J. A hypothesis for the structure and rheology of glutenin. *J. Sci. Food Agric.* **1968**, *19*, 617–623. [CrossRef]
156. Lindsay, M.P.; Skerritt, J.H. The glutenin macropolymer of wheat flour doughs: Structure–function perspectives. *Trends Food Sci. Technol.* **1999**, *10*, 247–253. [CrossRef]
157. Ewart, J.A. Calculated molecular weight distribution for glutenin. *J. Sci. Food Agric.* **1987**, *38*, 277–289. [CrossRef]

Review

Phenolic Compounds in Whole Grain Sorghum and Their Health Benefits

Jingwen Xu [1,*], Weiqun Wang [2] and Yong Zhao [1]

[1] College of Food Science and Technology, Shanghai Ocean University, Shanghai 201306, China; yzhao@shou.edu.cn

[2] Department of Food Nutrition Dietetics and Health, Kansas State University, Manhattan, KS 66506, USA; wwang@ksu.edu

* Correspondence: jw-xu@shou.edu.cn

Abstract: Sorghum grain (*Sorghum bicolor* L. Moench) is a staple food grown across the globe, and is mainly cultivated in the semi-arid regions of Africa and Asia. Recently, sorghum grain is increasingly utilized for human consumption, due to the gluten-free nature and potential phenolic-induced health benefits. Sorghum grain is rich in bioactive phenolic compounds, such as ferulic acid, gallic acid, vanillic acid, luteolin, and apigenin, 3-deoxyanthocyanidins (3-DXA), which are known to provide many health benefits, including antioxidant, anti-inflammatory, anti-proliferative, anti-diabetic, and anti-atherogenic activities. Given an increasing trend of sorghum consumption for humans, this article reviews the content and profile of phenolics in sorghum. It covers aspects of their health benefits and explores their mechanisms of action. The impact of thermal processing, such as boiling, steaming, roasting, and extrusion on sorghum phenolics is also discussed. Compelling data suggest the biological functions of sorghum phenolics, however, further investigations appear warrant to clarify the gap in the current research, and identify promising research topics in future.

Keywords: sorghum phenolics; antioxidant; anti-inflammatory; anti-proliferative; anti-diabetic; anti-atherogenic

Citation: Xu, J.; Wang, W.; Zhao, Y. Phenolic Compounds in Whole Grain Sorghum and Their Health Benefits. *Foods* **2021**, *10*, 1921. https://doi.org/ 10.3390/foods10081921

Academic Editor: Anthony Fardet

Received: 13 July 2021
Accepted: 11 August 2021
Published: 19 August 2021

1. Introduction

Sorghum (*Sorghum bicolor* L. Moench) is the fifth most produced cereal crop globally after wheat, maize, rice, and barley, and is the main cereal food consumed in the semi-arid regions of Africa and Asia, due to the high resistance to drought [1]. Sorghum grain possesses tannins, which are bitter tasting. Sorghum grain used to be utilized for animal feed and biofuel production, rather than human food in the United States. However, sorghum is the main cereal grain for the populations in sub-Saharan Africa. In the last decade, the US has selected less tannin variants, and there has been a growing interest of sorghum consumption due to the gluten-free nature. Celiac disease is an immune disorder which affects millions of people in the US, which results from the consumption of glutamine-rich cereal grains, such as wheat, barley, and rye. The gluten-free nature of sorghum shows its great potential as an alternative cereal grain for human consumption by eliminating the risk of celiac disease for celiac patients. Except for the gluten-free nature, sorghum grain possesses bioactive phenolic compounds, such as phenolic acids, flavonoids, and anthocyanins [2,3], which are known to be associated with reducing the risk of many chronic diseases, such as diabetes, obesity, cancer, and cardiovascular disease [4–7]. More studies have recently been focused on the processing and health-promoting effects of sorghum phenolics.

Sorghum grain is composed of pericarp, testa, endosperm, and germ from the outside to the inner. Testa is located between the pericarp and endosperm, which is unique in the sorghum grains and distinct from other cereal grains [8]. Sorghum grain contains pigmented pericarp (i.e., black, red, yellow, brown) and non-pigmented pericarp (i.e., white) [9]. According to Dykes et al. [10], the genes R and Y contribute to the pericarp color, for example, a white color is shown when Y is homozygous recessive; a yellow color is shown when R and Y are homozygous recessive and homozygous dominant; and a red color is shown when R and Y are dominant. Bioactive phenolic compounds are primarily located in the pericarp and testa, wherein they are bound to the non-starch polysaccharides, such as cellulose, hemicellulose, lignin, and pectin in cell wall [11]. Dykes et al. [12] demonstrated that factors, such as varieties and growing conditions, determined the content and the profile of phenolic compounds in sorghum grains. Generally, pigmented sorghum grains contain more content of phenolics than the white sorghum grains, due to the presence of pigmented anthocyanins [3,13–15].

Previously, review articles have been focused on the bioactive compounds in sorghum and their health benefits as potential food ingredients [16], and the processing technologies for reducing anti-nutritional factors in sorghum grain [17]. To date, there is no comprehensive review regarding the phenolics in sorghum grain and their biological functions, such as antioxidant activity, anti-inflammation activity, anti-cancer effect, anti-diabetic, and anti-atherogenic functions. Given the increasing consumption of sorghum phenolics, this review article will focus on the health-promoting effect of phenolic compounds in sorghum, with regards to the recent antioxidant activity, anti-inflammation activity, anti-proliferative effect, anti-diabetic, and anti-atherogenic functions, aimed to better understand the biological functions of phenolics in sorghum grains for improving human health. The effect of thermal processing on phenolics in sorghum grains, in terms of content and profile, will also be discussed. The current research gap will be identified, and promising research topics will be further recommended.

2. Phenolic Compounds in Sorghum Grain

Phenolic compounds, belonging to secondary metabolites, are well-known to naturally bio-synthesize in plants. Sorghum grain possesses many phenolic acids and flavonoids, wherein flavonoids can be further divided into flavonone, flavonol, anthocyanins, and condensed tannins, known as proanthocyanidins in sorghum. Caffeic acid, cinnamic acid, ferulic acid, gallic acid, salicylic acid, vanillic acid, and *p*-coumaric acid dominate the phenolic acids in sorghum grains (shown in Figure 1) [18,19]. The predominant flavonoids in sorghum include luteolin, apigenin, eriodictyol, and naringenin [12]. We previously reported that 3-deoxyanthocyanidins (3-DXA) was the predominant anthocyanins in sorghum grains, which mainly included luteolinidin, apigeninidin, 5-methoxyluteolinidin, and 7-methoxy apigeninidin. This was found through the identification of anthocyanins in 25 sorghum grains with various pigmented pericarps, including red, brown, yellow, and white pericarps [13]. Except for the predominant phenolic acid and flavonoids, stilbenoids and polyamines are also present in sorghum grains in a small amount. The content of silbenoid trans-piceid and trans-resveratrol of red sorghum is reported to be 0.4–1.0 mg/kg and 0.2 mg/kg [20]. Total phenolic content, total flavonoid content, and total anthocyanin content of sorghum varieties associated with pigmented pericarp sorghum grains are summarized in Table 1.

Conventionally, phenolic compounds in sorghum grain are mainly obtained through refluxing extraction, water extraction, maceration extraction, soxhlet extraction, and organic solvent extraction [3,12,17,19]. However, the extraction yield, content, and profile of phenolics in sorghum are varied between the different extraction solvents. For example, Devi et al. [23] reported that the acidified methanol extract of sorghum (red sorghum, collected in Tamil Nadu, India) bran polyphenols showed greater content of anthocyanins (4.7 mg/g) than methanol extract (1.95 mg/g) and acetone extract (1 mg/g). To the content of total flavonoids and phenolics of sorghum bran, acidified methanol extract was also higher than methanol extract and acetone extract.

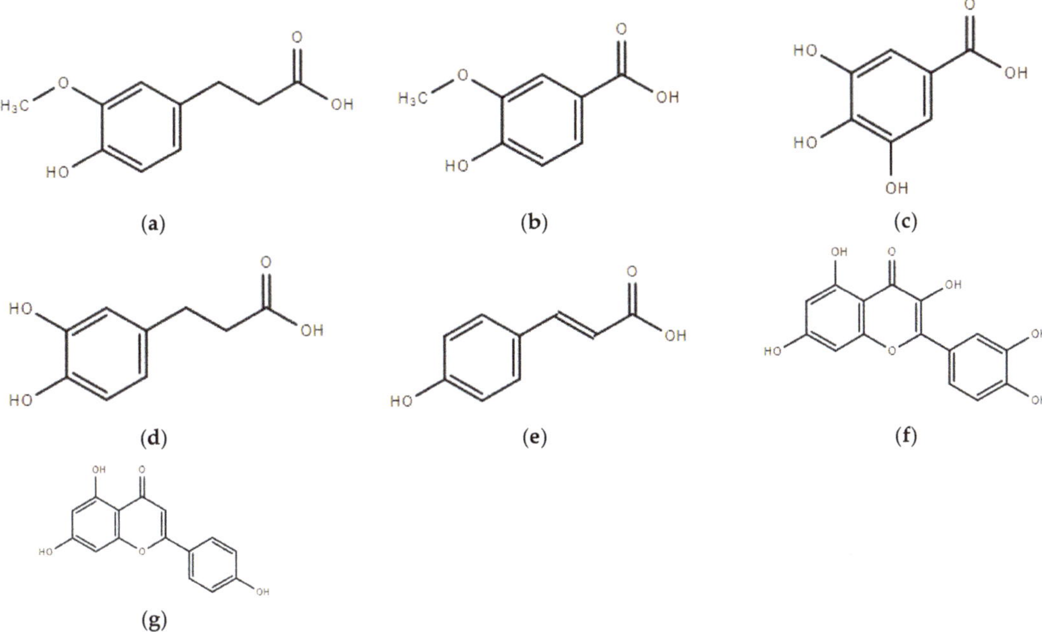

Figure 1. Typical phenolics in sorghum grains, (**a**) ferulic acid; (**b**) vanillic acid; (**c**) gallic acid; (**d**) caffeic acid; (**e**) *p*-coumaric acid; (**f**) luteolin; (**g**) apigenin.

Table 1. Total phenolics content (TPC), total flavonoids content (TFC), total anthocyanins content (TAC) of sorghum varieties associated with pigmented pericarps.

Sorghum Source	TPC (mg/g)	TFC (mg/g)	TAC (mg/g)	Reference
White pericarp	0.24–34.78 GAE	0.06–0.38 RE	0.02 CCE; 0.09 GAE	[2,21,22]
Yellow pericarp	-	-	Bran: 0.26–0.81 AE; Flour: 0.10–0.35 AE	[3]
Black pericarp	4.13–11.50 GAE	0–0.20 RE	3.02 GAE, 0.18 CCE	[2,21]
Brown pericarp	3.58 GAE; 1.74 FAE	1.39 CE	5.55 GAE;	[21]
Red pericarp	0.66–47.86 GAE	0–0.60 RE	0.41–0.60 GAE; 2.66–8.93 CCE	[21,22]

GAE: gallic acid equivalents; RE: rutin equivalents; CCE: cyanidin chloride equivalents; FAE: ferulic acid equivalents; CE: catechin equivalents; AE: apigenidin equivalents.

Nowadays, emerging technologies have been utilized for phenolics extraction from sorghum grain, aimed to improve the extraction yield and phenolics content. So far, ultrasound-assisted extraction [24], pulsed-electric field, accelerated solvent extraction [25], microwave-assisted extraction [26], and subcritical water extraction [27] have been reported for phenolic extraction. For example, the accelerated solvent extraction method at a temperature of 120 and 150 °C, by using solvent of 50% and 70% ethanol/water (*v*/*v*), could result in the content of phenolic compounds of black sorghum bran (A05028/RTx3362), which is up to 45 mg/g gallic acid equivalent of dry weight (gallic acid equivalent, dry weight) (GAE, dw) [25]. Luo et al. [24] reported that the polyphenolic content of red sorghum bran was 49.7 mg/g GAE (dw), through an ultrasound-assisted extraction method for 21 min of processing time, 53% ethanol of solvent, and 52:1 mL/g of solid–liquid ratio. Luo et al. [27] developed the subcritical water extraction method for extracting phenolics up to 47.2 mg/g GAE (dw) from sorghum, through optimized conditions, including 144.5 °C of temperature, 21 min of extraction, and 35 mL/g of solid–liquid ratio.

The identification, characterization, quantification, and qualification of phenolic compounds from sorghum rely on chromatographic techniques, such as high performance liquid chromatography (HPLC) combined with different detectors, including diode array detector (DAD) [13,27], UV-vis photodiode array detector (PDA) [21], tandem quadrupole detector mass spectroscopy (TQD-MS) [26], electron spray ionization (ESI), and atmospheric pressure chemical ionization (APCI) triple quadrupole MS [3,28].

3. Health Benefits and Potential Molecular Mechanisms of Phenolic Compounds in Sorghum Grain

3.1. Antioxidant Activity

Overproduction of reactive oxygen species (ROS) in the human body will potentially result in oxidative stress, which is implicated in the increasing risk of many chronic diseases, such as inflammation, diabetes, atherosclerosis, and cardiovascular disease [29]. Antioxidants from dietary polyphenols can scavenge free radicals for preventing chronic diseases [29]. This section will mainly focus on the antioxidant activity of phenolics isolated from sorghum grains through tests from both in vitro and in vivo.

Phenolic compounds from sorghum varieties have been reported to possess antioxidant activities, which are mainly characterized by scavenging the radicals of DPPH, ABTS, FRAP, and ORAC in vitro [2,15,30]. For example, the ORAC value of phenolics isolated from black sorghum (Shawaya) bran was 3.7 mmol Trolox equivalents/mg (TE/mg) [31]. The IC_{50} value of DPPH radical scavenging activity of eight brown sorghum genotypes (SOR 01, SOR 03, SOR 08, SOR 11, SOR 17, SOR 21, SOR 24, SOR 33) varied from 91.2 to 361.2 mg/mL, and the IC_{50} value of ABTS radical scavenging activity ranged from 203.4 to 352.6 mg/mL [28]. Brown pericarp sorghum (IS131C) was found to possess greater antioxidant activity than black sorghum (Shawya Short Black 1), red sorghum (Mr-Buster, Cracka), and white sorghum (Liberty) when compared the antioxidant activity assayed by ABTS, DPPH, and FRAP [15]. According to Xiong et al. (2021), varieties of IS31C and Shawya Short Black 1 showed higher values of ABTS, DPPH, and FRAP than varieties of Liberty, Mr-Buster, and Cracka [15].

3-DXA in sorghum has also been shown to reduce the oxidative stress in vitro, through modulating the defense system against oxidative stress and inducing NADH: quinone oxyreductase (NQO) activity [4,31]. Belonging to the phase II enzyme, NAD(P)H quinone reductase is known as a detoxifying enzyme, balancing the carcinogen-activating phase I enzymes. The 2−3 double bond in the C-ring of flavonoids and 3-deoxyflavonoids in sorghum is considered to induce NQO activity [4].

Human colorectal cancer Caco-2 cells and hepatocarcinoma HepG2 cells are two common cellular antioxidant activity testing models through MTT assay [7,15]. EC_{50} values of sorghum varieties (i.e., Liberty, Mr-Buster, Cracka, IS131C, Shawaya Short Black 1) varied from 0.4 to 127.1 mg/mL, which assessed the total antioxidant effect of both extracellular and intracellular environments [15]. The median effective concentration (EC_{50}) is the concentration of substance in an environmental medium expected to produce an effect in 50% of test organisms.

The antioxidant activity of sorghum phenolics in vivo has also been investigated through enzymatic activity, such as SOD and GPx. Lewis [32] studied the effect of a diet containing fiber from black sorghum (containing 3-DXA) and white sorghum (containing phenolic acids) on cellular antioxidant activity in rats. Black sorghum (3-DXA rich sorghum) resulted in an increase in superoxide dismutase activity (SOD) and a decrease in glutathione peroxidase (GPx) activity in normolipidemic rats [32]. SOD is a cellular antioxidant enzyme, which can catalyze the dismutation of superoxide anion to hydrogen peroxide, thus detoxifying oxygen and water by catalase or GPx [33]. GPx is an important antioxidant enzyme for reducing hydrogen and lipid peroxides. In addition, white sorghum (containing phenolic acids) increases the catalase activity (CAT) [32]. Catalase is known as a cellular hydrogen peroxide scavenger. Taken together, black sorghum (containing 3-DXA) and white sorghum (containing phenolic acids) show strong antioxidant activity in the rat model. However, Moraes et al. [34] reported diets containing sorghum flour (i.e., BRS 305, BRS 309, BRS 310) did not significantly influence the SOD level in normolipidemic rats. In another study conducted by Ajiboye et al. [35], phenolic extracts of red sorghum variety (obtained from Igbona market, Osogbo, Nigeria, 100 mg/kg body weight) increased the detoxifying enzymes in ROS, including SOD, CAT, GPx, glutathione reductase (GSH-Red), and glucose 6-phosphate dehydrogenase (*Glc* 6-PD) in rat liver. Taken together, the antioxidant activity of sorghum phenolic extracts in vivo is highly variable, and is dependent on the sorghum varieties, sorghum bran, sorghum flour, and whole grain of sorghum. Sorghum bran phenolic extracts showed greater antioxidant activity than sorghum flour, due to the presence of a higher content of phenolics in the bran. The antioxidant activity of sorghum phenolic extracts is summarized in Table 2.

Table 2. Antioxidant activities of phenolic extracts of sorghum grains.

Sorghum Source	Bioactive Extracts	Antioxidant Activity	Reference
Hongyingzi, Hongzhenzhu, Dongbei sorghum, Jiangsu sorghum, Jiliang 2 sorghum, Longza 11, black grain sorghum, white Longmi sorghum.	Caffeic acid, *p*-coumaric acid, ferulic acid, protocatechuic acid, luteolindin, apigeninidin, luteolin, apigenin, taxifolin, naringenin.	Antioxidant activities against DPPH and FRAP assays.	[2]
Tannin-containing sorghum varieties (Sumac, Hi-Tannin, Seredo, CR 35:5 × 2), non-tannin varieties (white variety, KARI-Mtama, red variety, ICSV-III), Mizzou, Tx430.	Condensed tannins, 3-DXA, phenolics.	Induced phase II detoxifying enzymes; anti-proliferative effect on esophageal, OE33, colon cancer cells.	[4]
Liberty, Mr-Buster, Cracker, IS131C, Shawaya Short Black 1.	Phenolic extracts.	Antioxidant activities against DPPH and FRAP assays; Anti-proliferative effect on Caco-2 cells.	[15]
Tx3362, Shawaya Black, Black PI Tall, Hyb 107, Hyb 115, Hyb 116, Hyb 117, Hyb 118.	Total phenolics, condensed tannins, flavan-4-ols, 3-DXA.	Antioxidant activities against DPPH and ABTS assays.	[5]

3.2. Anti-Inflammatory Effect

Inflammation refers to an immune response to cellular injury or infection by pathogens, and triggers many chronic diseases. Pro-inflammatory cytokines, such as interleukin 1, β (IL-1β), tumor necrosis factor (TNF-α), and interleukin 6 (IL-6) are known to be involved in the inflammation pathogenesis through various cellular and molecular pathways. Phytochemicals are reported to modulate inflammation by inhibiting pro-inflammatory enzymes [36,37]. Therefore, this section will focus on the effect of phenolic extracts from sorghum grains on inhibiting inflammation in vitro and in vivo.

Black sorghum bran phenolics extract (10% *w/v* in 50% ethanol) showed an inhibitory effect on TNF-α and IL-1β in lipopolysaccharide-stimulated peripheral blood mononuclear cells at dilutions of 1:100–1:200 and 1:100–1:400, respectively [38]. Hong et al. [39] showed that acidified ethanol extracts of sorghum (SC84MX, SC84KS, PI570481) at 50 mg gallic acid equivalent/mL inhibited nitric oxide (NO) production up to 72.45%, 68.32%, and 95.36%, respectively. The increase in the secretion of TNF-α and IL-6 of RAW 264.7 macrophages infected by the bacteria *Legionella pneumophila* was observed after the treatment of polyphenol extracts of sorghum (PI570481) (0.625 and 1.25 mg/mL) [39]. The mRNA expressions of IL-6 and IL-β of RAW 264.7 macrophage cells were significantly inhibited by the soluble phenolic extracts of the sorghum variety (Tong Za 117) at 300 to 500 mg/mL and 50 to 500 mg/mL, respectively, though a non-toxic mechanism [37].

In addition to the in vitro evaluation, the inhibitory effect of sorghum phenolic extracts on inflammation is also reported in vivo. Black sorghum bran phenolic extracts also showed an anti-inflammatory effect on an 12-*O*-tetradecanoylphorbol acetate (TPA) induced mouse ear model [38]. In addition, golden gelatinous sorghum extracts inhibited the expression levels of cyclooxygenase-2 and inducible nitric oxide synthase, through a TPA induced mice ear edema model [36]. Ritchie et al. [40] studied the inhibitory effect of diets containing 6% dietary fiber from sorghum brans, including black bran (high levels of 3-DXA), Sumac bran (high levels of condensed tannins and low levels of 3-DXA), and a combination of high-tannin bran and black bran on colon inflammation. Diets containing sorghum bran upregulated the colonocyte proliferation and gene expression of trefoil factor (Tff3), and transformed growth factor beta (Tgfb) after the inflammation induced by DSS [40]. Tff3 and Tgfb are known to repair lesions and maintain epithelial barrier integrity, which are both involved in cellular migration and suppression of apoptosis. The effect of extruded sorghum flour on inflammation and oxidative stress in high fat diet-fed rats was studied by de Sousa et al. [41]. A diet containing extruded sorghum flour increased the total antioxidant capacity of serum plasma and SOD level, but reduced the concentrations of p65 through NF-κB in liver and lipids peroxidation [41]. The anti-inflammatory effect of sorghum phenolic extracts is summarized in Table 3.

Table 3. Anti-inflammatory effect of phenolic extracts of sorghum grains.

Sorghum Source	Bioactive Extracts	Anti-Inflammatory Effect	Reference
Red sorghum.	Phenolics, flavonoids, anthocyanins.	Antioxidant activities against DPPH, FRAP, superoxide radical scavenging, hydroxyl radical scavenging assays, metal chelating, hydrogen peroxide.	[23]
Tong Za 117, Tong Za 141, Tong Za 142, Tong Za 143, Chi Za 109, Chi Za 101.	Ferulic acid, p-coumaric acid, caffeic acid, 3,4-dihydroxybenzoic acid, luteolinidin, apigeninidin, 5-methoxyluteolinidin, 7-methoxy apigeninidin.	Antioxidant activities against DPPH and ABTS assays; inhibitory effect on IL-6 and IL-1β.	[37]
Sumac, Mycogen 726, black sorghum, white sorghums.	Total phenolic extracts.	Inhibitory effect on IL-1β and TNF-α.	[38]
SC84MX, SC84KS, PI57048.	Phenolics, flavonoids, tannins, 3-DXA, anthocyanins.	Antioxidant activities against DPPH, ORAC and nitric oxide assays; inhibited cellular production of NO, IL-6, ROS.	[39]
PUI570481	Polyphenol extracts.	Inhibitory effect on IL-6 and TNF-α.	[42]
1-Terral REV 9924, 2-Pioneer 84P8D, 3-Dekalb DK-54-00, 4-FFR353, 5-DynaGro DG765B, 6-Pioneer 83P99, 7-Dekalb DK-51-01, 8-Terral REV 9782, 9-Terral REV 9562, 10-Terral REV9883.	Naringenin, eriodicytol, apigenin, luteolin, apigeninidin, luteolinidin.	Antioxidant activities against DPPH and NO assays; Inhibitory effect on OVCA cells.	[43]
White sorghum, red sorghum.	Gallic acid, protocatechuic acid, chlorogenic acid, caffeic acid, luteolinidin, apigeninidin, p-coumaric acid, flavanols, quercetin, hydroxycinnamic acid, 5-methoxy luteolinidin, 7-methoxy-luteolinidin, 5,7-dimethoxy-luteolinidin, 7-methoxy-apigeninidin, 5,7-dimethoxy-apigeninidin.	Antioxidant activities against ORAC and nitric oxide assays; inhibited cellular production of NO, IL-6, ROS.	[44]

So far, most studies have shown the anti-inflammatory effect of sorghum phenolic extracts. However, the underlying mechanism is not fully understood. In addition, studies have focused on the anti-inflammatory effect of phenolic extracts of sorghum grains, rather than the individual phenolic compounds. Therefore, little is known about which phenolic compound dominates to inhibit inflammation. Taken together, more studies are still warranted for further investigation regarding the the inhibitory effect of sorghum phenolics on inflammation.

3.3. Anti-Proliferative Effect

Cancer is a complex disease, involving the functioning oncogenes, de-functioning tumor suppressor genes, and tumor mutations caused by the endogenous and exogenous factors [45]. This section will focus on the effect of phenolic extracts from sorghum grains on cancer inhibition in vitro and in vivo.

Studies have reported sorghum phenolic extracts possessthe antioxidant activity, phase II enzyme induction, regulation of p53 gene, anti-proliferative effect on cancer cells, and induction of cancer cell apoptosis [4,7,46–48]. Quercetin, a flavonoid found in sorghum grain, has been reported to inhibit b-catenin signaling in SW480 colon cancer cells [49]. Luteolin, a predominant flavonoid in sorghum has also shown an anti-proliferative effect on human colorectal cancer HCT15 and CO115 cells, by harboring KRAS and BRAF activating mutations [50].

The IC_{50} values obtained through an MTT test of sorghum (i.e., KARI-Mtama, Mizzou, Tx430, Seredo, Sumac, Hi-tannin) phenolic extracts inhibiting HT-29 and OE33 cancer cell proliferation ranged from 54.8 to 389 mg/mL and 95.3 to 654 mg/mL, respectively [4]. Black sorghum 3-DXA extract has also been shown to have an inhibitory effect on HT-29 human colon cancer cells (IC_{50} = 180−557 mg/mL) [46]. Hargrove et al. [51] reported that phenolic extracts from sorghum (Sumac sorghum, black sorghum) bran inhibited the aromatase activity in vitro, which had IC_{50} values of 12.1 and 18.8 mg/mL, respectively. Suganyadevi et al. [52] reported that 3-deoxyanthocyanins in red sorghum bran induced apoptosis in breast cancer MCF 7 cells through stimulating the p53 gene and down-regulating the Bcl-2 gene. P53 gene is known to be responsible for the cell cycle arrest and apoptosis. Phenolic extracts of black pericarp sorghum have also been shown to inhibit the growth of human HepG2 cells and Caco-2 cells, through a cell cycle arrest at G2/M phase and an induction of apoptosis [7].

In addition to the sorghum bran, sorghum stalk phenolic extract has also been shown to have an inhibitory effect against colon cancer proliferation in vitro. Massey et al. [53] isolated the phenolics from the pith and dermal layer of sweet sorghum (i.e., Dale and M81E) stalk, and found that the dermal layer contained more content of phenolics than the pith for both varieties, especially the content of 3-DXA apigeninidin and luteolinidin. Phenolic extracts from the dermal layer of sorghum varieties Dale and M81E showed higher antioxidant activity than the pith assayed by ABTS [53]. The extract of dermal layer of sweet sorghum Dale inhibited the growth of colon cancer HCT116 cells and colon cancer stem cells (CCSCs), through modulating the gene p53 above 35 mg of gallic acid equivalent/mL [53].

The inhibitory effect of sorghum phenolic extracts on cancer has also been reported. Wu et al. [54] found an antioxidant activity of sorghum (Moench) procyanidins (150 mg/kg) against oxidative stress in a rat model, that significantly reversed the increase in malondi-aldehyde (MDA) level and decreased SOD and GPx in both liver homogenate and serum of rat induced by D-galactose. Sorghum procyanidins (100, 200, 400 mg/kg) inhibited the tumor growth and reduced tumor weight in C57BL/6J mice of lung cancer, and the inhibitory effect of tumor growth and weight was dose-dependent, through the suppression of vascular endothelial growth factor (VEGF) production in mice [54]. Hwanggeumchal sorghum phenolic extracts have also been shown to have an inhibitory effect on human breast cancer MDA-MB-231 cells and MC7 xenografts in mice, through modulating Jak/STAT pathways, hindering the STAT5b/IGF-1R and STAT3/VEGF pathways, and down-regulating the angiogenic factors, such as VEGF, VEGF-R2, and cell cycle regulators such as cyclin D, cyclin E, and pRb [55]. In addition, Hwanggeumchal sorghum extracts also induced the apoptosis of MDA-MB-231 cells arrested at G1 phase. The anti-proliferative effect of sorghum phenolic extracts is summarized in Table 4.

Table 4. Anti-proliferative effect of phenolic extracts of sorghum grains.

Sorghum Source	Bioactive Extracts	Anti-Proliferative Effect	Reference
Black sorghum varieties (Macia, Sumac, PI152653, PI152687, PI193073, PI1329694, PI1559733, PI1559855, PI1568282, PI1570366, PI1570481, PI1570484, PI1570819, PI1570889, PI1570993).	Total phenolic extracts.	Anti-proliferative effect on HepG2 and Caco-2 cells: induction G1/S cell cycle arrest, activation of p53.	[48]
Red sorghum	3-DXA extracts.	Inhibitory effect on MCF7 cancer cells through up-regulating p53 and down-regulating Bcl-2 genes.	[52]
Dale, M81E	Vanillic acid, p-coumaric acid, ferulic acid, caffeic acid, apigeninidin, luteolinidin, malvidin-3-O-glucoside, apigenin, luteolin, trans-resveratrol, luteoferol.	Inhibitory effect on HCT116 and colon cancer stem cells through activating p53 gene.	[53]
Hwanggeumchal sorghum.	Total polyphenol extracts.	Anti-proliferative effect on MDA-MB 231 and MC7 cells: down-regulating VEGF, VEGF-R2, cyclin D, cyclin E, pRb and up-regulating p53.	[55]
TX430, Sumac.	Total phenolic extracts.	Anti-proliferative effect on HepG2 and HCT15 cells.	[56]

To date, the anti-proliferative effect of sorghum phenolic extracts has been investigated in regards to the effective dose and underlying mechanism. However, most studies have been focused on the complex phenolic extracts of sorghum, rather than the individual phenolic. Hence, little is known about which phenolic plays a leading role in the anti-proliferative effect on cancer. Thus, more investigations are still needed to understand the anti-proliferative effect of individual phenolic from sorghum grain.

3.4. Anti-Diabetic Effect

Diabetes is one of the most challenging chronic diseases worldwide. The insulin resistance and pancreatic b-cell dysfunction result in the hyperglycemia and abnormal carbohydrate metabolism, further leading to type 2 diabetes (T2D). Sorghum phenolic extracts have been found to effectively inhibit diabetes, through reducing serum glucose, total cholesterol, and triglycerides [6,57,58]. This section will focus on the anti-diabetic effect of sorghum phenolic extracts in vitro and in vivo.

Chung et al. [57] found that phenolic extracts of Hwanggeumchal sorghum (250 and 500 mg/kg for 14 days) could significantly reduce the serum glucose, total cholesterol, triglycerides, urea, uric acid, creatinine, aspartate amino transferase, and alanine amino transferase in streptozotocin-induced diabetic rats. In addition, phenolic extracts of Hwanggeumchal sorghum at 250 mg/kg also resulted in an increase in serum insulin in diabetic rats, but not in normal rats [57]. However, the mechanism of the anti-diabetic effect of sorghum phenolics was not discussed. Phenolic extracts of Hwanggeumchal sorghum (0.5% and 1% addition to dietary intake) given to high fat diet-fed rats resulted in a significant reduction in perirenal fat, total and low-density lipoprotein cholesterol (LDL-cholesterol), triglycerides, and glucose [59]. The hypoglycemic effect of sorghum extracts was considered to be associated with the regulation of PPAR-g-mediated metabolism in rats. The anti-diabetic effect of sorghum extract was also evaluated in diabetic rats induced by streptozotocin, and results showed sorghum extracts (0.4 and 0.6 g/kg) decreased the expression of phosphoenolpyruvate carboxykinase and the phosphor-p38/p38 ratio, but did not affect the glucose transporter 4 translocation and the phosphor-Akt/Akt ratio [60]. Similarly, Wu et al. [58] also reported that feeding sorghum red pigments (200 mg/kg body weight) to diabetic mice induced by glucose reversed glucose tolerance and serum levels of triglycerides, total and LDL-cholesterol, and ameliorated lipid metabolism. In addition, the feeding of sorghum red pigments (200 mg/kg body weight) to diabetic mice reduced body weight by 26.5%, compared to the diabetic mice [58]. However, the mechanism of the anti-diabetic effect of sorghum phenolics was not shown.

In addition to the phenolic extracts from sorghum grain, sorghum flour has also been shown to modulate adiposity and inflammation in obese rats fed with a high fat diet. Tested diets (replacement of 50% cellulose and 100% of corn starch by sorghum flour, and replacement of 100% cellulose and 100% of corn starch by sorghum flour in obese diet) lowered the percentage of adiposity, fatty acid synthase gene expression, TNF-α, blood levels of glucose, and adipocyte hypertrophy in rats [6].

In addition, polyphenolics and anthocyanins have been reported to inhibit the starch digestive enzymes, such as α-amylase and α-glucosidase, thus retarding starch digestibility and lowering the value of glucose index (GI), which is also considered to possess the anti-diabetic effect [28,47]. The IC_{50} values of the inhibitory effect of proanthocyanidins from Sumac sorghum and black sorghum bran phenolic extracts on α-amylase were reported to be 1.4 and 11.4 mg/mL, respectively [51]. The effect of red sorghum phenolic extract on pancreatic lipase inhibition, α-amylase activity, and α-glucosidase inhibitory activity was studied by Irondi et al. [61], and they found that IC_{50} values were 12.72 ± 1.13, 16.93 ± 1.08, and 10.78 ± 0.63 mg/mL, respectively. IC_{50} values of brown sorghum genotypes (SOR 01, SOR 03, SOR 08, SOR 11, SOR 17, SOR 21, SOR 24, SOR 33) on α-glucosidase and α-amylase inhibition were reported to be 14.7 to 61.0 mg/mL and 10.6 to 852.6 mg/mL, respectively [28]. The anti-diabetic effect of sorghum phenolic extracts is summarized in Table 5.

Table 5. Anti-diabetic and anti-atherogenic effect of phenolic extracts of sorghum grains.

Sorghum Source	Bioactive Extracts	Anti-Diabetic and Anti-Atherogenic Effects	Reference
Brown sorghum varieties (SOR 01, SOR 03, SOR 08, SOR 11, SOR 17, SOR 21, SOR 24, SOR 33)	Gallic acid, chlorogenic acid, caffeic acid, ellagic acid, p-coumaric acid, quercetin, luteolin, apigenin.	Inhibitory effect on α-amylase and α-glucosidase activities.	[28]
Hwanggeumchal sorghum.	Phenolic extracts.	Reduced the serum glucose, total cholesterol, triglycerides, urea, uric acid, creatinine.	[57]
KNICS-579	Polyphenol extracts.	Reduced the concentration of triglycerides, total LDL-cholesterol and glucose.	[60]
Red sorghum	Total phenolic extracts.	Antioxidant activity against ABTS, DPPH, FRAP assays; Inhibitory effect on pancreatic lipase, α-amylase and α-glucosidase activities.	[61]

To date, studies have shown the anti-diabetic effect of sorghum phenolics, including the reduction in serum glucose, decrease in total cholesterol and triglycerides, and inhibition of α-glucosidase and α-amylase activity. However, the underlying mechanisms of the inhibitory effect of sorghum phenolics on diabetes have not been fully studied. Therefore, more research is warranted for further investigation of the role of sorghum phenolics in diabetes inhibition.

3.5. Anti-Atherogenic Effect

Atherosclerosis is a major factor in the development of coronary heart disease. Blood vessel walls turn thicker in the development of atherosclerotic lesions, thus affecting blood circulation. Many risk factors are involved in the induction of heart disease, such as hypercholesterolemia, hypertension, cigarette smoking, and diabetes [62]. Moreover, hypercholesterolemia is associated with an increased level of LDL-cholesterol. Hence, hypercholesterolemia is a major risk factor for the development of cardiovascular disease. An increase in total cholesterol, triglycerides, and LDL-cholesterol are all considered to be associated with an increased risk of atherosclerosis and cardiovascular diseases, whereas high-density lipoprotein cholesterol (HDL-lipoprotein) is considered to be associated with the reduced risk of atherosclerosis and cardiovascular disease [62].

The anti-atherogenic effect of sorghum on high fat diet-fed mice was studied by Shen et al. [22], and they found that the diet containing sorghum reduced serum cholesterol by 24.47% and triglyceride by 32.72%, and increased HDL-cholesterol by 27.27% compared to the high fat diet group. In addition, the sorghum feed also increased SOD and GPx activities in the serum, compared to high fat diet mice [22]. A high fat diet might result in the generation of ROS, which can be controlled through enzymatic defense mechanisms, such as SOD, GPx, and CAT. Phenolics in sorghum possess a strong antioxidant capacity, thus increasing the activities of enzymes, SOD and GSH-Px, for reducing the oxidative effect. To date, little is known about the anti-atherogenic effect of sorghum phenolic extracts, as well as the underlying mechanism both in vitro and in vivo. Therefore, more studies are needed to better understand the anti-atherogenic effect of sorghum phenolics.

4. Effect of Processing on Phenolic Compounds in Sorghum

Sorghum grain is a good alternative to cereal grains, due to it containing nutritional and health-promoting factors. However, tannins, known as bitter-tasting compounds, have negative effects on human consumption, although they can provide biological functions for improving human health. So far, to our knowledge, sorghum grain is thermally cooked, such as through boiling, cooking, nixtamalization (alkaline cooking), extrusion, roasting, and steaming prior to human consumption. Tannins and polyphenols are considered as anti-nutritional factors which affect the digestion of starch and protein. Thermal processing could potentially reduce the content of tannins and phenolic compounds [63–65]. Therefore, the biological functions of phenolic compounds will also be affected by thermal processing, due to heat sensitivity. Thus, this section will focus on the effect of thermal processing on phenolic compounds in sorghum varieties. In addition, the current research gap and future research topics will also be discussed.

The reduced content of tannins in sorghum cultivars (Wadakar, low b-glucan type II non-tannin sorghum; Tabat-C, low b-glucan type I non-tannin sorghum; new in bread line Tabat-NL, high b-glucan type I non-tannin sorghum) has been reported through cooking (20 min) and boiling also decreased the content of phytate and polyphenols [64]. In addition, Hamad et al. [64] also reported that boiling resulted in an increase in digested starch, rapid digestible starch, hydrolysis index, and estimated GI, indicating that boiling eliminated anti-nutritional factors, such as tannins and polyphenols, to further improve starch degradation [64]. Luzardo-Ocampo et al. [66] reported that cooking and nixtamalization (10 g Ca(OH)$_2$/kg flour, 94 °C for 40 min) significantly reduced the content of condensed tannins, total phenolics, and flavonoids in the digestible fraction white sorghum (Tortillas y Pan), whereas it increased the content of total phenolics and flavonoids in the digestible fraction of red sorghum (Niquel) compared to the raw sorghum [66]. The variance in the content of condensed tannins, total phenolics, and flavonoids in white and red sorghums was believed to result from the different degree of cooking and nixtamalization in the undigested form, mouth, stomach, digestible fraction, and non-digestible fraction [66].

Pressured steam cooking and air-dried flaking also resulted in the decrease in content of phytate and tannins, from 69.87% to 93.73% and 19.49% to 46.05%, respectively, in sorghum varieties of IS8237C, Liberty, and Alpha [67]. Xiong et al. [64] showed that steaming (100 °C for 50 min) and roasting (150 °C for 60 min) increased the content of total phenolics, total flavonoids, and condensed tannins to a different extent in non-tannin white color sorghum (Liberty). Steaming and roasting can damage the cellular structure of cereal grains, thus releasing the bound phenolic compounds, and increasing the content of phenolics after thermal processing [68,69]. Extrusion cooking has also been shown to decrease the content of total phenols and total flavonoids of sorghum genotypes SC319, B.DLO357, and SC391 [70].

To date, more studies have been focused on the effect of thermal processing on nutrients digestion in sorghum grains, rather than on the phenolics regarding the content and biological functions after thermal processing. Therefore, more research is recommended to better understand the effect of phenolics in sorghum grains, with regards to the content, profile, and biological activities through thermal treatments.

5. Conclusions and Outlook

This review discussed the phenolic compounds in sorghum grain, in terms of the extraction method, profile, and biological functions both in vitro and in vivo. Representative phenolic compounds in sorghum are ferulic acid, caffeic acid, gallic acid, luteolin, apigenin, 3-DXA, and others. Bioactive phenolic compounds possess many biological functions, such as an antioxidant activity, anti-inflammatory effect, anti-proliferative effect, anti-diabetic, and anti-atherogenic effects. However, the underlying mechanisms regarding the inhibitory effect of sorghum phenolics on inflammation, diabetes, and atherosclerosis remain unclear. Therefore, more studies are warranted for a better understanding of the molecular mechanisms involved in the inhibitory effect of sorghum phenolics on inflammation, diabetes, and atherosclerosis.

To date, sorghum grain has been served as a cereal grain alternative, or has replaced wheat or other cereal grains in innovative bakery products for human consumption. The thermal processing could potentially reduce the content of tannins and phytate, which are the anti-nutritional factors for improving nutrient digestion. However, the loss of biological functions of phenolics and tannins in sorghum has also been reported. Although steaming and roasting processes increased the content of total phenolics and flavonoids due to the release of bound phenolics, most studies showed a reduction in phenolics in sorghum grains after thermal treatment. Therefore, moderate processing is urgently required for maintaining the content of phenolics and their biological activities, and for reducing the anti-nutritional factors in sorghum grains to improve nutrient digestion.

Author Contributions: Conceptualization, J.X. and W.W.; writing—review and editing, J.X., W.W. and Y.Z.; project administration, J.X.; funding acquisition, W.W. All authors have read and agreed to the published version of the manuscript.

Funding: This research was funded by USDA hatch grant KS18HA1012 (Contribution Number 22-006-J from the Kansas Agricultural Experiment Station).

Institutional Review Board Statement: Not applicable.

Informed Consent Statement: Not applicable.

Data Availability Statement: Not applicable.

Conflicts of Interest: The authors declare no conflict of interest.

References

1. FAO. FAOSTAT. Available online: http://www.fao.org/faostat/en/#data/QC (accessed on 30 March 2019).
2. Shen, S.; Huang, R.; Li, C.; Wu, W.; Chen, H.; Shi, J.; Chen, S.; Ye, X. Phenolic compositions and antioxidant activities differ significantly among sorghum grains with different applications. *Molecules* **2018**, *23*, 1023. [CrossRef]
3. Kumari, P.K.; Umakanth, A.V.; Narsaiah, T.B.; Uma, A. Exploring anthocyanins, antioxidant capacity and a-glucosidase inhibition in bran and flour extracts of selected sorghum genotypes. *Food Biosci.* **2021**, *41*, 100979. [CrossRef]
4. Awika, J.M.; Yang, L.; Browning, J.D.; Faraj, A. Comparative Antioxidant, Antiproliferative and Phase II Enzyme Inducing Potential of Sorghum (*Sorghum bicolor*) Varieties. *LWT-Food Sci. Tech.* **2009**, *42*, 1041–1046. [CrossRef]

5. Dykes, L.; Ronney, W.L.; Rooney, L.W. Evaluation of phenolics and antioxidantactivity of black sorghum hybrids. *J. Cereal Sci.* **2013**, *58*, 278–283. [CrossRef]

6. Arbex, P.M.; de Moreira, M.E.C.; Toledo, R.C.L.; de Morais Cardoso, L.; Pinheiro-Sant'ana, H.M.; dos Benjamin, L.A.; Licursi, L.; Carvalho, C.W.P.; Queiroz, V.A.V.; Martino, H.S.D. Extruded Sorghum Flour (*Sorghum bicolor* L.) Modulate Adiposity and Inflammation in High Fat Diet-Induced Obese Rats. *J. Funct. Foods* **2018**, *42*, 346–355. [CrossRef]

7. Chen, X.; Shen, J.; Xu, J.; Herald, T.; Smolensky, D.; Perumal, R.; Wang, W. Sorghum phenolic compounds are associated with cell growth inhibition through cell cycle arrest and apoptosis in human hepatocarcinoma and colorectal adenocarcinoma cells. *Foods* **2021**, *10*, 993. [CrossRef] [PubMed]

8. Earp, C.F.; McDonough, C.M.; Rooney, L.W. Microscopy of pericarp development in the caryopsis of *Sorghum bicolor* (L.) Moench. *J. Cereal Sci.* **2004**, *39*, 21–27. [CrossRef]

9. Dykes, L.; Rooney, L.W.; Waniska, R.D.; Rooney, W.L. Phenolic compounds and antioxidant activity of sorghum grains of varying genotypes. *J. Agric. Food Chem.* **2005**, *53*, 6813–6818. [CrossRef] [PubMed]

10. Dykes, L.; Seitz, M.L.; Rooney, L.W.; Rooney, L.W. Flavonoid composition of red sorghum genotypes. *Food Chem.* **2009**, *116*, 313–317. [CrossRef]

11. de Cardoso, L.M.; Pinheiro, S.S.; de Carvalho, C.W.P.; Queiroz, V.A.V.; de Menezes, C.B.; Moreira, A.V.B.; de Barros, F.A.R.; Awika, J.M.; Martino, H.S.D.; Pinheiro-Sant'Ana, H.M. Phenolic Compounds Profile in Sorghum Processed by Extrusion Cooking and Dry Heat in a Conventional Oven. *J. Cereal Sci.* **2015**, *65*, 220–226. [CrossRef]

12. Dykes, L.; Peterson, G.C.; Rooney, W.L.; Rooney, L.W. Flavonoid composition of lemon-yellow sorghum genotypes. *Food Chem.* **2011**, *127*, 173–179. [CrossRef]

13. Su, X.; Rhodes, D.; Xu, J.; Chen, X.; Davis, H.; Wang, D.; Herald, T.J.; Wang, W. Phenotypic diversity of anthocyanins in sorghum accessions with various pericarp pigments. *J. Nutr. Food Sci.* **2017**, *7*. [CrossRef]

14. Xiong, Y.; Zhang, P.; Warner, R.D.; Shen, S.; Johnson, S.; Fang, Z. Comprehensive Profiling of Phenolic Compounds by HPLC-DAD-ESI-QTOF-MS/MS to Reveal Their Location and Form of Presence in Different Sorghum Grain Genotypes. *Food Res. Int.* **2020**, *137*, 109671. [CrossRef] [PubMed]

15. Xiong, Y.; Damasceno Teixeira, T.V.; Zhang, P.; Warner, R.D.; Shen, S.; Fang, Z. Cellular Antioxidant Activities of Phenolic Extracts from Five Sorghum Grain Genotypes. *Food Biosci.* **2021**, *41*, 101068. [CrossRef]

16. Girard, A.L.; Awika, J.M. Sorghum Polyphenols and Other Bioactive Components as Functional and Health Promoting Food Ingredients. *J. Cereal Sci.* **2018**, *84*, 112–124. [CrossRef]

17. Rashwan, A.K.; Yones, H.A.; Karim, N.; Taha, E.M.; Chen, W. Potential Processing Technologies for Developing Sorghum-Based Food Products: An Update and Comprehensive Review. *Trends Food Sci. Tech.* **2021**, *110*, 168–182. [CrossRef]

18. Billa, E.; Koullas, D.P.; Monties, B.; Koubios, E.G. Structure and composition of sweet sorghum stalk components. *Ind. Crops Prod.* **1997**, *6*, 297–302. [CrossRef]

19. Awika, J.M.; Rooney, L.W.; Waniska, R.D. Anthocyanins from Black Sorghum and Their Antioxidant Properties. *Food Chem.* **2004**, *90*, 293–301. [CrossRef]

20. Bröhan, M.; Jerković, V.; Collin, S. Potentiality of red sorghum for producing stilbenoid-enriched beers with high antioxidant activity. *J. Agric. Food Chem.* **2011**, *59*, 4088–4094. [CrossRef]

21. Rao, S.; Santhakumar, A.B.; Chinkwo, K.A.; Wu, G.; Johnson, S.K.; Blanchard, C.L. Characterization of Phenolic Compounds and Antioxidant Activity in Sorghum Grains. *J. Cereal Sci.* **2018**, *84*, 103–111. [CrossRef]

22. Shen, Y.; Song, X.; Chen, Y.; Li, L.; Sun, J.; Huang, C.; Ou, S.; Zhang, H. Effects of sorghum, purple rice and rhubarb rice on lipids status and antioxidant capacity in mice fed a high-fat diet. *J. Funct. Foods* **2017**, *37*, 103–111. [CrossRef]

23. Devi, P.S.; Kumar, M.S.; Das, S.M. DNA damage protecting activity and free radical scavenging activity of anthocyanins from red sorghum (*Sorghum bicolor*) bran. *Biotech. Res. Int.* **2012**, *2012*, 258787. [CrossRef] [PubMed]

24. Luo, X.; Cui, J.; Zhang, H.; Duan, Y.; Zhang, D.; Cai, M.; Chen, G. Ultrasound Assisted Extraction of Polyphenolic Compounds from Red Sorghum (*Sorghum bicolor* L.) Bran and Their Biological Activities and Polyphenolic Compositions. *Ind. Crop. Prod.* **2018**, *112*, 296–304. [CrossRef]

25. Barros, F.; Dykes, L.; Awika, J.M.; Rooney, L.W. Accelerated Solvent Extraction of Phenolic Compounds from Sorghum Brans. *J. Cereal Sci.* **2013**, *58*, 305–312. [CrossRef]

26. Herrman, D.A.; Brantsen, J.F.; Ravisankar, S.; Lee, K.M.; Awika, J.M. Stability of 3-deoxyanthocynin pigement structure relative to anthocyanins from grains under microwave assisted extraction. *Food Chem.* **2020**, *333*, 127494. [CrossRef]

27. Luo, X.; Cui, J.; Zhang, H.; Duan, Y. Subcritical Water Extraction of Polyphenolic Compounds from Sorghum (*Sorghum bicolor* L.) Bran and Their Biological Activities. *Food Chem.* **2018**, *262*, 14–20. [CrossRef] [PubMed]

28. Ofosu, F.K.; Elahi, F.; Daliri, E.B.-M.; Tyagi, A.; Chen, X.Q.; Chelliah, R.; Kim, J.-H.; Han, S.-I.; Oh, D.-H. UHPLC-ESI-QTOF-MS/MS Characterization, Antioxidant and Antidiabetic Properties of Sorghum Grains. *Food Chem.* **2021**, *337*, 127788. [CrossRef]

29. Zhang, Y.J.; Gan, R.Y.; Li, S.; Zhou, Y.; Li, A.N.; Xu, D.P.; Li, H.B. Antioxidant phytochemicals for the prevention and treatment of chronic diseases. *Molecules* **2015**, *20*, 21138–21156. [CrossRef]

30. Wu, G.; Shen, Y.; Qi, Y.; Zhang, H.; Wang, L.; Qian, H.; Qi, X.; Li, Y.; Johnson, S.K. Improvement of in vitro and cellular antioxidant properties of Chinese steamed bread through sorghum addition. *LWT-Food Sci. Tech.* **2018**, *91*, 77–83. [CrossRef]

31. González-Montilla, F.M.; Chávez-Santoscoy, R.A.; Gutiérrez-Uribe, J.A.; Serna-Saldivar, S.O. Isolation and Identification of Phase II Enzyme Inductors Obtained from Black Shawaya Sorghum [*Sorghum bicolor* (L.) Moench] Bran. *J. Cereal Sci.* **2012**, *55*, 126–131. [CrossRef]

32. Lewis, J.B. Effects of Bran from Sorghum Grains Containing Different Classes and Levels of Bioactive Compounds in Colon Carcinogenesis. Master's Thesis, Texas A&M University, College Station, TA, USA, 2008.

33. Ighodaro, O.M.; Akinloye, O.A. First line defense antioxidants-superoxide dismutase (SOD), catalase (CAT) and glutathione peroxidase (GPX): Their fundamental role in the entire antioxidant defense grid. *Alex. J. Med.* **2017**, *54*, 287–293.

34. Moraes, E.A.; Natal, D.I.G.; Queiroz, V.A.V.; Schaffert, R.E.; Cecon, P.R.; de Paula, S.O.; dos Anjos Benjamim, L.; Reibeiro, S.M.R.; Martino, H.S.D. Sorghum genotype may reduce low-grade inflammatory response and oxidative stress and maintains jejunum morphology of rats fed a hyperlipidic diet. *Food Res. Int.* **2012**, *49*, 553–559. [CrossRef]

35. Ajiboye, T.O.; Komolafe, Y.O.; Oloyede, O.B.; Ogunbode, S.M.; Adeoye, M.D.; Abdulsalami, I.O.; Nurudeen, Q.Q. Polyphenolic extract of *Sorghum bicolor* grains reactive oxygen species detoxification in N-nitrosodiethylamine-treated rats. *Food Sci. Hum. Wellness* **2013**, *2*, 39–45. [CrossRef]

36. Shim, T.J.; Kim, T.M.; Jang, K.C.; Ko, J.Y.; Kim, D.J. Toxicological evaluation and anti-inflammatory activity of a golden gelatinous sorghum bran extract. *Biosci. Biotech. Biochem.* **2013**, *77*, 697–705. [CrossRef]

37. Li, M.; Xu, T.; Zheng, W.; Gao, B.; Zhu, H.; Xu, R.; Deng, H.; Wang, B.; Wu, Y.; Sun, X.; et al. Triacylglycerols compositions, soluble and bound phenolics of red sorghums, and their radical scavenging and anti-inflammatory activities. *Food Chem.* **2021**, *340*, 128123. [CrossRef] [PubMed]

38. Burdette, A.; Garner, P.L.; Mayer, E.P.; Hargrove, J.L.; Hartle, D.K.; Greenspan, P. Anti-inflammatory activity of select sorghum (*Sorghum bicolor*) brans. *J. Med. Food* **2010**, *13*, 879–887. [CrossRef]

39. Hong, S.; Pangloli, P.; Perumal, R.; Cox, S.; Noronha, L.E.; Dia, V.P.; Smolensky, D.A. Comparative study on phenolic content, antioxidant activity and anti-inflammatory capacity of aqueous and ethanolic extracts of sorghum in lipopolyshaccaride-induced RAW264.7 macrophages. *Antioxidants* **2020**, *9*, 1297. [CrossRef] [PubMed]

40. Ritchie, L.; Taddeo, S.S.; Weeks, B.R.; Carroll, R.J.; Dykes, L.; Rooney, L.W.; Turner, N.D. Impact of novel sorghum bran diets on DSS-induced colitis. *Nutrients* **2017**, *9*, 330. [CrossRef]

41. de Sousa, A.R.; de Castro Moreira, M.E.; Grancieri, M.; Toledo, R.C.L.; de Oliveira Araújo, F.; Mantovani, H.C.; Queiroz, V.A.V.; Martino, H.S.D. Extruded Sorghum (*Sorghum bicolor* L.) Improves Gut Microbiota, Reduces Inflammation, and Oxidative Stress in Obese Rats Fed a High-Fat Diet. *J. Funct. Foods* **2019**, *58*, 282–291. [CrossRef]

42. Gilchrist, A.K.; Smolensky, D.; Ngwaga, T.; Chauhan, D.; Cox, S.; Perumal, R.; Noronha, L.E.; Rhames, S.R. High-polyphenol extracts from *Sorghum bicolor* attenuate replication of Legionella pneumophila within RW 264.7 macrophages. *FEMS Microb. Lett.* **2020**, *367*. [CrossRef] [PubMed]

43. Dia, P.; Pangloli, P.; Jones, L.; McClure, A.; Patel, A. Phytochemical concentrations and biological activities of *Sorghum bicolor* alcoholic extracts. *Food Funct.* **2016**, *7*, 3410–3420. [CrossRef] [PubMed]

44. Carbonneau, M.-A.; Cisse, M.; Mora-Soumille, N.; Dairi, S.; Rosa, M.; Michel, F.; Lauret, C.; Cristol, J.-P.; Dangles, O. Antioxidant Properties of 3-Deoxyanthocyanidins and Polyphenolic Extracts from Côte d'Ivoire's Red and White Sorghums Assessed by ORAC and in vitro LDL Oxidisability Tests. *Food Chem.* **2014**, *145*, 701–709. [CrossRef] [PubMed]

45. Thoma, C.R.; Zimmermann, M.; Agarkova, I.; Kelm, J.M.; Krek, W. 3D cell culture systems modeling tumor growth determinants in cancer target discovery. *Adv. Drug Deliv. Rev.* **2014**, *69*, 29–41. [CrossRef] [PubMed]

46. Yang, L.; Browning, J.D.; Awika, J.M. Sorghum 3-deoxyanthocyanins possess strong phase II enzyme inducer activity and cancer cell growth inhibition properties. *J. Agric. Food Chem.* **2009**, *57*, 1797–1804. [CrossRef] [PubMed]

47. Kunyanga, C.N.; Imungi, J.K.; Okoh, M.W.; Biesalki, H.K. Total phenolic content, antioxidant and antidiabetic properties of methanolic extract of raw and traditionally processed Kenyan indigenous food ingredients. *LWT-Food Sci. Tech.* **2012**, *45*, 269–276. [CrossRef]

48. Smolensky, D.; Rhodes, D.; McVey, D.S.; Fawver, Z.; Perumal, R.; Herald, T.; Noronha, L. High-Polyphenol Sorghum Bran Extract Inhibits Cancer Cell Growth Through ROS Induction, Cell Cycle Arrest, and Apoptosis. *J. Med. Food* **2018**, *21*, 990–998. [CrossRef] [PubMed]

49. Park, C.H.; Chang, J.Y.; Hahm, E.R.; Park, S.; Kim, H.-K.; Yang, C.H. Quercetin, a Potent Inhibitor against b-Catenin/Tcf Signaling in SW480 Colon Cancer Cells. *Biochem. Biophys. Res. Commun.* **2005**, *328*, 227–234. [CrossRef]

50. Xavier, C.P.R.; Lima, C.F.; Preto, A.; Seruca, R.; Fernandes-Ferreira, M.; Pereira-Wilson, C. Luteolin, Quercetin and Ursolic Acid Are Potent Inhibitors of Proliferation and Inducers of Apoptosis in Both KRAS and BRAF Mutated Human Colorectal Cancer Cells. *Cancer Lett.* **2009**, *281*, 162–170. [CrossRef]

51. Hargrove, J.L.; Greenspan, P.; Hartle, D.K.; Dowd, C. Inhibition of Aromatase and α-Amylase by Flavonoids and Proanthocyanidins from *Sorghum bicolor* Bran Extracts. *J. Med. Food* **2011**, *14*, 799–807. [CrossRef] [PubMed]

52. Suganyadevi, P.; Saravanakumar, K.M.; Mohandas, S. The Antiproliferative Activity of 3-Deoxyanthocyanins Extracted from Red Sorghum (*Sorghum bicolor*) Bran through P53-Dependent and Bcl-2 Gene Expression in Breast Cancer Cell Line. *Life Sci.* **2013**, *92*, 379–382. [CrossRef]

53. Massey, A.R.; Reddivari, L.; Vanamala, J. The Dermal Layer of Sweet Sorghum (*Sorghum bicolor*) Stalk, a Byproduct of Biofuel Production and Source of Unique 3-Deoxyanthocyanidins, Has More Antiproliferative and Proapoptotic Activity than the Pith in P53 Variants of HCT116 and Colon Cancer Stem Cells. *J. Agric. Food Chem.* **2014**, *62*, 3150–3159. [CrossRef] [PubMed]

54. Wu, L.; Huang, Z.; Qin, P.; Yao, Y.; Meng, X.; Zou, J.; Zhu, K.; Ren, G. Chemical characterization of a procyanidin rich extract from sorghum bran and its effect on oxidative stress and tumor inhibition in vivo. *J. Agric. Food Chem.* **2011**, *59*, 8609–8615. [CrossRef]

55. Park, J.H.; Darvin, P.; Lim, E.J.; Joung, Y.H.; Hong, D.Y.; Park, E.U.; Park, S.H.; Choi, S.K.; Moon, E.-S.; Cho, B.W.; et al. Hwanggeumchal Sorghum Induces Cell Cycle Arrest, and Suppresses Tumor Growth and Metastasis through Jak2/STAT Pathways in Breast Cancer Xenografts. *PLoS ONE* **2012**, *7*, e40531. [CrossRef]

56. Cox, S.; Noronba, L.; Herald, T.; Bean, S.; Lee, S.H.; Perumal, E.; Wang, W.; Smolensky, D. Evaluation of ethanol-based extraction conditions of sorghum bran bioactive compounds with downstream anti-proliferative properties in human cancer cells. *Heliyon* **2019**, *5*, e01589. [CrossRef] [PubMed]

57. Chung, I.-M.; Kim, E.-H.; Yeo, M.-A.; Kim, S.-J.; Seo, M.; Moon, H.-I. Antidiabetic Effects of Three Korean Sorghum Phenolic Extracts in Normal and Streptozotocin-Induced Diabetic Rats. *Food Res. Int.* **2011**, *44*, 127–132. [CrossRef]

58. Wu, L.; Liu, Y.; Qin, Y.; Wang, L.; Wu, Z. HPLC-ESI-qTOF-MS/MS characterization, antioxidant activities and inhibitory ability of digestive enzymes with molecular docking analysis of various parts of raspberry (*Rubus ideaus* L.). *Antioxidants* **2019**, *8*, 274. [CrossRef]

59. Park, J.H.; Lee, S.H.; Chung, I.-M.; Park, Y. Sorghum Extract Exerts an Anti-Diabetic Effect by Improving Insulin Sensitivity via PPAR-γ in Mice Fed a High-Fat Diet. *Nutr. Res. Pract.* **2012**, *6*, 322. [CrossRef]

60. Kim, J.; Park, Y. Anti-diabetic effect of sorghum extract on hepatic gluconeogenesis of streptozotoxin-induced diabetic rats. *Nutr. Met.* **2012**, *9*, 106. [CrossRef] [PubMed]

61. Irondi, E.A.; Adegoke, B.M.; Effion, E.S.; Oyewo, S.O.; Alamu, E.O.; Boligon, A.A. Enzymes inhibitory property, antioxidant activity and phenolics profile of raw and roasted red sorghum grains in vitro. *Food Sci. Hum. Wellness* **2019**, *8*, 142–148. [CrossRef]

62. Gobbo, L.C.D.; Falk, M.C.; Feldman, R.; Lewis, K.; Mozaffarian, D. Effects of tree nuts on blood lipids, apolipoproteins, and blood pressure: Systematic review, meta-analysis, and dose-response of 61 controlled intervention trials. *Am. J. Clin. Nutr.* **2015**, *102*, 1347–1356. [CrossRef]

63. Gaytán-Martínez, M.; Cabrera-Ramírez, Á.H.; Morales-Sánchez, E.; Ramírez-Jiménez, A.K.; Cruz-Ramírez, J.; Campos-Vega, R.; Velazquez, G.; Loarca-Piña, G.; Mendoza, S. Effect of Nixtamalization Process on the Content and Composition of Phenolic Compounds and Antioxidant Activity of Two Sorghums Varieties. *J. Cereal Sci.* **2017**, *77*, 1–8. [CrossRef]

64. Hamad, S.A.A.; Mustafa, I.; Magboul, B.I.; Qasem, A.A.; Ahmed, I.A.M. Nutritional quality of raw and cooked flours of a high b-glucan sorghum inbred line. *J. Cereal Sci.* **2019**, *90*, 102857. [CrossRef]

65. Xiong, Y.; Zhang, P.; Luo, J.; Johnson, S.; Fang, Z. Effect of Processing on the Phenolic Contents, Antioxidant Activity and Volatile Compounds of Sorghum Grain Tea. *J. Cereal Sci.* **2019**, *85*, 6–14. [CrossRef]

66. Luzardo-Ocampo, I.; Ramírez-Jiménez, A.K.; Cabrera-Ramírez, Á.H.; Rodríguez-Castillo, N.; Campos-Vega, R.; Loarca-Piña, G.; Gaytán-Martínez, M. Impact of Cooking and Nixtamalization on the Bioaccessibility and Antioxidant Capacity of Phenolic Compounds from Two Sorghum Varieties. *Food Chem.* **2020**, *309*, 125684. [CrossRef] [PubMed]

67. Wu, G.; Ashton, J.; Simic, A.; Fang, Z.; Johnson, S.K. Mineral availability is modified by tannin and phytate content in sorghum flaked breakfast cereals. *Food Res. Int.* **2017**, *103*, 509–514. [CrossRef]

68. Randhir, R.; Kwon, Y.-I.; Shetty, K. Effect of thermal processing on phenolics, antioxidant activity and health-relevant functionality of select grain sprouts ad seedlings. *Innov. Food Sci. Emerg. Technol.* **2008**, *9*, 355–364. [CrossRef]

69. Wu, L.; Huang, Z.; Qin, P.; Ren, G. Effects of processing on phytochemical profiles and biological activities for production of sorghum tea. *Food Res. Int.* **2013**, *53*, 685–687. [CrossRef]

70. de Morais Cardoso, L.; Pinheiro, S.S.; Martino, H.S.D.; Pinheiro-Sant'Ana, H.M. Sorghum (*Sorghum bicolor* I.): Nutrients, bioactive compounds, and potential impact on human health. *Crit. Rev. Food Sci. Nutr.* **2015**, *57*, 372–390. [CrossRef]

MDPI
St. Alban-Anlage 66
4052 Basel
Switzerland
Tel. +41 61 683 77 34
Fax +41 61 302 89 18
www.mdpi.com

Foods Editorial Office
E-mail: foods@mdpi.com
www.mdpi.com/journal/foods